LAND, PEOPLE AND ECONOMY OF PAKISTAN

The book unravels the complexity of Pakistan's physical terrain, from the Arabian Sea coast to the Himalayan heights, focusing on its bio-climatic environment. While throwing light on the evolution of societal organisation and the nature and intensity of economic enterprises, it discusses the human response to the natural milieu and the way natural resources have been harnessed.

This volume discusses the distribution of natural resources and deciphers the evolutionary development of manufacturing, transport and trade, the main non-agricultural enterprises, and looks at the potential for their growth. It provides an overview of the country's economy and its fluctuations through different political regimes. It also investigates the country's diverse social, linguistic and ethnic composition. A habitat study incorporating villages and towns and their numerical and spatial distribution and also the hierarchical organisation of rural society have received a special focus in the book.

The book, with its qualitative and quantitative information, will be useful to students, teachers and researchers studying physical geology, climatology, hydrology, soil science, environmental studies, archaeology, sociology, economics, demography, area studies and regional development studies. It will also be an indispensable companion to professionals in floral-landscaping, habitat study, administrators and policymakers.

K.R. Dikshit is a former Professor of Geography (now retired) at the University of Pune, India. He has written several books, such as *Geography of Gujarat* (1970), *Contribution to Indian Geography: Geomorphology* (ed.1983), *Maharashtra in Maps* (1986), *Environment, Forest Ecology and Man in the Western Ghats: The Case of Mahabaleshwar Plateau* (1993) and *North-East India: Land, People and Economy* (2014) as well as many research papers.

Jutta K. Dikshit has been teaching geography at the Department of Geography at the University of Pune, India. She studied geography and German language and literature at the Johannes Gutenberg University, Mainz (Germany) and obtained her doctoral degree from the Sorbonne, Paris. Before coming to India, she worked as a lecturer at the University of Saarbrucken for a number of years. Her research papers are mainly in the field of physical geography. She is the editor of the book *The Urban Fringe of Indian Cities* (2011), and she is co-author of the book *North-East India: Land, People and Economy* (2014).

LAND, PEOPLE AND ECONOMY OF PAKISTAN

A Geographic Perspective

K.R. Dikshit and Jutta Dikshit

Routledge
Taylor & Francis Group
LONDON AND NEW YORK

Designed cover image: @getty

First published 2025
by Routledge
4 Park Square, Milton Park, Abingdon, Oxon OX14 4RN

and by Routledge
605 Third Avenue, New York, NY 10158

Routledge is an imprint of the Taylor & Francis Group, an informa business

British Library Cataloguing-in-Publication Data
A catalogue record for this book is available from the British Library

ISBN: 978-1-032-83596-9 (hbk)
ISBN: 978-1-032-83595-2 (pbk)
ISBN: 978-1-003-51014-7 (ebk)

DOI: 10.4324/9781003510147

Typeset in Sabon
by Deanta Global Publishing Services, Chennai, India

CONTENTS

FIGURES

TABLES

PREFACE

The decision to write a book on Pakistan was neither sudden nor a thoughtless adventure, embarked on, by the authors. The idea of writing about the Geography of Pakistan was embedded in our idea of writing a book on South Asia, a region attracting global attention as a potential economic power and a strategically important region of the world. But, while beginning to draw an outline of the book, it dawned on us, that we shall be doing injustice to each of the countries (the three core countries, India, Pakistan and Bangladesh, the three land-locked mountainous countries, Nepal, Bhutan and Afghanistan and the two island countries, Sri Lanka and Maldives). We, then, considered it more appropriate to write about any one country, and we chose Pakistan.

We settled on the theme after a prolonged discussion with other geographer colleagues and some preliminary investigation of the state of knowledge of Pakistan, among our professional colleagues and students of geography. There appeared to us a sheer lack of interest and a pathetic ignorance of even the basic geographical facts of Pakistan. This further strengthened our resolve to produce a geography text on Pakistan.

Our choice of Pakistan, in preference to other countries of South Asia, was also guided by some other important considerations. First, Pakistan is one of the three most important countries – India, Pakistan and Bangladesh – of South Asia, with a common parentage, a large country, next only to India in terms of area and population. Pakistan has a very variegated relief ranging from the sea level of the Arabian Sea in the south to the perpetually snow-covered Himalayan heights. Formed either by diastrophic movements or sculpted by the Indus and its tributaries, these geomorphic features are as fascinating as they are challenging and have attracted geographers from across the world, especially those interested in mountain studies. We wanted to share the knowledge generated by those savants and as authors also liked to better our understanding of Pakistan and add to the existing knowledge, howsoever infinitely little it may be. The humanity in Pakistan, though following a common faith, Islam, which is also the state religion, presented a far greater diversity in space. With their historical antecedents, ethnic affiliation and linguistic and social heritage, each of the provinces in the country appeared different from others and presented a challenge in deciphering their characteristic and unique features.

We started on a very determined and an enthusiastic high note, but the initial euphoria gradually waned and we were faced with the reality of writing on a large complex region of which we had only a perfunctory knowledge.

We began collecting references and writings very assiduously from all possible sources and books, articles, research papers from periodicals and even travel accounts of visitors to Pakistan. Our daily reading of the Dawn, the most independent newspaper of Pakistan, kept us posted not only with the happenings in Pakistan but even with serious discussions in research institutions like PIDE (Pakistan Institute of Development Economics) reported in the journal. Informative articles written in the Dawn by scholars like Professor Pervez Hood Bhoy and journalists like the late Irfan Husain and many others were additional sources of our information and knowledge and gave us a cross-section of opinions about Pakistan. Besides, our contact with the academics, especially from the universities in India specialising in international studies, like the JNU (Jawahar Lal Nehru University, New Delhi) and Delhi University, and frequent discussions with them, equipped us with an insight into different aspects of a complex Pakistani society.

We worked at our own pace, and it took almost nine years to complete the book. During the course of our work, many events took place that were not to our liking. The souring of relations between the two neighbouring countries, India and Pakistan, disrupted our thinking as well as our work. But we moved on, avoiding political controversies that were, anyway, beyond the scope of this book.

During the course of writing and preparing the manuscript the authors realised that it would be too bulky to include all aspects of the geography of Pakistan in one volume, and hence we thought of bringing out two separate titles, one incorporating the General or Thematic Geography of Pakistan and the other discussing the regional aspects of the Geography of Pakistan.

We, as authors, have done our best to provide an authentic, unbiased and comprehensive account of the land, people and economy of Pakistan and hope that the readers find it worth the toil we have put in.

Jutta Dikshit and K.R. Dikshit

ACKNOWLEDGEMENTS

In writing this book, the authors have received inspiration, encouragement and help from several sources and individuals. It is not possible to acknowledge their help and thank them individually. But, we are indebted to certain individuals so deeply that we carry a moral obligation to at least thank them. Also, certain institutions and libraries have been of great help. The individuals who helped us with some not-easily-available books and maps are many. We acknowledge their help and thank them collectively.

To start with, we would like to acknowledge the moral support and constant encouragement received from Professor Gopal Krishan of Chandigarh, who showed great interest in our work and inspired us with his own example of commitment to work and the scholarly works that he has produced. But more importantly, he took a deep interest in many aspects of the geography of Pakistan, specifically of Punjab. He had a far better understanding of the society of Punjab than we could ever imagine, and enlightened us with his knowledge of Punjab and Punjabis, during our frequent discussions. We profited greatly from the discussions and his suggestions that ameliorated our understanding. Our heartfelt thanks and gratefulness to Professor Gopal Krishan! The late Professor Surya Kant of Chandigarh, another distinguished professor, and a samaritan in many respects, came to our rescue whenever we had difficulty in procuring some information. He often sent to us the photocopy of the paper we wanted, and in one case, he sent a photocopy of an entire book. We are deeply grateful to him and cannot thank him enough. Professor Charles C. Benninger, a Harvard-educated, distinguished architect of international repute and a family friend of ours, was of great help in securing some books not easily available, from the USA, which he offered us as a gift. We sincerely thank him for his help and generosity.

Our son, Richard Varuna Dikshit, who kept in touch with the progress of our work, procured for us several valuable maps of Pakistan from external sources besides some rare books. Thank you Richie for your interest in our work and help in many ways.

We thank sincerely Professor Sachchidanand Sinha of JNU for showing interest in our book and even recommending us to the publishers. We value his trust in us and the encouragement he gave. We also must thank Professor Nina Singh who showed interest in the publication of the book. Professor R.B. Bhagat of the International Centre of Population Studies, Bombay, was very helpful in securing access to the library of the Institute, from where we sourced the population data including the migration data, from India to Pakistan, following the 1947 partition of India. We stayed at their guest house and enjoyed the hospitality of the Institute for a couple of days. We are thankful to the Institute, its Director Professor Ram and Professor Bhagat in equal measure. Professor Debendra Nayak of JNU, Shilllong made regular enquiries about the progress of the book and encouraged with his curiosity about the book. We thank Debendra for his encouragement.

A couple of people who formed an informal team to help us finish the book deserve special mention and thanks. The first and the most important is the late Professor R.C. Tiwari, who has been some kind of an informal editor for us. He went through almost the entire script and made the job of the revision of the text easier. He will remain for us a partner in the progress of our work. Another person in this category is Dr Abhinav Kurkute, a lecturer in a local college, who helped us with the cartographic work. All the maps and diagrams in the book carry the imprint of Dr Kurkute`s skill in map-making. We acknowledge the help rendered by him.

Among the institutions, we would like to thank the Survey of India for authenticating the international boundary between India and Pakistan. Among the libraries, we are grateful to the library and librarians of Gokhale Institute of Politics and Economics, the Deccan College post-Graduate Research Institute and the library of the Savitribai Phule Pune University. We would also like to acknowledge the help we received from the Central Government libraries, i.e., Shastri Bhavan Library and the library of the Indian Council of World Affairs, New Delhi.

Finally, we thank Mr Amit Kumar, Senior Commissioning Editor and through him Routledge India for bringing out the book in a neat and organised manner.

In writing the book, the authors have received help from many other individuals and institutions. It is not possible to thank them individually. Therefore, apologetically though, we thank all the institutions and individuals who helped us in writing this book selectively.

K.R. Dikshit and Jutta Dikshit

1

INTRODUCTION

In writing the introduction to the present volume, the authors considered it appropriate to present it in three sections. The first section provides a glimpse of the earlier writings – largely books – on the geography of Pakistan. It has to be borne in mind that much that was written on India, before 1947, included the present area of Pakistan, as the latter was a part of British India. Thus, the present review includes the geographical writings on India that appeared before Pakistan came into existence in 1947, as well as the books that appeared during the post-1947 decades. The second section gives, in bare outlines, an introduction to the geography of Pakistan, and the third section acquaints the readers with the organisation of the book.

The reasons behind this unconventional organisation where a brief introduction of the books finds a place in the introduction of the book are twofold. First, the young age of the nation, barely 75 years, as Pakistan appeared on the map of the world as a distinct independent nation only in 1947. And, second, a turbulent existence, subjected to unforeseen political upheavals. The severest of these adverse events was the secession of the Eastern Pakistan province from its parent country, Pakistan, and the emergence of a separate independent nation, Bangladesh, resulting in the loss of one of its limbs. Hence, a brief geographical outline, including a glimpse of its history as a nation and the bare details of its territory, population, economy and some hints of its contemporary politico-economic situation appear in order.

1.1 A Glimpse of Earlier Writings

Readers of the present volume may wonder if there are no pre-existing works on the geography of Pakistan. As a newly emerged nation state, Pakistan

DOI: 10.4324/9781003510147-1

has been the focus of attention of not only the governments of other countries, especially the great powers but equally of intellectuals and academics who found an uncharted territory to explore and write about. A disproportionately large volume of literature exists on Pakistan. Besides a large number of research papers and articles, there exist a countless number of books varying in content and size. There are books on Pakistan's history, politics, economy, foreign affairs, defence and strategic affairs, not ignoring the India–Pakistan relations, besides a number of biographies of army generals and politicians. The most numerous are the books on political and economic crisis in Pakistan. Yet, books on geography of Pakistan, as expected, have been far and few between. Here, a few publications, largely books, that discussed the geography of Pakistan, or of the areas that later formed part a of Pakistan territory, are reviewed.

Reference of earliest Writings: Leaving aside the authors like al-Beruni (11th century) who made some passing reference to places or physical features of Hindustan in his famous treatise, *Kitab-ul-Hind*,[1] or the late 16th century work of Abul Fazl, *Ain-e-Akbari*, especially the third part, which deals with administration and carries an inventory like description of places, there was hardly any serious geographical writing till the British appeared on the scene.

One of the early writings by a British author was the book *The Ancient Geography of India*[2] authored by Alexander Cunningham, the first Director General of the Archaeological Survey of India. Not a geography book in the sense we understand, the book finds the corresponding names of places with the ancient place names as recorded during the Buddhist period or as mentioned in the travels of the Chinese Pilgrim, Hiuen Tsang. To this end, he divided India into five regions (Northern, Western, Central, Southern and Eastern). The Western and Northern regions, together, of Cunningham cover the entire area of Pakistan, extending even into Afghanistan. To give an example, he found the correspondence of ancient *Kaofu* with present Afghanistan, *Kapisine* with Kafiristan, Parashpura with Peshawar and Taki with Punjab plain; occasionally, these original names are so twisted that they appear artificial. The book, nonetheless, remained a perpetual source of information for archaeologists and is still referred widely by those interested in India's antiquity.

The actual geographical work, by the colonial rulers, started with surveying the entire country, fixing the boundaries of administrative divisions, right up to the village level, and initiating the exercise of gazetteer writing, in a hierarchical order, from the Imperial Gazetteers for the entire country to provincial and district series. These gazetteers were, for long, the main source of geographical information on any part of British India as well as the princely states. The first of the series, *the Imperial Gazetteer of India*[3] edited by Sir W.W. Hunter, appeared in 1881, followed by a revised and enlarged edition, containing 26 volumes, including the first four volumes titled *Descriptive, Historical, Economic and Administrative*, followed by an alphabetic compendium of description of the provinces, towns, districts and *taluka* places and other physical features, like mountains, rivers, etc., contained in the next 21 volumes, ending with *an Atlas of the Indian empire*, being the 26th volume of the series. Published by Oxford University Press, these Gazetteers ran into several editions (1908, 1909 and 1931) and remained a principal reliable source of information on many aspects of the geography of British India. The volumes carry reliable general and statistical information, besides a substantial geographical account of the regions (provinces and districts) required for administrative purposes.

Besides these sources of geographical information, the first commissioned book that discussed the geography of the areas presently forming a part of Pakistan, during the colonial period, was one by James Douie (1916) *The Punjab, North-west frontier Province and Kashmir*[4] written as a part of the "provincial geographies of India" series under the editorship of Sir T.H. Holland. A well-written book, with an inventory-like account, yet profusely illustrated with maps and photographs, containing almost all aspects of the geography of the provinces, it was, perhaps, the first book in English that threw some light on the geography of the areas, now forming a part of Pakistan. The chapter on Punjab is especially interesting. With an elaborate description of the land management and the irrigation system with canal network, it gives a clear picture of the Punjab of the early 20th century.

1.1.1 Books by Geographers

One of the earliest geography textbooks written by a geologist-geographer, Sir Dudley Stamp *Asia: A Regional and Economic Geography*[5] (Methuen, 1929) carries a section about India. Dedicated to his wife, "in the memory of the bullock cart days of Burma and Irrawaddy Nights" and running into 720 pages, the book devotes 200 pages to the Indian Empire. Note the title of the chapter "Indian Empire", not just India, as in the case of Turkey or Japan. Smacks of colonial arrogance! The book provides a simplified account of Punjab, North-West Frontiers and even Balochistan.

Mention must be made of Maneck B. Pithawala, Director of Post-graduate Research in Geography and Geology, Dean of the science Faculty, University of Sindh, who wrote several papers and some short books, over a period two decades, starting with *A Geographical Analysis of lower Indus Basin* (1937) and ending with physical and economic geography of Sind, the lower Indus basin (1959), he largely wrote on Sind and on the river Indus. A list of some of his publications is appended at the end of the chapter.

The first ever comprehensive geographic text, in fact, too comprehensive, on India and Pakistan, was written by O.H.K. Spate (Methuen, 1954).

Titled, *India and Pakistan: A General and Regional Geography*[6] with a chapter on Ceylon by B.H. Farmer, the book runs into 1,000 pages. Besides being a comprehensive geography of India and Pakistan, the text is a masterpiece of lucid prose writing and geographical interpretation, seldom found in other geography texts. Spate`s book enjoyed a long lease of popularity, went into several reprints, and in addition, a revised edition (1972) with A.T.A. Learmonth, as a co-author. Pakistan claims more than its share in the contents of the book. Though a bit outdated, it is still a good read and known for its lucid style, remains unsurpassed as a regional text. Almost, at the same time as the publication of Spate`s India and Pakistan, or a tad earlier, another thematic book – Kingsley Davis: *The Population of India and Pakistan*[7] appeared on the scene. Being more of a sociological interpretation of the region's demography, the book was the first of its kind in the population study of the sub-continent.

1.1.2 Geographical Literature on Pakistan after 1947 – the Year of Pakistan's Birth

It must be noted that not long after Pakistan appeared as a nation state (1947) on the map of the world, Pakistani geographers produced several geographic texts on their country. The first in the line was Nafis Ahmad who justifying the creation of Pakistan wrote *The Basis of Pakistan.*[8] Soon, he followed it with the publication of another book *Economic Geography of East Pakistan.*[9] It has to be understood that Professor Nafis Ahmad was a Professor at Dacca University and hence concentrated only on East Pakistan, now Bangladesh. On the Western side, Lahore-based Professor Kazi Said-Uddin Ahmad, Professor Emeritus, University of the Punjab, wrote *A Geography of Pakistan.*[10] Originally published by the University of California (1964), the book went into a second editions and was published by the Oxford University Press (Karachi–Oxford University Press, 1968, 268p.). It has to be noted that the book running into 268 pages dealt with the Pakistan of 1968, having a much larger area that included the present-day Pakistan and Bangladesh. Despite a generalised approach, the book became very popular in the absence of any moderate size textbook about Pakistan. Following the popularity of this book with students, another similar book, *A Geography of Pakistan*[11] was written by K.U. Kureshi. The book went into a revised and enlarged edition in 1988 by National Book Service, 1988.[12] The book, essentially meant as a text for students of Pakistan studies, was widely consulted by even mainstream geography students.

Another book that did not attract much attention was *The Political Geography of Pakistan* written by Pakistani–Canadian A. Tayyeb (1966). Titled *Pakistan: A Political Geography*,[13] it is one of the rare scholarly texts that did not receive the attention it deserved. Part of the reason is that it is loaded with an all too familiar theoretical discussion of the evolution of political geography. The book still remains a significant contribution to the political geography of Pakistan of its early historical phase.

These books written by Pakistani authors had for their company another book authored by B.L.C. Johnson. Running into 214 pages, it gave a very sketchy account of Pakistan and its geography. Titled just *Pakistan*,[14] the book was published in 1979, as one of the Heinemann educational books series, London. It may be mentioned that Professor Johnson wrote similar books on India and Bangladesh. The book was finally placed online by Cambridge University Press in 2011.

The most popular and widely read author on the geography of Pakistan is Fazle Karim Khan who first wrote *A Geography of Pakistan: Environment, People and Economy.*[15] A revised and updated edition of the book with a slightly modified title *Pakistan: Geography, Economy and People*[16] appeared in 2015. An O` level text, endorsed by the University of Cambridge International Examinations and profusely illustrated with coloured maps, diagrams and photographs with a number of summary boxes and questions at the end of each chapter, the book is presently a standard geography text read and consulted by most students of geography in Pakistan. Incidentally, the book completely ignores the regions and regional description and interpretation, as if Pakistan is a homogeneous mass of land. The inclusion of even simplified, regional descriptions of the contrasting regions, from the Indus Delta to the Himalayan heights, would have certainly enhanced the value of the book, as a regional geography text.

On the background of what existed earlier, the present authors decided to produce a complete geography text on Pakistan with updated figures and accounts. The present volume *Land, People and Economy of Pakistan: A Geographic Perspective* carries a thematic approach. An outline of the book will be found in the third part of the introduction.

1.2 Geography of Pakistan

Pakistan (official name – Islamic Republic of Pakistan) made its appearance on the political map of the world on August 14, 1947. This followed the independence of India from the British rule; the division of the country into two independent territorial units, India and Pakistan, and the transfer of power from the British Crown to a newly born independent Pakistan. On August 14, 1947, Lord Louis Mountbatten, the last Vice-Roy of India, representing the British Crown, handed over a sovereign independent Pakistan to Quaid-i-Azam Muhammad Ali Jinnah, who was sworn in as the first Governor General of Pakistan at a ceremonial function at Karachi. And, Pakistan, as a nation state, came into existence.

Pakistan, with an area of 796,096 km^2 and a population of 241.49 million (2023)[17] is the second largest country in South Asia, in terms of area as well as population. It is the fifth most populous country in the world, after China, India, USA, and Indonesia, in that order and is also the second most populous Islamic country, next only to Indonesia. Ninety-eight percent of its population follows Islam as its faith. A small minority, roughly 2 percent, consists of a handful of Hindus, some Christians and an insignificant number of followers of other faiths like Zoroastrianism, Buddhism and Ahmadis.

Today's Pakistan is not what it was at its birth in 1947. Pakistan, then, consisted of two non-contiguous territorial units: Western Pakistan consisting of four provincial units, Punjab, Sindh, Balochistan and North-West Frontier Province and East Pakistan, consisting of a single provincial unit, known as East Pakistan. In 1971, the country was split into two independent nations, Pakistan as it exists today, and Bangladesh, the erstwhile provincial unit of East Pakistan. East Pakistan, the present Bangladesh, separated from Pakistan, after remaining a part of the parent country, for almost a quarter of a century (August 1947 to December 1971).

1.2.1 The Separation of East Pakistan and the Emergence of a New State – Bangladesh

It is difficult to elaborate on the complexity of the geo-political situation that finally ended in the separation of East Pakistan and the creation of an independent sovereign Bangladesh. One may, however, point to some aspects of the geographical reality that led to the division of Pakistan. First, the two units were 1,600 km apart, separated by an intervening stretch of the Indian territory. Administratively, this was not an efficient spatial organisation. But, more importantly, it was the specific regional interest of East Pakistan that led to its separation from the parent state.[18,19] East Pakistan was distinctly different from its western counterpart, in ethnicity, language and culture, nurtured over the millennia in response to the milieu of the rain-soaked, tropical deltaic terrain. The people in East Pakistan had a passionate love for Bengali, their mother tongue, a far cry from the Punjabi and Urdu-speaking West Pakistan. The stress of the artificial co-existence of the two peoples together, different in all aspects, held together only by a common belief system and religion, was too much to bear for long. Added to this was the perception in East Pakistan of being exploited and treated as a colony. There could be many reasons that triggered the separatist movement in East Pakistan, but the perception of being treated as an underdog was always there that erupted in protests, revolts, violence and finally separation from West Pakistan and the emergence of a new nation state, Bangladesh, in December 1971.

TABLE 1.1 Area and Population of Pakistan before and after 1971

Territorial Units	*(1951)*		*Territorial Units*	*(2017)*	
	Area in km²	*Population*		*Area in km²*	*Population*
Pakistan including East Pakistan	944,632	75,842,000	Pakistan	796,096	207,654,626
East Bengal	140,171	42,063,000	Bangladesh	147,565	172,594,319

(Bangladesh after 1971).
Source: Census of Pakistan, 1951 and 2017.

Thus, while Pakistan had a much larger area (943,643 km²) at its birth in 1947, it was reduced to 84 percent of its original area and almost half of its population after the loss of East Pakistan territory. The changes in the political regimes and administrative systems that occurred during the intervening period (1947–2017) are too cumbersome to recount, and, in any case, beyond the scope of this book.

A quick look at the area and population of Pakistan (including East Pakistan) as recorded in 1951 Census, and of contemporary Pakistan, with the provinces, as recorded in 2017, are given below (Table 1.1).

1.2.2 Administrative Units of Contemporary Pakistan

Officially known as the "Islamic Republic of Pakistan", Pakistan is a federal Republic, with four provinces, a Federally Administered Tribal Area (FATA) and a Federal Capital Territory, Islamabad. The area and population of the constituent units are given below (Table 1.2). The FATA is no longer a separate administrative unit and was merged and included in Khyber Pakhtunkhwa, through a constitutional amendment, in 2018.

1.2.3 Some Basic Geographical Features

Locationally, Pakistan is situated between India on its East and Afghanistan and Iran on its West. It shares its northern boundary with the narrow Wakhan corridor of Afghanistan and the Xinjiang province of China. Bound between two internationally recognised boundary lines, the Durand line on the west, separating it from Afghanistan and the Radcliffe line on the east, making borders with India, Pakistan occupies a west-east distance of approximately 1,200 km, from 60°–56'E longitude, the extreme western point of Balochistan, to 73° East longitude along 29–30 North Latitude, and north-south stretch of roughly 1,500 km, from the Rann of Kutch, at approximately 24° North latitude, to the northernmost point of Khyber–Pakhtunkhwa at 37° N. Latitude. The country has an international border

TABLE 1.2 Administrative Units, Area and Population of Pakistan

Administrative Units	*Area in km²*	*Population*	*Share of Area (in percentage)*	*Share of Population (in %)*
Pakistan	796,096	207,684,626	100.00	100.00
Balochistan	347,190	12,335,129	43.60	5.93
Khyber Pakhtunkhwa	74,521	30,508,920	9.36	14.69
Punjab	205,345	109,889,655	25.79	52.96
Sindh	140,914	47,854,510	17.70	23.94
FATA	27,220	4,993,044	3.42	2.40
FCA (Islamabad)	906	2,003,368	0.11	0.96

Source: Bureau of Statistics, Census Organization, Govt. of Pakistan, Census 2017, Table 1.2 Area, Population by Sex, Sex Ratio, Population Density, Urban Proportion, Household Size and Annual Growth Rate

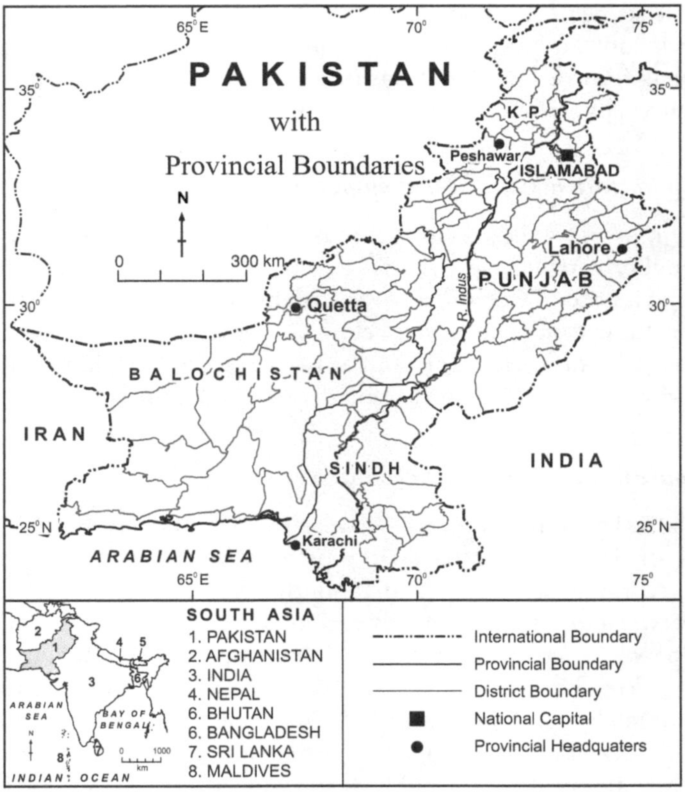

FIGURE 1.1 Pakistan and Its Provinces.

Source: survey of Pakistan 2020, Administrative Map of Pakistan, Survey of Pakistan Press, Rawalpindi.

of over 6,700 km. This is shared with India, its eastern neighbour (>2,900 km), Afghanistan (>2,400 km), China (over 500 km) and Iran (more than 900 km). Pakistan has a coastline of over a thousand kilometres (1,046 km), along the Arabian Sea.

Almost, the entire territory of Pakistan falls in the catchment of the Indus, the river that has a transnational character. Its tributaries are an important source of water for Pakistan, as much as they are for India and Afghanistan. For Pakistan, Indus is its lifeblood that sustains its economic and cultural life. The mountainous North, the ranges of the Karakoram and the Hindu Kush, with a large number of high peaks, perpetually snow-covered, and a series of glaciers at higher altitudes, are equally important to Pakistan and serve as a huge storehouse of water, released during summers when the country needs it most. The glacial and snow-covered mountains as a source of water are as important for Pakistan, as the monsoon rains are for India.

With just 25 percent of the country's area under active cropping (Net Sown Area) every year and a burgeoning population that has been growing at a mean annual growth of 2.5 percent, and a fertility rate of 3 percent, Pakistan faces an imbalance between its population and agricultural resources.

1.2.4 Governance and the Government

Pakistan is a Parliamentary Democracy, governed by a constitution which provides for a bi-cameral legislature, a government headed by the Prime minister, a President as head of the State, and a Judiciary headed by an apex body, the Supreme Court of Pakistan. The Parliament consists of a Lower House, the National Assembly of Pakistan and an Upper House, the Senate. The National Assembly has a total strength of 336 members of which 266 members are elected directly and the remaining 70 members represent the two reserved constituencies, 60 for women and 10 for the non-Muslim minority community. The reservation for women in an Islamic republic is, indeed, very progressive and reflects the far-sightedness and democratic values of the makers of Pakistan's Constitution.

1.3 Organisation of the Book

The present text, *Land, People and Economy of Pakistan: A Geographic Perspective*, divided into three parts, approaches the subject thematically. The first part, organised in chapters, discusses different aspects of Pakistan's physical environment, reflected in its bio-climatic characteristics, organisation of the drainage network and relief characteristics. The second part covers the people, their population and socio-cultural organisation, including the ethnolinguistic groups and their spatial distribution. In the third part, there is a discussion of Pakistan's natural resources and their utilisation, seen in the agricultural organisation in the country, besides mineral resources and industries.

The penultimate chapter discusses the expansion of the transport network and the changing volume and direction of trade. The last chapter gives briefly the contemporary economic crises in which Pakistan finds itself.

Maps and diagrams appear in the book as illustrations to impart clarity to the written text. Maps have also been drawn to show the physical characteristics of a region or to display, in outline, some specific facts about a region. The maps in the present book carry an official version of the international boundary between India and Pakistan, authenticated by the Survey of India. This may not totally agree with the official version of the international boundary in Pakistan. But, this is what results from a disagreement about the areas or the boundaries between two neighbouring countries. Each requires the stamp of approval from its own authorised agency and hence the difference.

1.3.1 Data Procurement and Analysis

It may be emphasised that much of the data on all aspects of Pakistan, unless classified, is available online. Every ministry has its publications which are easily accessible, may it be the Annual Reports, the Five Year Plans, Census Reports, the Annual Economic Surveys and even the Statistical Abstracts – all are available online. Even the provincial reports appear online. Part of the data is also accessed from the publications of The United Nations, World Bank or other international sources. The authors have taken care to acknowledge the source wherever necessary.

1.3.2 Reference Material

There is copious literature available on Pakistan, more on politics, defence, strategic and international relations than on mundane themes like transport or rural settlements. The authors have made sincere efforts to consult as many publications as possible from books and periodicals, more from periodicals than books, and referred to a variety of views and perceptions. It may be mentioned, with appreciation, that most of the national-level journals of Pakistan permit free access to their journals. This is very helpful to researchers and the present authors have benefitted greatly by consulting relevant research material contained in these journals.

The maps, an important tool of geographers, have been obtained by the present authors from a variety of sources. Besides the current international sheets which are easily available, the authors have used very extensively the old editions of the quarter-inch sheets of the Survey of India (1:250,000) for location and physical features. Besides, very accurate maps on a scale of 1:1,300,000, and more commonly on a 1:2,000,000 scale, are available from international sources and markets. Besides these, Google provides very good satellite images of any part of the country, with multiple layers that can be intelligently consulted. The Survey of Pakistan has published a series of maps which are not permitted to be downloaded but can be consulted. This is especially so with administrative maps. Agencies like WAPDA (Water and Power Development Authority) have their own maps and flow charts, as much as the Irrigation Department of Pakistan has its own map of the network of canals in the country.

In writing the book, the authors have strived to remain non-judgemental, so far as the governance of the country and the policy matters of the state are concerned. Yet, it is likely that one may find some observations made by the authors, presenting a new perspective based on their own perception, at odds with a general understanding. To that extent, we expect the indulgence of the readers.

The authors, as can be judged from the text, have put in honest efforts over several years with a passion that combined hard work and concern for authenticity. They would feel rewarded if the book is read by a large number of readers especially geographers, for whom the book is primarily intended.

Notes

1 Al Biruni (1030) *Kitab-ul-Hind,* translated by Edward C. Sachau as 'India-An Account of the Religion, Philosophy, Literature, Geography, Chronology, Astronomy, Customs, Laws and Astronomy of India (English edition with notes and indices) in two volumes, Kegan Paul, Trench, Trubner & Co., London, 1910.
2 Cunningham, Alexander (1870) *The Ancient Geography of India: The Buddhist Period,* (London-1871) Reprinted Indological Book House, Varanasi
3 Hunter, (Sir) W. W. (1881) *Imperial Gazetteers of India, in Multiple Series,* Government of India, Oxford University Press, Oxford.
4 Douie, James (1916) *The Punjab, North-West Frontier Province and Kashmir,* Cambridge University Press, Cambridge, 373 p., classified as a rare book
5 Stamp Sir Dudley (1929) *Asia: An Economic and Regional Geography,* Methuen Publishing, London.
6 Spate, O. H. K. (1954) *India and Pakistan, A General and Regional Geography,* Methuen Publishing, London.
7 Davis, Kingsley (1951) *The Population of India and Pakistan,* Princeton University Press, Princeton.
8 Ahmad, Nafis (1947) *The Basis of Pakistan,* Thacker Spink & Co., Calcutta.
9 Ahmad, Nafis (1958) *Economic Geography of East Pakistan,* Oxford University Press, London and Karachi.
10 Kazi, Said-Uddin Ahmad (1964) *A Geography of Pakista*n. Originally published from the University of California, the book went into a second edition, and was published by (Oxford University Press, Karachi, 1968).
11 Kureshi, K. U. (1977) *A Geography of Pakistan,* Oxford University Press, London and Karachi.

12 Kureshi, K. U. (1988) *A Geography of Pakistan* (Revised edition), National Book Service, Karachi.
13 Tayyeb, A. (1966) *Pakistan: A Political Geography,* Oxford University Press, Karachi.
14 Johnson, B.L. C (1979) *Pakistan*, Heinemann Educational Books Series, 214 p. (The book was finally placed on-line by Cambridge University Press, London in 2011).
15 Khan, Fazle Karim (1991) *A Geography of Pakistan: Environment, People and Economy*, Oxford University Press, Karachi.
16 Khan, Fazle Karim (2015) *Pakistan: Geography, Economy and People*, Oxford University Press, Karachi.
17 Pakistan Bureau of Statistics (2023) *Digital Census*, Government of Pakistan, Islamabad.
18 Ali, Tariq (1983) *Can Pakistan Survive? The Death of a State*, Penguin Books Ltd., London.
19 Cohen, S. P. (2011) *The Future of Pakistan,* Oxford University Press, Karachi.

PART I

Physical Geography of Pakistan

2

PHYSICAL SETTING AND RELIEF FEATURES OF PAKISTAN

Preamble: Physical features like mountains, hills, plateaus, plains and rivers came into existence much before the contemporary national boundaries were drawn; and as such, they don't respect these national or international boundaries. Hence, not infrequently, these major landforms straddle over several countries and cut across national boundaries. The Andes, the Alps and the Himalayas extend across several countries. In the case of Pakistan, the mountains, like the Himalayas, the Karakoram and the Hindu Kush, do not abruptly end at the international borders, and they could be described as much a part of the physical features of Pakistan as of some neighbouring countries.[1] Some of the areas characterised by significant physical features, especially under the shadow of the Himalayas, are under a "de facto" control of Pakistan while some others are under the control and administration of India. Therefore, without any prejudice to the claims and counterclaims of contending parties about the extension of their rule or sovereignty over a specific area of Kashmir, some of the prominent physical features, lying in the area administered by Pakistan, are discussed in this chapter.

2.1 Physical Build of Pakistan

The physical build and relief features of Pakistan, as indeed of any other region of the world, are the product of its lithology, structural base and tectonic movements on the one hand, and the work of sub-aerial processes of weathering, erosion and sedimentation on the other (the well-known Davisian dictum of structure, process and stage). The relief features of Pakistan are discussed under two sections: an introduction to geology,

DOI: 10.4324/9781003510147-3

structure and tectonic movements that form the base of the existing relief; and the relief features of Pakistan.

2.1.1 Geology, Structure and Tectonic Movements of Pakistan[2]

Much of South Asia, especially the Indian subcontinent, comprising India, Pakistan, Nepal, Bhutan and Bangladesh, has a unified geological history, as the evolution of its geological structure is linked with the history and movement of the Indian plate, which extends from the Chaman Fault, marking the western extremity on Pakistan–Afghanistan border in the West, to Sagaing Fault in the East, separating the Indian plate from the Myanmar Plateau. Between these two extremities, the movement of the Indian plate that carries the territories of India and Pakistan has impacted the structure and geology of these countries, depending on their disposition vis-à-vis the north-moving Indian plate.

2.1.2 The Movement of the Indian Plate and the Himalayan Orogeny

Once a constituent unit of the large Gondwanaland, which consisted of Latin America, Africa, Madagascar, Australia and India, the Indian plate underwent many prolonged episodes of sedimentation, tectonic movements and volcanism from the Proterozoic (about 2000 million years ago) till mid-Mesozoic around 130 million years ago, when the Indian plate drifted apart and started moving northward and came into contact with the Asian plate. As the Indian plate moved northward at a fast rate of 15 to 20 cm/year, the seas enveloping the continent from the north, east and west continued receiving an enormous amount of sedimentation. The sea lying between the Indian and the Eurasian plate, widely known as the *Tethys*, was getting narrowed, and accumulated a huge amount of sediments. As the Indian plate came into contact with the Eurasian plate around 55 million years ago, during the Eocene period, the collision produced a variety of tectonic features in the frontal zone. The continued movement of the Indian plate northward and its subduction under the Asian plate caused excessive compression, leading to the development of a series of thrust faults, throwing huge masses of not only meta-sedimentary and meta-sediments but even highly metamorphosed crystalline rocks. This was the birth of the great Himalayas stretching for over 2500 km from east to west and occupying a width of 400 km.

The movement of the plate northward also led to the closing of the Tethys, marked by what is commonly known as the Indus–Tsangpo Suture Zone (I-TS). Occasionally, the subducting Indian plate penetrated deep, even below the crust to reach the mantle, causing excessive temperature and melting of rocks. The eruption of this material on the surface, in a fairly large area, created volcanic island arcs, or simply island arcs. Such volcanoes throw up fragments of basic rocks from the ocean crust, such as ophiolite or ophiolite melange. In the case of Pakistan, the northern most part of the region carries the well-known Kohistan island arc, covering the Gilgit and Chilas region. On the eastern side lies the intrusive Karakoram batholith, composed of granite and granitoid rocks.

The Himalayas are confined to the south of the Indus–Tsangpo Suture (ITS). The territory beyond the ITS to the north is known as the Trans-Himalayan Zone.

2.1.3 Simplified Geology and Structure of Pakistan

In the present text, a more simplified account of the geology and structure of Pakistan, based on the Geological Map of Pakistan prepared by the Geological Survey of Pakistan, is adopted for the description of geology and structure of Pakistan. According to this map, as mentioned above, there are four geological divisions of Pakistan (Figure 2.1).

1. **The Northern Montane Area** – As the name suggests, it is the northern mountainous part of the country.
2. **The Indus Basin** – Basin, here, implies the basin of sedimentation as opposed to a drainage basin. The Indus basin consists of a number of

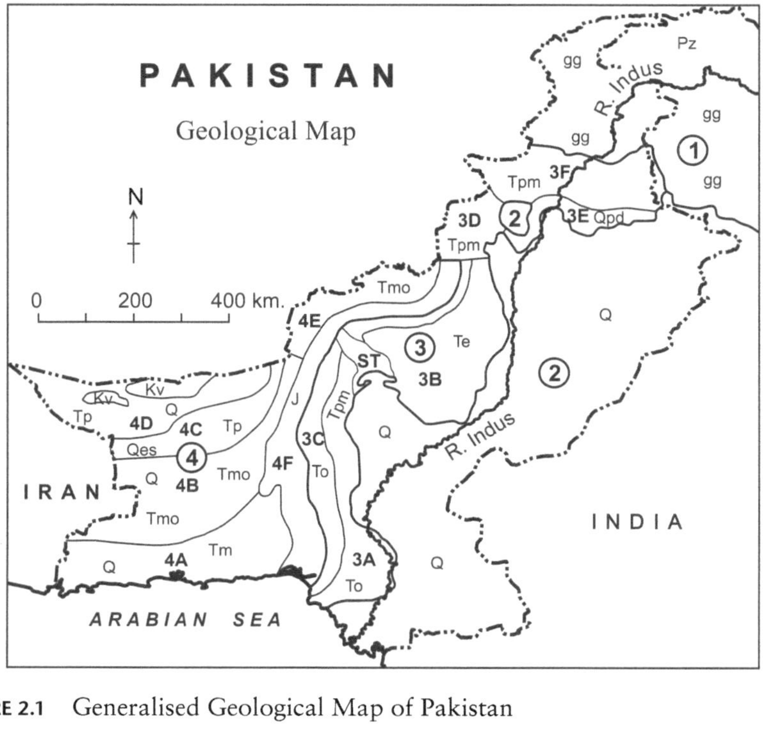

FIGURE 2.1 Generalised Geological Map of Pakistan

provinces. These are a. Kirthar province, b. Sulaiman province, c. Axial belt, d. Waziristan, e. Trans Indus and Salt Range and f. Kohat–Potwar province.

3. **Indus Plain** – This consists of Quaternary deposits and recent and sub-recent alluvium occupying the plains of Punjab and Sindh.
4. **Balochistan Basin** – This consists of a. South Makran belt, b. North Makran Belt, c. Ras Koh belt, d. Chagai belt e. Zhob belt and f. Axial belt.

2.1.4 The Northern Montane Area

This mountainous zone lies beyond 34°30' North Latitude. Starting from the north of Islamabad, this mountainous division has the following geological units. From South to North, these are a. The Siwaliks, b. The Lesser Himalayas, c. The Greater Himalayas, d. The Tethys Himalayas, e. The Indus–Tsangpo narrow zone and f. Trans-Himalayan zone. It may be mentioned that not all the units mentioned above occur in all the longitudinal cross-sections of the Himalayas. All these units are separated by the low angle reverse faults, commonly known as thrusts, of different magnitude and intensity.

2.1.4.1 The Siwalks in Pakistan

The Siwaliks are the foothills of the Himalayas, usually composed of sandstone and conglomerates with abundant fossils of mammals like elephants, marked by the Himalayan Frontal Fault on the south and the main boundary fault on the North that separates it from the Lesser Himalayas. The Siwaliks seldom rise above 1,000 m beyond which the Lesser Himalayas start. Developed during the last phase of the Himalayan orogeny, these are composed of the eroded sediments of the Himalayas and uplifted about a million years ago.

2.1.4.2 The Lesser Himalayas

The Himalayan zone between 1,000 and 3,000 m is known as the Lesser Himalayas. This is the zone where most of the hill stations are located. The Lesser Himalayas are largely composed of sedimentary and metamorphic rocks like shale, quartzite and marble with occasional patches of Cambrian and pre-Cambrian rocks. The Lesser Himalayas in Pakistan are best represented by Murree Hills, bound on the north by the Main Central Thrust (MCT). It is in this zone that one of the Himalayan Syntaxis, the Hazara–Kashmir Syntaxis occurs. In this case, the entire mass of the Lesser Himalayas is dragged north, making an acute bend in the outline of the

Himalayas. This is usually explained by a rotational shift of the Indian plate in an anti-clock wise direction.

2.1.4.3 The Greater Himalayas

This is the highest part of the Himalayas, ranging in height from 3,000 m to 8,000 m. The Greater Himalayas are separated from the Lesser Himalayas by the Main Central Thrust, on the South, and on the northern side, it is separated from the Trans Himalayan zone by a Trans Himalayan Fault (THF). This is the highest zone of Pakistan, composed of igneous and metamorphic rocks, chiefly granite, gneiss and schist. The greater Himalayas have a non-continuous existence. In Pakistan, these are represented by a mountain complex of Changra Mt (6830 m), Nanga Parbat (8126 m) and some other ranges and peaks.

2.1.4.4 A Brief Description of Nanga Parbat

Nanga Parbat (35°14'N and 74°35' E) is the westernmost limit of the Greater Himalayan chain and is the ninth highest peak in the world and home to the famous Rakhiot Glacier. The mountain has very steep slopes like a chiselled pyramid and rises 7,000 m over a distance of 25 km from the Indus valley. It is a great attraction, as well as a challenge, to the mountaineers. The mountain has claimed the life of several climbers. It was successfully scaled for the first time in 1953 by an Austrian climber Hermann Buhl. This first ascent by an Austrian has gone into folklores and is the subject of mountaineering history.

It may be noted that Nanga Parbat has attracted not only Austrian and German climbers but even scientists studying mountains and high-altitude areas. Among the scientists to study Nanga Parbat and the life and society in the shadow of the Nanga Parbat in recent years are Marcus Nuesser and Susane Schmidt of South Asian Institute, Heidelberg, who studied socio-hydrological interactions in the North-western Himalayas. Though they concentrated their study on socio-hydrological interaction, they have left no one in doubt that the glaciers of Nanga Parbat are receding. Talking about Chungphare Glacier, they observe – A comparison of the sketch map by Schlagintweit and the topographic map produced by the German Himalaya Expedition shows that the Chungphare glacier retreated by about 600 m between 1856 and 1934. Since then, it has further retreated by about 600 m between 1934 and 2013, as can be derived juxtaposing the historical topographical and contemporary satellite imagery ... In 2013, the Chungphare glacier was 11.7 km long, covered a total area of 24.6 km^2, and terminated at an altitude of 2,920 m, as derived from remote sensing data.[3] This is alarming as the glaciers with their huge storage of water are the lifeline of Indus.

Tethys Himalaya: North of the Greater Himalaya, but lower than the main Himalaya, is the Tethys Himalaya which carries the compacted and metamorphosed sediments of the Tethys sea starting from the Cambrian to the Eocene period. The Tethys Himalaya also contains igneous rocks, largely granites and granitoid rocks which when exposed in large areas, make dome-shaped structures or batholiths. The Tethys Himalaya could be exemplified with the Deosai mountains, appearing like a high altitude plateau, south-east of Nanga Parbat.

The Indus–Tsangpo Suture Zone: This is the narrow zone, marked, in part, by Indus in the western part, and represents the northernmost zone of contact between the Indian and the Asian plate. This is also the linear area where the Tethys Sea finally disappeared. The zone has deep sea sediments and even ocean floor volcanic rocks. The river Indus on the west and the Tsangpo on the east, not only represent the contact zone of the Indian and the Asian plate, they also mark the northern limit of the Himalayas, beyond which to the north lies the Trans Himalayan zone.

The High Trans Himalayan Zone: This is the most significant part of the northern mountainous zone of Pakistan, lying beyond the Himalayas, and extending even beyond the Pakistan border to the north. The region carries considerable high level terrain, many ranges and the mountain peaks. Extending from the Gilgit-Rakaposhi mountains on the west to Karakoram on the east, this is a very extensive zone with glaciated terrain and a large number of streams joining Indus from the north. The region has many peaks with more than 7,000 m in height. Geologically, the region is divided into two parts, the western part occupied by Kohistan Arc and the Eastern part usually termed as Karakoram batholith.

The Kohistan Arc is the volcanic terrain in the western part of North Pakistan where the Indian plate subducting under the Asian plate has gone deep enough to reach the earth's mantle, causing melting of the rocks and the ejection of these partly molten material on to the surface through volcanic fissures. solidifying as volcanic rocks. The most commonly occurring rocks in all the Island Arcs are the ophiolite or ophiolite melange. The Kohistan Arc covers the Gilgit region and further west including Chitral.

1. **Northern Montane Area**
   gg – Igneous, metamorphic and mafic intrusive rocks
   Pz – Paleozoic rocks mostly metamorphic and volcanic rocks
2. **Indus Plain**
   Q – Recent and subrecent alluvium; stream deposits, sand deposits, deltaic and tidal deposits
3. **Indus Basin**
   3A Kirthar Province
      To – Oligocene Nari formation mostly sandstone and limestone
      Tpm – Pliocene and Miocene shale, sandstone and conglomerate

3B – Sulaiman Province
  Te – Eocene limestone and shale
  To – Oligocene sandstone and limestone
3C – Axial Belt
  To – Oligocene sandstone and limestone
3D – Waziristan
  Tpm – Pliocene and Miocene siltstone, shale sandstone and limestone
3E – Trans-Indus and Salt Ranges
  Qpd – Piedmont deposits
3F Kohat Potwar Province
  Tpm – Pliocene and Miocene siltstone, shale sandstone and limestone

4. **Balochistan Basin**
   4A – South Makaran Belt
     Q – Recent and subrecent alluvium and extrusive mud
     Tm – Miocene siltstone, shale sandstone and limestone
   4B – North Makaran Belt
     Q – Recent and subrecent alluvium
     Tmo – Miocene and Oligocene shale, sandstone and limestone
   4C – Ras Koh Belt
     Qes – Eolian sand
     Tp – Paleocene shale, sandstone and conglomerate
   4D – Chagai Belt
     Q – Recent and subrecent alluvium and Eolian sand
     Tp – Paleocene shale, sandstone and conglomerate
     Kv – Cretaceous limestone, shale, sandstone, conglomerate and volcanic rocks
   4E – Zhob Belt
     Tmo – Miocene and Oligocene shale, sandstone and limestone

2.1.5 The Indus Basin

It must be mentioned that the Indus Basin here refers to a basin, an arm of the sea, where sedimentation has taken place in the previous geological periods. As described by the geological survey of Pakistan, the Indus sedimentation basin covers the Sulaiman–Kirthar zone, the axial zone of these mountains, Balochistan, South Waziristan and Kohat–Potwar region. The sedimentary beds of Tertiary origin, dating back to Palaeocene to Pliocene and even Quaternary, with varying thickness ranging from a few hundred metres to over 2,000 m have been subjected to folding and thrusts, during the period of Himalayan orogeny, giving rise in the process to Sulaiman–Kirthar mountains. The western margin of this unit is riddled with many faults, probably associated with the great Chaman Fault.

2.1.5.1 The Kirthat–Sulaiman Range including South Waziristan

Structurally these geological units are alike. They are all marked by tightly folded beds of rocks, ranging from Jurassic to Quaternary, with a dominance of sedimentary rocks from almost all periods of Cenozoic, ranging from Palaeocene to Pliocene epoch. This 800 km long mountain chain running from the Arabian Sea coast to South Waziristan in Khyber Pakhtunkhwa is composed of sedimentary rocks formed during different periods of the entire Cenozoic Era. Most of the rocks are sandstone, limestone and occasionally shale. The axial linear zone of this region is formed of Jurassic limestone and embedded shales.

2.1.6 The Indus Plain

As the name suggests, the Indus Plain is formed with the sediments of the Quaternary Era. The plain covering much of Punjab and Sindh is covered with recent alluvium as well as the terraces accruing at different heights. The alluvium is not uniform and varies in texture from clayey loam to sandy loan depending in the disposition of the land. The finest alluvium occurs in the flood plains of the rivers.

2.1.7 Balochistan Basin

Much of Balochistan, like the Indus Basin, is composed of sedimentary rocks from different epochs of Tertiary Era. The rocks are usually sandstone, limestone and shale. In the northern part of Balochistan, in Chagai hills and Ras koh belt, there are intrusive mafic rocks containing dark-coloured minerals like pyroxene and mica. These rocks contain 40 to 45 percent silica. Mafic intrusive rocks in the form of dykes and sills are found in Chagai Hills. Intrusive rocks also occur in the southern axial belt like the Hindubagh area.

2.2 Relief Features of Pakistan

The physical build of a country and the associated relief features refer to all such geomorphic expressions that were created by nature, either through sub-surface diastrophic forces or through sub-aerial processes of weathering, erosion or deposition. Most of these surface features were sculpted by nature long before man appeared on this planet.

The Pakistani territory has a varied topography, with relief ranging from the sea level to some of the highest mountain ranges of the world; from the Arabian Sea coast to the second highest mountain range of the world, the Mount Godwin Austen (K_2) reaching a height of 8,611 m. A generally accepted view in Pakistan is that about three-fourths of Pakistan's area, including Pakistan-ruled Kashmir and Gilgit--Baltistan, is occupied either by mountainous terrains or desert-like plateaus, and only one-fourth is formed of sedimentary basins of the Indus and its tributaries, appearing as plains in Punjab and Sindh.

The broad category of mountainous relief includes the high Himalayan ranges including the Nanga Parbat, the northern ranges of Chitral and Kohistan, bordering the Wakhan Corridor of Afghanistan, the frontier hills of Khyber Pakhtunkhwa and the outer ranges of the Himalayas and the sub-Himalayan hills. In addition, the northsouth aligned Sulaiman and the Kirthar Ranges separating Balochistan from Punjab and Sindh, and the Makran Ranges, overlooking the Arabian Sea coast, are the other mountainous features of Pakistan. Besides, there are localised hills and ranges like the Salt Range of Punjab or the Brahui and Harbol Hills in Balochistan. The plateau region includes the extensive mid-altitude (roughly 2,000 m ASL) plateau of North Balochistan (Figure 2.2).

For elaborating the relief features, the following divisions and sub-divisions appear appropriate, though any region can be further sub-divided into many sub-regions and yet smaller units. The relief features of Pakistan are

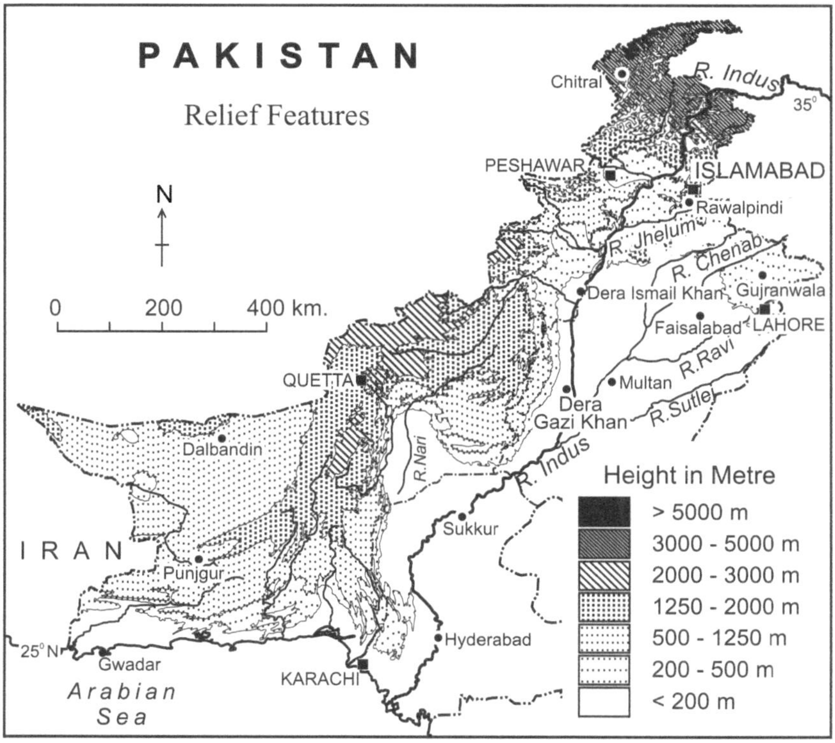

FIGURE 2.2 Relief Features of Pakistan

described by grouping them into the following six physiographic divisions of Pakistan. Each division represents an assemblage of some distinct relief features. These divisions are further sub-divided, as follows:

1. **The northern and north-eastern mountains and mountainous relief, including the Karakoram, the Himalayas and the Trans-Himalayan region**
   a. High altitude northern mountainous region north of 35°N latitude, from Chitral on the west to Karakoram Range on the east
   b. The north-western mountainous rim, forming the boundary between Afghanistan and Pakistan, and extending from Dir (35°12' N and 71°51' E) in the north to Chaman (30°55' N to 65°30' E) in the south on Afghanistan–Balochistan border
   c. Longitudinally aligned Sulaiman–Kirthar complex of hill ranges
2. **The transitional zone of sub-mountain region – lower ranges, hills and plateaus**
   a. The sub-Himalayan region including the Himalayan foothills and the Salt Range
   b. The Potwar Plateau
3. **The mid-altitude plateau of Balochistan**
   a. Chagai Hills complex in the north
   b. Raskoh residual complex
   c. The Kharan Sandy Desert
   d. Makran–Siahan Central Hill Range divide
   e. Central Makran Range
   f. The Turbat–Dasht Valley
   g. Coastal Makran Range and Makran Coast
4. **The Indus Plains and other minor plains like Peshawar Basin**
   a. The Punjab Plains – A series of doabs
   b. The Sind–Sagar Doab or the Thal Desert
   c. The Sindh alluvial crescent
5. **The Eastern Deserts**
   a. The Cholistan Desert (in Punjab)
   b. The Thar Desert (in Sindh)
6. **The Indus Delta**

2.2.1 The High Altitude Northern Mountain Region

2.2.1.1 The Himalayan Arc

The most scenic mountainous region of Pakistan, occupying the extreme northern part of its territory, extends from Afghanistan border in the west to Xinjang province of China in the east, occupying an area of roughly 100,000 km². This high mountainous area, lying south of the Wakhan Corridor of Afghanistan, is an over-arching complex, resembling an umbrella over Pakistan, formed by Hindukush Ranges on the west and the Karakoram Range on the east. The Northern Mountain Region is the highest region of Pakistan with the Muzdagh–Karakoram Range forming the northern most snow-covered mountain, followed by Baltistan traversed by the Indus and its tributaries. It includes the administrative units Chitral, Gilgit and Baltistan in the north and upper Dir, Northern Swat and Kohistan in the south. Not far to the north is the Pamir, commonly known as the roof of the world, lying in the Badakhsan province of Tajikistan, with its highest peak reaching 7,495 m ASL.

Almost the entire area rises above 4,500 m ASL contour, and the west-east arcuately aligned snow-covered ranges are bound by 5,500 m contours. These perpetually snow-covered ranges carry glaciers, which abound in the area especially in the eastern part.

The area can be divided into three parts:

1. Chitral – The western most part.
2. The Gilgit–Baltisan Mountainous Terrain: The Gilgit–Baltistan region was earlier known as Northern Areas, but is now called Gilgit–Baltistan or simply GB.
3. The Mid-Eastern Trans-Himalayan Mountainous Region

1. *Chitral* – The north-western extremity of Pakistan, south of the narrow Wakhan Corridor of Afghanistan, coinciding with Chitral district, is marked by two northeast-southwest aligned sub-parallel ranges, separated by the intervening valley of Yarkhan–Chitral. Apparently, it is an old glacial valley, extending from Barowghil (Broghil) Pass (3,777 m) in the northeast to Chitral town (1,495 m) in the southwest. Not wider than 15–20 km in its middle stretches, the valley sides are concave and rise to a height of over 3,000 m ASL in a space of 10 km. Bordering the valley on the two sides are high mountain ranges rising to heights ranging between 6,000 to 7,500 m. On the extreme northwest is the Hindukush range that forms the border between Pakistan and the Wakhan Corridor of Afghanistan and reaches its highest point in Tirich Mir (7,708 m) on its south-western extremity. The Hindukush Range has a number of interconnected peaks, separated by transverse tributaries of the river Chitral. From southwest to northeast, representing some of the highest points on the Hindukush, is a linear cluster of four glaciated peaks. These are the Tirich Mir (7,708 m), the highest in the region, Istor-o-Nal (7,403 m), Nashak (Noshaq 7,492 m) and Saraghrar (7,349 m).
2. *T*he Gilgit–Baltistan Mountainous Terrain – East of Chitral Basin is the high altitude terrain, commonly known as the Gilgit–Baltistan.

This mountainous territory is divided into two unequal parts by the river Hunza, the easternmost tributary of Gilgit and the Karakoram Highway. The western part is known as Gilgit, occupied and drained by the river Gilgit, and the higher and larger eastern part, drained by the Indus, flowing through the centre of the region, is known as Baltistan.

a. *The Ghizar–Gilgit–Diamir Mountainous Region*

Geologically recognised as Kohistan Arc, an intra-oceanic island arc, bounded in the south by the main mantle thrust and the Indus suture, this highland plateau, lying to the west of the integrated drainage line formed by the segments of a number of rivers, is the Highland of Ghizar–Gilgit and Diamir. With accordant surfaces at 5,000 m ASL and patches of perpetual snow, the area is traversed by west-east flowing Gilgit and its tributaries. This mountainous high altitude plateau is enclosed on the north and west by a line of snow-covered ranges and peaks, ranging from over 6,000 m ASL in the west to 7,785 m. ASL, as in the Batura Glacier at around 74°30' E longitude.

This mountainous region is covered by the high altitude Gilgit Basin, draining into the Indus. The crest line of the divide, between Chitral and Gilgit Valleys, carries discontinuous patches of snow-covered peaks, the highest of which, in the mid-southern section, is Buni Zom (6,541 m). The river Gilgit represents the northwest extension of the Indus suture that runs right across through Tibet, Ladakh and Baltistan. The Gilgit and the Indus flowing in opposite, yet convergent directions, the former from northwest to southeast and the latter from southeast to northwest, are fully aligned, and occupy the same tectonic depression. The river Gilgit flows through the centre of its basin with a number of tributaries joining it from the peripheral ranges, varying in height from 4,000 to 6,000 m ASl. The snow-melt from these snow-covered ranges is the source of perennial flow in Gilgit and its tributaries. Among the peripheral ranges is the famous Shandrup–Hundrup National Park at approximately 3,800 m. Gilgit region is the summer sports ground for Pakistan, with a polo ground close to the National Park, at Shandur top. Not too far from the Shandur Pass (4,765m) that connects Chitral with Gilgit. The Shandur top is the scene of annual polo matches, more specifically between Gilgit and Chitral.

Some of the important northern tributaries of Gilgit before it joins the Indus near Jaglot, are, from west to east, Yasin, Ishkoman and Hunza. And, between the latter two, close to the international border, are the two glaciers, the Batura Glacier (7,762 m) and the Baltor Glacier (7,143 m). The river Hunza and its tributaries Shimashal and Ghunjerab (also written as Khunjerab), pierce deep into the glaciated terrain further south.

b. *Baltistan Mountainous Region*

The high altitude mountainous region east of the Karakoram Highway is administratively known as Baltistan. The entire Baltistan area, under the shadow of the Karakoram Range, is traversed by the Indus, flowing from southeast to northwest. The alignment of the Indus continues even beyond its southward turn and continues in its tributary Gilgit. The river occupies what the geologists call the Indus suture zone, the zone where the Indian plate, closes in the Asian plate, sub-ducts under the latter, closes the Tethys and the depression so formed is occupied by Indus. The Indus enters the Baltistan region at a point marked 34°40' N latitude and 76°30' E longitude and flows in a north-westerly direction for a distance of about 250 km before it abruptly turns south. In its course, it collects the waters of its tributary Shyok flowing in a southwest-northeast direction, sub-parallel to the Indus and joining the latter at Gol. Further west, it is joined by Shigar at Skardu. Both these tributaries, Shyok and Shigar, descend from the northern heights of the Karakoram complex and bring considerable volume of water. The tributaries from the south are not very significant, as they descend from the Deosai Plateau which is neither snow-covered nor does it receive a heavy rainfall. Besides, the rivers originating on the southern slope of the Deosai Range flow southward and join the Jhelum.

The Indus valley, except at the points of confluence of its tributaries, is often narrow, not exceeding 5 km. The valley widens a little, extending upstream along the Shigar, wide and large enough to provide for an airport at Skardu (2317 m ASL). The Indus is the main river in the region and collects the entire snowmelt and precipitation through its tributaries or sub-tributaries from all over the region.

3. The Mid-Eastern Trans-Himalayan Mountainous Region The extreme northern part of Baltistan, along the international border with China, is the highest and most mountainous region of Pakistan. Extending between 74°30' E and 77° E longitude, this 350–400 km long, northwest-southeast oriented highland, has a series of well-defined ranges and valleys from north to south. These are from north to south:
    a. The Great Karakoram – with its ranges and peaks, glaciated landscape and the existing glaciers
    b. The Rakaposhi Range and the Harmosh Massif
    c. The Indus–Shyok Valley occupying the Indus suture
    d. The southernmost, high altitude Deosai Plateau with its radial drainage

a. *The Great Karakoram Range*

This trans-Himalayan mountain range, dividing the Asian plate from the Indian plate is often referred to as a large batholith. But, most commonly, it is known as the Karakoram Range, home to a large number of sub-ranges, peaks and glaciers. The Karakoram Range shelters the second highest peak of the world K2 (Mount Godwin Austin) with a height of 8611 m ASL, followed by a number of lower peaks. There are dozens of peaks registering a height of over 7,000 m. Some of these rise to a height of over 8,000 m; they are

1.	K2 (Mt Godwin Austin)	8,611 m	35°53' N and 76°31' E
2.	Nanga Parbat	8,126 m	35°14' N and 74°35' E
3.	Ghasherbrum I	8,060 m	35°43' N and 76°42' E
4.	Broad Peak	8,047 m	35°48' N and 76°34' E
5.	Ghasherbrum II	8,035 m	35°46' N and 76°39' E

The Nanga Parbat is not strictly a part of the Karakoram Range, but lying in the same mountainous region, is included in the category of high peaks. Besides, there are many peaks with heights rising above 7,000 m. The four highest peaks in the Karakoram complex (K2, Ghasherbrum I, Ghasherbrum II and Broad Peak) form a cluster huddled in a linear distance of not more than 50 km. So also, is the case in the western part of the Karakoram Range (around 36°20' N and 75°30' E), where there is cluster of huge peaks and glaciers, the highest being the Disteghil Sar at 7,885 m.

The Karakoram Range and the associated glaciers make one of the largest glacial fields of the world, next only to the polar regions. These glaciers are a permanent source of water for the Indus and its tributaries. Much of the discharge of the Indus in summer is derived from the glaciers associated with these northern mountain ranges, especially the Karakoram and its glacial field.

b. *Rakaposhi Range and Harmosh Massif*

Between the high Karakoram and the Indus are placed a number of secondary ranges, occasionally rising above 7,000 m. These ranges could be considered an offshoot of the Karakoram Range, which in its western part is split into a number of ranges, all following the same alignment as the Karakoram, and detached from each other by the tributaries and sub-tributaries of Indus. The Rakaposhi Range, with its three main peaks, (Rakaposhi – 7,788 m, Diran peak – 7,257 m and Malubiting – 7,458 m) is a sub-Karakoram Range, separated from the main Karakoram Range by a number of glacial valleys that discharge their water during summers in the Indus (36°10' N and 74°30' E). An eastern extension of the Rakaposhi Massif forms the Harmosh Range, detached from the former by a low-lying Harmosh Pass and Harmosh Lake. The entire area lies above the 6,000 m contour, descending steeply in spurs to the south, to finally meet the Indus.

The Karakoram mountain complex is home to a large number of glaciers, as the snow line in the region ranges from 5,500 m to 6,000 m, covering extensive areas under permanent snow. Most of these glaciers, occupying higher reaches, have an alignment parallel to the Karakoram Range; only a few move transverse to the range. Some of the largest glaciers in the world occur in this high altitude zone, the largest of which is the Siachin Glacier, outside the Pakistan Administered Territory, and presently under the control of India. Its length is estimated to be over 72 km, forming the source of the southeast flowing Nubra River, a tributary of Shyok, which, in turn, is the tributary of the Indus.

Holding a large mass of ice and glaciers, in one region, next only to Artic and Antarctic areas, Karakoram region is the rock and ice wall of great magnitude that adds substantially, through its melt water, to the perennial flow of the Indus.

c. *Deosai Plateau*

The region south of the Indus, between 75° and 76° E longitude, parallel to the Indus and serving as a divide between the Indus and its tributary Jhelum, is the Deosai plateau. The central part of the plateau with a west-east running ridge, that attains a maximum height of 5,700 m, is higher. Sloping gradually to the east, the plateau is also the source region of a number of tributaries which, flowing in a south-easterly direction, join the Indus. Devoid of perpetual snow with a height of around 4,000 to 4,500 m ASL and spot heights lingering around 5,250 m ASL, the relative flatness of the area gives an impression of a plateau. It is grass covered during summers providing high altitude pastures to local sheep herders and graziers. The plateau carries some natural lakes, the most important of which is the Satpara Lake.

d. *Nanga Parbat Complex*

Located in the south central part of the Gilgit–Baltistan mountainous region, not too far from the Indus loop, is the Nanga Parbat complex. With its peak towering 8126 m ASL, the ninth highest in the world, this high residual mountain has a northeast-southwest linear alignment and is surrounded by the tributaries of the Indus, locally known as *Gahs*, in the east and west, and by the river Indus to its northwest. With its precise location (35°14' N and 74°-35' E), the peak is surrounded by glaciers on all sides. It has the Bazhin Glacier on its south-eastern flank, the Chungphar Glacier in the northeast, and the Diamir and Rupal Glaciers in the south, a valley

> glacier occupying the source of Rupal Gah. The greatest single free-falling precipitation cliff in the world (4,500 m) is the Rupal face of Nanga Parbat, along its south-eastern flank (Kazmi and Quasijan, 1997). The peak of Nanga Parbat can be seen from the Indus and the Karakoram Highway to the west. The awesome view of Nanga Parbat "*is quite overwhelming in its magnitude: it is in fact one of the grandest spectacles that nature offers to the gaze of man*" (Burrard and Hayden, 1933).[4]

From Hunza in the north to Indus–Kohistan in the south, this mountainous region, largely snow-covered in the Karakoram region and detached snow-covered peaks and high mountains, often bare, devoid of vegetation, form a challenging as well as enchanting landscape to the mountaineers. Because of the region representing a high altitude and a snow bound ecosystem, the Government of Pakistan has notified two protected areas, viz., the Deosai National Park about 25 km southeast of Skardu and the Central Karakoram Natural Park about 75 km northwest of Skardu.

Besides the glaciers, there are several lakes – and there are a dozen of them – in the high altitude region of Gilgit–Baltistan, the most important of which is the Satpara Lake near Skardu. Most of these lakes seem to be scooped hollows of the former glacial period, subsequently filled with water, never drying up because of lower evaporation at the higher altitude and fed also during the summer from the melt waters of adjacent glaciers and snow fields.

Despite its awe-inspiring appearance, appearing impregnable, there are a few passes that traverse the Karakoram Range. The most important is the Khunjerab Pass (4,733 m ASL) that links Pakistan with the Xinjiang province of China, through the Karakoram Highway. The only other important pass is the Karakoram Pass in the eastern sector in the Indian territory at 5,575 m ASL (approximately 35°20' N and 77°45' E longitude), open only in summer. In the western part, west of 75° E longitude, the Hindukush Ranges forming roughly the northern frontier of Pakistan are breached in several points to enter into the Wakhan Corridor of Afghanistan. From east to west, these are as follows: a. Mingleke La (4,709 m ASL), b. Chillenji La (5,247 m ASL), c. Barowghil La (3,777 m ASL) and d. Anash La Dolah Pass (4,555m ASL) in the extreme west.

Besides, there are a number of regional or local passes crossing the divides between the adjacent valleys or river basins, for instance, between the Chitral and Gilgit Basins, there are several passes facilitating trans-regional movement.

The Drainage Characteristics in the Region – The principal river of the region is Indus, and all other rivers in this mountainous region, are tributaries or sub-tributaries of Indus. The Indus as well as several of its tributaries guided in their alignment by thrusts in the trans-Himalayan region. The Indus occupies the Indus suture in its upper course. The two main tributaries of Indus, in the south-eastern part, are the Shyok and Shigar, both right bank tributaries, the former flowing in a southeast-northwest direction and the latter joining the Indus from the northwest, near Skardu, All the rivers including the Indus flow in a relatively narrow valleys. As the Indus, after traversing through the barren and rocky terrain of Ladakh, enters into the high mountainous region, overlooked by the snow-covered towering heights of Karakoram, it develops a relatively narrow cross profile. In some sections, it makes a gorge, as near Kachura, some 25–30 km northwest of Skardu, and then further west at Shengus and Sassi, the latter in the Indus loop where it takes a southward turn. By comparison, the Gilgit and its tributaries, joining the Indus from the west, have wider valleys permitting human settlements and transport routes along the rivers.

2.2.2 The North-Western Mountain Rim

From the Chitral Highland in the north to Chaman, on the Pakistan–Afghanistan border, about 140 km northwest of Quetta, is an unbroken chain of mountains, coinciding with the western margin of the Indian plate. This mountainous linear belt, occupying the boundary between Pakistan and Afghanistan, is a semi-detached series of hills and ranges of a relatively lower altitude. The extreme northern boundary is formed by the Hindukush, and its offshoots further south make the boundary between Pakistan and Afghanistan. Except in the extreme northern part, nowhere does the mountainous border show heights exceeding 4,000 m. The northern part of this mountainous belt is formed by the Chitral Highlands, lying above 5,000 m, with occasional glaciated peaks reaching heights above 7,000 m ASL. Drained by Chitral and its tributaries that are flowing between Hindukush on the west and Swat Kohistan on the east, the mountainous border loses height as one moves south in Kafiristan and Bajaur Hills.

The border lands are further lowered towards the south, under the impact of the river Kabul, that originating in Afghanistan, breaches the border heights and develops its own flood plain. The river, by a general lowering of the heights, also makes way for the Khyber Pass at 1,070 m ASL.

The border mountainous region, south of Peshawar, is marked by greater heights and very rough, and dissected relief. This area falls in erstwhile FATA (Federally Administered Tribal Areas) of Pakistan. Marking the international border west of Peshawar is the east-west aligned Safed Koh, the highest range of this belt rising to a height of 4,755 m ASL at Sakharin. The Safed Koh and the associated mountains are the source of a number of small streams including Kurram, the largest tributary of Indus in the area. The Kurram makes a triangular plain in the area with its apex at Parachinar.

The mountains are usually lower than 2000 m in the southern part, represented by South Waziristan and the mountains of Balochistan further south where the Toba Kakar Range is the principal mountain. North of Zhob – Quetta axis, Toba Kakar complex has a number of parallel and sub-parallel hills all aligned northeast-southwest, and divided by valleys, which are occupied by non-perennial streams. The highest point of these ranges is 3,332 m ASL, close to what was earlier Hindubagh, now changed into Muslimbagh. The general height of the Toba Kakar Ranges is about 2,500 m with occasional higher peaks.

The land, between the Toba Kakar Range on the west and the Indus on the east, is occupied by the valley of Zhob on the west and the Sulaiman Range in the east. The Zhob Valley provides the site and sets the alignment of the Zhob–Quetta Highway and railway line. The Zhob Valley is a wide synclinal depression marked on its eastern part by the Sulaiman Range.

2.2.2.1 The Sulaiman Range

Geologically, Sulaiman is similar to any mountain system on the margin of the Indian plate, resulting from its subduction under the Central Asian plate, with the marine sediments on its margin thrown into folds and thrusts, giving a typical arrangement of geological cross-section from east to west. In the case of Sulaiman, the easternmost part is marked by the Indus foredeep, followed westward by the folds and thrusts of Sulaiman Range proper, followed further west by the hills composed of molasse of the post-orogen phase and by flysch partly involved in the thrust process. Furthermost, the cross-section is marked by the appearance of mafic and ultramafic rocks with volcanic intrusion and ophiolite. The entire Sulaiman Range has predominance of limestone, inter-bedded with shales and occasional sandstone, all varying with marine facies and some sandstone and conglomerate in the eastern part, more like the post-orogenic molasse.

The Sulaiman Range, running almost parallel and overlooking the Indus valley in the east is a north-south aligned mountain system, extending between 29° N and 32° N latitude. With increasing height from south to north, the range has several peaks rising to over 3000 m ASL. With an asymmetrical east-west cross profile, the mountain has a steep face to the Indus valley on the east and a gradual descent to the Zhob depression on the west. In its south-north orientation, the range gradually increases in height from around 1500 m in the south to 3500 m ASL in the north, on the border with FATA just south of 32° N latitude. The highest point of Sulaiman Range is Takht-i-Sulaiman (3,487 m ASL) in the extreme north.

Other important peaks include Fort Munro (1,939 m ASL), From Fort Munro northward, the Sulaiman Range rises gradually with several peaks rising above 2400 m ASL, like one near Sami Aghbarg at 2,453 m ASL.

The mountain range stands as an effective water divide, between the Indus on the east and the rest of Balochistan on the west. Except the river Zhob which, flowing northeast ward, crosses the border of Balochistan and joins the river Gomal, much of the trans-Sulaiman region, to the west, does not drain eastward into the Indus, but ends in the dry desert valleys or finds its outlet southward into Kachhi Plain.

2.2.2.2 The Quetta Hill Complex

The highest hill complex of Balochistan, about 60 km from the border with Afghanistan, is the Quetta Hill complex with a number of peaks rising above 3,000 m ASL, including the famous hill resort of Ziarat (3,389 m ASL) where Qaid-e-Azam Mohd Ali Jinnah spent the last few days of his life. Quetta (1,589 m ASL), located at the base of an escarpment, occupies a pass surrounded by several adjacent hills like the isolated Chiltan Hills.

Intervening between much of Balochistan and its northern mountainous region is the Sibi Basin, also known as the Kachhi Plain. It is a triangular lowland, extending from its apex at Sibi to the Indus Plain. It is an extension of the Indus Plain with a number of parallel or sub-parallel ephemeral streams, running south-eastward and finding their way into the Indus, or lost in the arid landscape of the region. The plain extends as far to the south-east as Jacobabad. This is a desert like lowland, the height ranging from 150 m ASL near the lower part to 700 m near the apex.

2.2.3 Longitudinally Aligned SulaimanKirthar Complex of Hill Ranges

2.2.3.1 Kirthar-Brahui Range

South of Quetta, the eastern fringe of Balochistan is marked by a series of north-south extending linear ranges. The northern most is the Central Brahui Range with its western extension the Harboi Hills, which also extends north-south almost parallel to the Central Brahui Range. The Brahui Range is named after the famous Brahui tribe, reportedly speaking a dialect that resembles the languages of the Dravidian family. Both the ranges, Central Brahui and Harboi, converge in a knot in the north, attaining a height of 3185 m ASL near Quetta. Quetta, in fact, occupies a strategic location (30°15’ N and 66°53’ E), at the foot of the convergence of three mountains, the Sulaiman, the Central Brahui and the Toba Kakar Ranges, surrounded by high peaks, and across the Bolan Pass, further west.

2.2.3.2 The Kirthar-Pab Ranges

Kirthar is the eastern of these twin ranges that form the border between Sindh and Balochistan. West of Kirthar is the Pab Range also known as the

Mor Range. In fact, the Kirthar – Pab Ranges are part of one and the same orogenic belt separated from each other by synclinal depression occupied by north-south flowing streams like Hub (hub darya) and Porali, the former joining the Arabian Sea west of Karachi and the later falling into Miani bay, passing through a lagoon of the same name.

The Kirthar Range, like its counterpart the Sulaiman Range, further north, is part of a whole mountain chain that has emerged following the sub-duction of the Indo-Pakistan plate, under the Iran–Afghan plate, throwing up 3000 m deep pile of late Mesozoic and Tertiary sediments ranging in age from Triassic to Pliocene into a series of folds and thrusts. The highest point in Kirthar Range is 2171 m ASL, whereas the Pab Range, to the west, has a maximum height of 2316 m, about 30 km south of the important Baloch town Khuzdar.

2.2.3.3 The Passes in the Mountain System

The north-western mountainous ranges, lying on the border between Pakistan and Afghanistan, have a number of passes. From north to south these are:

1. **Lawari La** – 3,118 m ASL, connecting Dir and Chitral Valleys
2. **Nama Pass** – 3,000 m ASL (35°01'N and 71°30'E latitude), connects Dir district with Do Bandi on the Afghanistan side.
3. **Khyber Pass** – It is the most important pass that has been historically the gateway to Hindoostan. Traversed by invaders, conquerors and travellers, from Central and Western Asia, the pass was the entry point to the Indus Valley and the vast land that is India. It has been the scene of invading armies marching through it. It has also been the site of many battles fought between the invading and the defending armies. The pass connects Peshawar, a strategically important frontier town in Pakistan, with Kabul, the capital of Afghanistan.

   The pass is formed by the headwaters of two streams, one flowing north-east to join the river Kabul and the other flowing south and southeast towards Jamrud, causing a breach in the Khyber Range, an offshoot of the Safed Koh Range. Locationally, the centre of the pass is at 34°06'N and 71°05'E) and the highest point is 1066 m ASL, between Jamrud, a few kilomotres west of Peshawar and Landi Kotal, further west. It begins 17 km west of Peshawar and runs for 53 km, till an important point on the western side of the pass, Dakka, is reached. The route through the pass rises gradually from Peshawar (330 m ASL) and reaches the highest point (1,066 m ASL) midway between Jamrud and Landi Kotal before it descends towards Kabul (Figure 2.3).

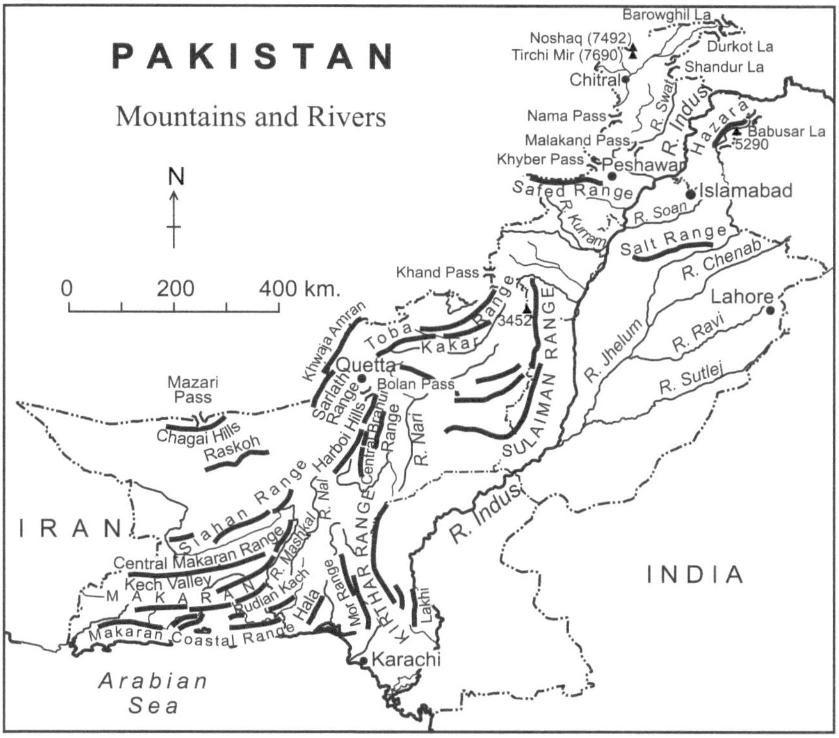

FIGURE 2.3 Mountains and Rivers of Pakistan

4. **Khand Pass** – 1,500 m ASL (32°15'N and 69°28'E), an important pass through which caravans were travelling from Afghanistan over the centuries to Pakistan reaching Dera Ismail Khan over the Gomal Pass in the Sulaiman Range. Even today, this is the pass that is used as passage between FATA and Afghanistan by the tribes of South Waziristan.
5. **Khojak Pass** (30°55'N and 66°25'E). This pass is about 15 km southeast of the border town Chaman, the last road-rail station on the coupled road and rail route. This route joins Sukkur, Shikarpur and Jacobabad in Sindh, with Quetta in Balochistan, the capital of Balochistan, and further northwest passing over the Khojak Pass to Chaman, the Pakistani city on the international border with Afghanistan, and leading further to Kandahar, the capital of Kandahar province of Afghanistan. As the Khyber Pass is important in providing the principal land passage between Peshawar and Kabul, the Khojak Pass is the gateway to Afghanistan through Khandahar. Quetta – Chaman – Kandahar is a significant axis, a southern route, for the movement of goods and people between Pakistan and Afghanistan.

6. **Mazari Pass** (29°25'N and 64°3'E) – At a mid-point on the northern boundary of Balochistan, on the northern margin of the Chagai Hills, the Mazari Pass is a convenient crossing to provide access to the Helmand Valley and further inside Afghanistan to the grand southern route of Afghanistan that connects, Kabul, Ghazni, Kandhahar and finally Herat further west.

2.2.4 The Sub-Mountainous Transitional Zone

2.2.4.1 Sub-Himalayan Region

South of Chitral, on the extreme north-west, Gilgit west of the Indus, and a much larger territory of Baltistan further east, is the zone of sub-mountainous relief, ranging in height from 1,500 to 3,500 m ASL, with occasional peaks reaching 5,000 m ASL on the tops of the divides away from the river valleys. This area, separated from the higher mountainous zone by an east-west stretch of the Indus, before it finally assumes a north-south course, is traditionally recognised as Kohistan, the western part of which is known as Swat–Kohistan. This mid-mountainous zone is drained by the north-south flowing Indus and its principal tributary, the Swat and the sub-tributary Panjkora in the west and the Kaghan Valley in the east. The upper reaches of the rivers – all being tributary or sub-tributary of the Indus – have extended their re-entrants piercing the 3,500 m contour that roughly separates the higher mountain zone from the sub-mountainous region. The region, between 3,500 m contour on the north and the 1,500 m contour in the south, is the Sub-Mountainous transitional region. This is the belt which is home to several hill stations like Murree, Nowshera, Abbottabad or Mingora. This zone also offers suitable sites for irrigation projects like Tarbela on Indus or Mangla on the Jhelum. This is also the belt of highest rainfall, with places like Murree (2,127 m ASL) receiving 1789 mm rainfall, the highest in Pakistan, Dir (1,370 m ASL) receiving 1469 mm rains and Balakot in the Kaghan Valley (981 m ASL) having 1,769 mm of rain annually. This has led to the growth of some forests in the region. This is the zone of high pastures attracting graziers in summers. The snow melt of the higher peaks has given rise to some permanent lakes, like Saiful Muluk Lake in the Kaghan Valley and the Mahadand Lake in the source region of the river Swat.

2.2.4.2 The Transitional Zone between the Himalayan Foothills and the Indus Plain

This narrow transitional zone is the piedmont where the land abruptly drops from 1,500 m ASL to 755 m ASL. Starting from Murree and Rawalpindi – Islamabad twin city in the east, to Mingora in the Swat valley in the west, this is a transitional zone where most of the tributaries of the Indus debouch from the higher hills. Acknowledged as the most habitable, this is the region with many ancient relicts of Buddhism like the *stupas* in Potwar area and relicts of Buddhist monasteries in the Swat Valley.

This zone has two distinct physical units: 1. The Potwar Plateau and 2. The Salt Range

The ***Potwar Plateau*** is formed almost exclusively of Tertiary rocks, especially Miocene and Pliocene shale and sandstone, with its eastward extension into Murree formation. The plateau lies between Kala Chitta on the north and the NE-SW trending Salt Range on its south. The Salt Range, unlike Potwar plateau, carries a cross-section of formation from the pre-Cambrian to Tertiary and reaches a height of 1,000 to 1,500 m ASL, just north of the river Jhelum.

The Potwar Plateau is filled with non-oceanic sediments of the Neogen period and has a height of 500 m ASL. With a fill of Quaternary alluvium and loess, the plateau is greatly ravined by the river Soan, a tributary of the Indus and its tributaries. The twin city of Islamabad–Rawalpindi, lies on the northern fringe of the plateau, at the foot of Kala Chitta Range overlooked by the Margalla Hills.

The ***Salt Range*** forming the southern limit of the Punjab Plain, separated from the latter by an escarpment, is the rump of an old pre-Cambrian formation carrying on its back several superimposed formations of different geological periods. The actual saline series belong to a pre-Cambrian period. Extending northeast-southwest for roughly 200 km, the Salt Range attains a height of 1522 m at Sakesar. In its western part, the range is traversed by a number of highways. Known for its salt mines, the range has the second largest salt mine in the world. Following the general slope, from the escarpment face in the south to the north, the drainage is directed northward.

The piedmont linear zone between 1,000 and 500 m ASL, extending northwest from Mangla on Jhelum, along the Grand Trunk Road, is climatically the most salubrious zone of Pakistan. A cultural axis of Pakistan, coinciding with the ancient northern route *Uttarapath*, the zone carries a string of towns like the ancient city of Taxila, Attock, Nowshera and Peshawar, along the Grand Trunk Road. The twin cities of Islamabad–Rawalpindi located in the area add to the importance of this zone. Abutted by the Margalla Hills and enclosing the Potwar Plateau in its fold, the zone has all the political and military power of Pakistan. The strategic advantage of this zone must have governed, and weighted heavily, on the choice of capital site of Islamabad.

2.2.5 Western Balochistan Ranges

West of the north-south Quetta – Khuzdar – Bela axis, Balochistan is marked by a series of arcuate ranges with convexity to the south, alternating by semi-arid and arid valleys and elongated depressions with inland lakes providing centripetal drainage for the ephemeral or even some of the regular streams.

From south to north, the arrangement of ranges and the intervening valleys is as follows.

N
8. Chagai Hills, the northernmost hill complex of Balochistan
7. Desertic Dalbandin – Nokkundi – Taman lowland
6. Raskoh Range
5. The North Central Sandy Desert of Balochistan and Siahan separated by the Rakhshan River
4. The Central Makran Ranges
3. The Kech Valley with several inland lakes
2. The Makran Coastal Range
1. The Makran Narrow Coastal Lowland

S

The above arrangement of a cross-section of hill ranges and valleys is a south-north sequence of relief along 64° E longitude from the Arabian Sea coast in the south to the Mazari Pass on the northern margin of the Chagai Hills.

These ranges and other physical features are the result of subduction of the Arabian Sea plate or Armorian plate that has caused the thrust and fold of a series of parallel ridges and sub-parallel ridges. These arcuate ridges are formed in tertiary sediments with the oldest rocks in Chagai complex with some intrusion in mafic rocks in the form of sills and dykes.

Makran Coastal Lowland – It is a 20 to 40 km wide strip of lowland, with widening inland profusion in the mouth of the river Dashi on Iran border. With several lagoons and bay-heads, this is the most densely populated part of Balochistan with several towns, fishing settlements and four airports. The important coastal towns and ports are Ormara in the east, Pasni in the centre, Gwadar further west, with a naval port developed by China between the Gwadar East and Gwadar West Bay, and Jiwani, the westernmost coastal town of Balochistan. The coast is backed by the Coastal Makran Range, formed in Miocene sediments.

North of the coastal Makran Range, lying between the coastal Makran Range and the Central Makran Ranges is the Kech Valley, prominent only in the western part, formed by joining together a number of ephemeral streams further east where the valley appears as a depression with seasonal flow. The route from Bela in east Balochistan follows this depression connecting important towns, like Awaran, Turbat, Mand and Gazdan on Iran border. Most of these ranges have the tertiary rocks ranging in age from Oligocene to Pliocene.

Between the Siahan Range that occupies the central part of West Balochistan and the Raskoh Range further north is the 150–200 km wide

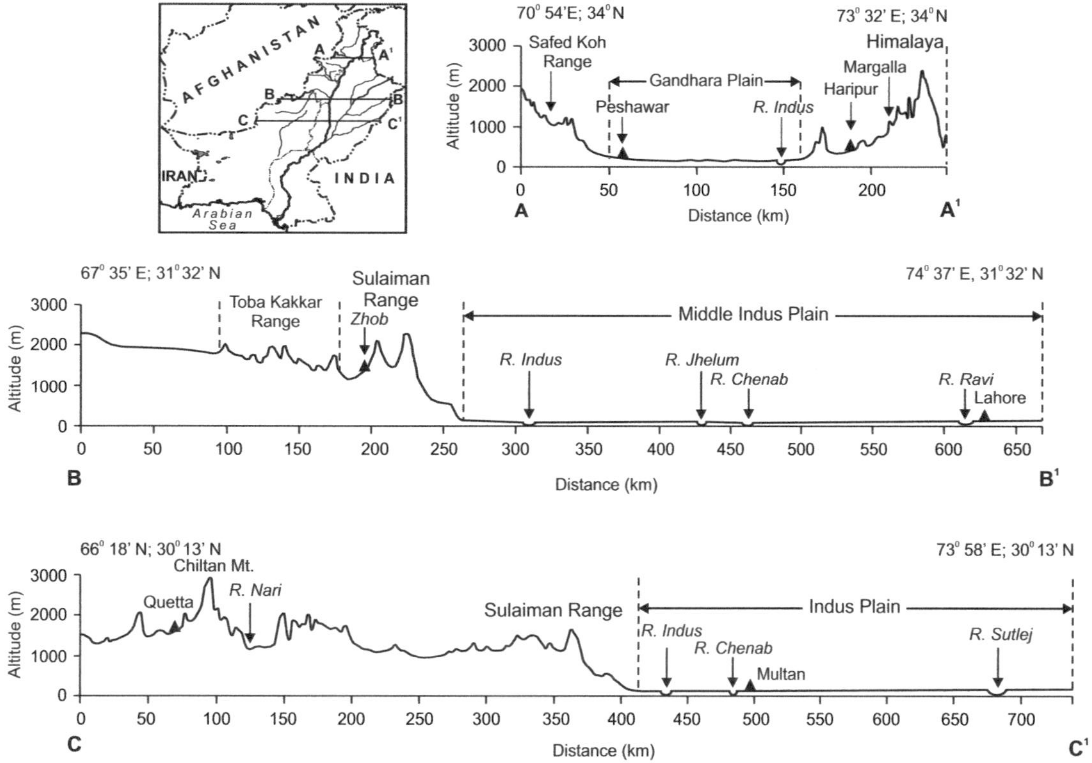

FIGURE 2.4 West - East Profiles: 1. Safed Koh Range to Margalla Hills, 2. Toba Kakar Range to R. Ravi, 3. Quetta – Sulaiman Range – Indus Plain

north south stretch of the Kharan Sandy Desert. Its lowest point is below 500 m ASL in Hamun-i-Mashkhel Salt Lake, in the extreme western part of the region, where the ephemeral streams from surrounding areas converge. More appropriately, it is the typical playa of a desert.

The Raskoh and Chagai Hills, though detached from each other, carry some patches of Cretaceous rocks, ultramafic granodiorite with sills and dykes and other volcanic intrusions. These features, typical of a suture zone are associated with southern extension of the Chaman Fault that separates the Indo-Pakistani plate from the Afghan sub-plate.

Chagai Hills occur on the centre of the northern boundary of Balochistan that separates the Khandahar and Helmand provinces of Afghanistan from the former. The highest point of the Chagai Hill complex, composed of ultra-volcanic rocks, is the Malik Nara Point (2,462 m ASL) though the general elevation of the hills is around 1,500 m ASL. Devoid of greenery, often appearing like a stony waste, the desolateness is broken only by the appearance of date palm groves close to water courses.

2.2.6 The Indus Plain

Between the Indo-Pakistan international boundary, close to Lahore on the east and Mianwali and Dera Ismail Khan on the west, both on the bank of Salt Range in the north and the Indus delta, on the Arabian Sea coast in the south, is the Great Indus Plain, drained by Indus and its five tributaries, Sutlej, Beas, Ravi, Chenab and Jhelum. This plain covers much of Punjab and Sindh with some hilly and desert margins. With an area of roughly 300,000km^2, the Indus plain is the human and economic core of Pakistan

One may divide it in a number of units.

1. Punjab plain
2. Sindh plain
3. Nal – Thal Parkar dry region
4. Indus Delta

2.2.6.1 The Punjab Plain

As the name suggests, the plain is formed by the alluvium of the Indus and its five tributaries. The Indus, the principal river, is the westernmost of all the rivers, forming a natural limit of Punjab. The other rivers, Jhelum, Chenab, Ravi and Sutlej, in a sequential order from west to east, emerge from the Himalayas beyond the national boundary, and flow south-westward and join Chenab, which finally joins the Indus as Panjnad near Mithankot, a town about 75 km north of Punjab–Sindh border. The Punjab Plain has a triangular configuration with its base in the north and the apex to the south, on the Punjab–Sindh provincial boundary.

This fertile alluvial plain has a couple of desert or desert like tracts. One such arid tract is the Sind–Sagar doab, also known as the Thal Desert. Interposed between Jhelum and Indus, it covers the districts of Bhakkar and Leiah. The Jhang–Bhakkar east–west route traverses through the centre of this dry tract. The other dry area is the Cholistan Desert, spread across the south-eastern part of Punjab. It formed a part of the erstwhile princely state of Bahawalpur, having a common border with the erstwhile princely state of Bikaner, now in India.

The Cholistan Desert is, in fact, a western extension of the Great Thar desert of India. The canals from Sutlej could benefit the area only partially, and much of it has remained a desert.

The Punjab Plain, covering much of the province, has a very gentle incline from north to south, almost imperceptible to any traveller. From Sargodha (187 m ASL) in the north to Multan (122 m ASL) in the south, a distance of around 300 km, the drop in height is very insignificant and the slope works out to around 25 cm a kilometre. This imperceptible gentle slope is responsible for the spread of an alluvium mantle in the entire plain, which consist of a series of *doabs*. In the east-west cross profile of the Punjab Plain, extending over 300 km, ranging in height from 250 m ASL in the west, near Dera Ismail Khan, to 215 m ASL on its eastern margin near Lahore, there is not much variation in relief. The height of the plain lingers around 180 m ASL, rising towards the piedmont region, as in the case of Sialkot, situated at a height of over 255 m ASL. It must be noted that Lahore, Gujranwala, Mandi Bahauddin and Jhelum in the northern part of the plain have a height of 215 m, 227 m, 252 m and 287 m ASL, respectively. These are influenced by their location in the piedmont zone and are characterised by some piedmont deposits besides the river alluvium of the adjacent rivers.

The western limit of the plain is formed by the river Indus, which, with its meanders and braided channels, runs almost parallel to western bordering mountain ranges, especially the Sulaiman and Kirthar Ranges, making hardly a 30–40 km wide flood plain to its west, developed into the piedmont zone. The highest point of this narrow flood plain is 200 m ASL.

The centre of the Punjab Plain is occupied by Sargodha and Faisalabad districts, having a height of 185 and 187 m ASL, respectively. Naturally, the cross profile dips a few metres as one approaches a river, but on an average the northern half of the Punjab Plain has an average height of 185 m ASL. In its southern part, the plain narrows down its width to approximately 200 km which shrinks further to approximately 100 km as the river crosses the Punjab–Sindh interprovincial boundary at 28°26' N latitude, about 65 km west of Rahim Yar Khan. The narrow plain before the Indus enters Sindh; it has an average height of less than 100 m ASL, which decreases further as the river passes through Sindh.

Geomorphologically, the Indus Plain is a deep basin with 6,000 to 7,000 m of sedimentary deposits ranging from late Mesozoic–Jurassic to Neogene and Tertiary period, all subjected to compression and folding. The Quaternary deposits of clay, silt and conglomerate with a thickness of about 500 m is overlain by a veneer of loose sandy and silty loam of contemporary phase. The alluvial plain consists of a contemporary flood plain, lower and upper terraces and occasionally a remnant of a very old concretionary deposit. The most extensive surface of the older or upper terrace, also called *chung formation*, is usually composed of loess-covered older alluvium. The middle terraces are less extensive and are hardly noticed in the area of low relief. The braided streams, especially the Indus and Chenab, occupy considerable width, sometimes as much as 10 km, and leave many river islands, which are also sites of agriculture once the flood plain recedes.

The Punjab Plain with an area of roughly 150,000 km² with its uninterrupted stretch of alluvial plain is the principal element in the agricultural landscape of Punjab and mainstay of the agricultural economy of Pakistan.

2.2.6.2 The Sindh Plain

Sandwiched between the Thar and Cholistan Deserts on the southeast and the Sulaiman and Kirthar Ranges in the northwest, the Indus Plain shrinks to 70 km in width at 29°30' N latitude. its southeast extremity marked by the city of Sadiqabad in Punjab and Mirpur Mathelo in Sindh. The Sindh Plain is an S-shaped 200 km wide alluvial strip that broadens in the delta region. It is bordered on the east by the extension of the Cholistan Desert in the north and the Thar Desert in the south. North–south, the plain extends for over 500 km, from the Arabian Sea coast in the south to the Sindh–Punjab boundary in the north. On the west, the plain is confronted by the heights of the Kirthar Range. About one-third of the Indus Plain on the east is desert, devoid of irrigation and area of low agricultural productivity.

From north to south, the plain slopes imperceptibly and the height difference between Jacobabad (28°18' N and 68°28' E) and Hyderabad (25°25'N and 68°25'E), a distance of roughly 300 km, is 27 m, showing a fall of 9 m per 100 km. The shape of the plain follows the course of the river Indus, 60–70 km on either side of the river, beyond which is the desert on the east and the foothills of the north–south aligned Kirthar Range on the west. In terms of relief, the Sindh Plain is almost featureless, the monotony relieved only by some lakes, cut-offs of the river, of which there are many, and the network of canals, taken out largely from Sukkur and Kotri Barrages.

In a regional cross-section from west to east, the piedmont zone along the Kirthar Range with isolated hillocks show a higher altitude that flattens as one moves east. The centre of the plain has a height of less than 100 m,

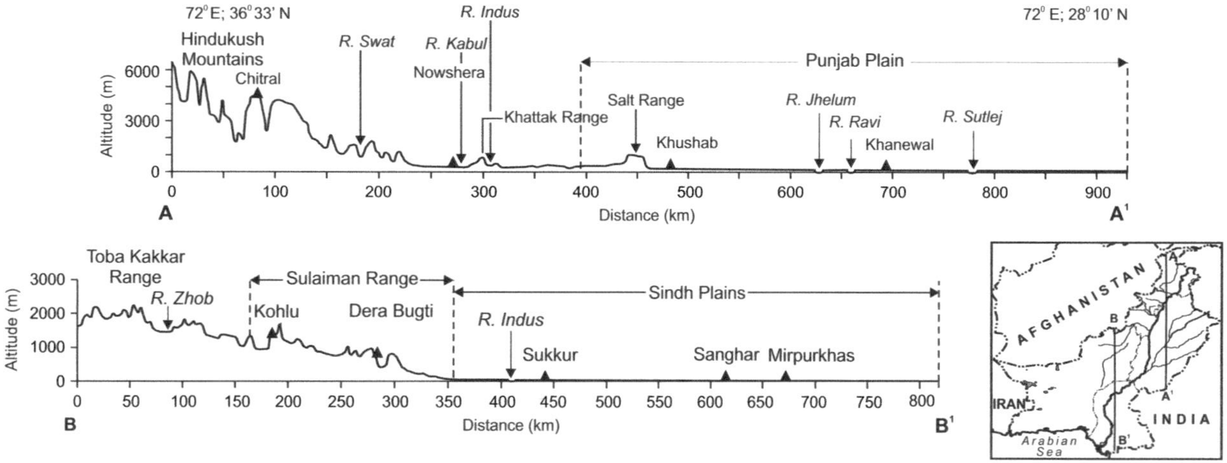

FIGURE 2.5 North – South profiles: 1 Hindu Kush – Chitral – Punjab Plain, Toba Kakar Range – Sulaiman Range – Sindh Plain

varying from north to south. In the northern part, the plain shows a height of around 65–70 m, as seen in the case of Rohri (66m ASL) and Sukkur having a height of 68.5 m ASL. The height declines southward as seen in the case of Hyderabad (28 m ASL) and Mirpur Khas having a height of just 15 m ASL.

The Indus has undergone lateral shifts during historic and even pre-historic times and has left behind many cut-offs, water bodies like ox-bow lakes, and even parallel channels some of which have regular streams. The most prominent example is a parallel stream called Wahar Nadi, west of Indus, and east of Dadu city. It appears that in the latest phase of lateral migration, the river Indus has shifted eastward, believably following a neo-tectonic movement that may have lifted up the western edge of the Indus Plain, along with Kirthar Ranges, forcing the river to shift laterally eastward leaving many parallel streams that join the Indus on both ends.

The Indus channel in Sindh is subject to excessive silting, not only because of gentle relief, but more because of a steep decline in its discharge as the result of the building of several barrages and many inundation canals. There is a great piling of lag deposits, after the floods recede. on the banks giving rise to natural levees and low embankments. The cross profile, therefore, shows increasing height as one approaches the river, which declines in the flood plain, further away from the stream on either side. It is more like dykes or manmade embankments. The situation, to some extent, resembles the Hwang Ho, though the discharge in the Indus in Sindh is much less and a high flood discharge is diverted in the canals or in the channels called *wahs*, especially designed for inundation. Many of these *wahs* (inundation channels) appear like distributaries. The configuration of the plain, having lower level on either side of the river, facilitates canal irrigation allowing easy flow through gravity.

The river Indus has large river islands in Sindh, some of them have turned out to be agriculturally very productive. An island over 200 km long, in the neighbourhood of Shikarpur and Larkana, enclosed between the Indus and western Nara is a case in point. The plain is also dotted with lakes, large and small. The largest freshwater body is Manchar, about 40 km southwest of Dadu (26°31' N and 69°47' E), a district headquarter in Central Sindh. During the rainy season, the lake is 35 km in length and covers an area of over 500 km^2. In the southern part of the plain, closer to the Indus Delta, the lakes turn brackish and a number of salt lakes, locally known as *dhands*, dot the landscape. These were once the inlets of the sea, turned into a closed lake as a result of the pro-gradation of the delta, like Khinjhar dhand, Sunahari dhand and Jhol dhand.

The braided channels of the river occupy a width varying from 12 to 15 km, a green belt marked by the growth of *Acacia arabica*, lining the belt on either side. Beyond the belt is the irrigated tract; and further east, is the semi-arid desert area traversed from north to south by the Nara canal, an excavated waterway following the course of the old Nara river, possibly an ancient channel of the Indus. This canal runs from Sukkur barrage in the north to Thalparkar district in the south, merging finally with the Jamrao Canal. It is the largest canal in Pakistan running for over 350 km. With the addition of the Jamrao Canal, it measures 800 km in length, and finally joins the river Puran, another old channel of Indus, before discharging itself into the sea. The Nara Game Reserve in Khairpur district owes its existence to Eastern Nara Canal. Similar to the Eastern Nara Canal, there is also a West Nara Canal that runs parallel to the Indus and feeds the Manchar Lake.

The Nara Canals, excavated usually along the deserted course of the Indus, are a very important element in the physical landscape of Sindh, These could be used as drainage canals in lowland, irrigation canals when needed, and could be allowed to develop wetland in its course. It could also exercise a moderating influence on the temperature conditions of Sindh during summer.

2.2.6.3 The Thar Desert

About a third of the total width of Sindh, occupying the eastern part, is semi-arid and arid, a western extension of the Thar Desert of India. Higher than the axial region of Sindh along the Indus, the desert belt extends all along the eastern border region of Sindh and South Punjab. The Punjab section of the eastern desert belt is known as the Cholistan Desert, a part of the old princely state of Bahawalpur, while the southern section in Sindh is part of the Thar desert. The desert landscape rises in height as one moves eastward. While Chhar (25°31' N and 69°67' E) located east of Eastern Nara Canal has a height of 06 m ASL, the rise in the land is spectacular with 120 m ASL on Indian border, 80 km to the east. The southern part of the Thar Desert in Sindh is much higher, some places rising to over 200 m ASL. Close to the Eastern Nara Canal, the land is dotted with a series of water bodies, the hollows filled with the seepage from the canal. Further east, it is desert, called *Registan* with sand dunes prominent in the landscape. Sometimes, these dunes are quite long and with shrubs get stabilised.

2.2.6.4 The Arabian Sea Coast of Sindh

The coastal area of Sindh could be divided into three parts: The eastern part bordering the Rann of Kutch of India, the central part marked by the Indus Delta and the rocky western part with Cape Monze (202 m ASL), forming the southern extremity of Pab Hills and the port city of Karachi.

The eastern part of the coast overlooking the Rann of Kutch is a lowland dotted all along, from east to west, by numerous salt lakes that turn dry during the summer but are flooded during the rainy season. Unlike the Indus Delta, the Kutch coast is much higher, the extreme eastern part rising to over 350 m ASL, as in the case of Nagar Parkar. A large part of the eastern coast of Sindh is neither inundated nor subject to deep tidal ingression.

The ***Indus Delta,*** with its head at Thatta, is a large area of tidal creeks and flats. The delta, with a lobate periphery and a bird-foot configuration, is marked by a number of distributaries of Indus. The main channel of the river joins the sea at a point about 100 km southeast of Karachi (24°48'N and 68°59'E). East of the mouth of Indus is a treeless coastal regions with creeks, mudflats, low islands, that get submerged during high tides. West of the Indus mouth, the land rises and there are not many tidal flats, though there are coastal islands suggesting the existence of a former delta.

The land of Pakistan has a variety of physical features, and in terms of relief, it extends from the sea-level in Sindh and Balochistan to the greatest heights of Karakoram, with K2 (Mount Godwin Austin), the highest part of the territory, rising to 8611 m ASL. Over two-thirds of Pakistan territory is mountainous, that includes the entire northern frontier region of Chitral, Gilgit and Baltistan and the sub-mountainous Siwalik Ranges running from Murree in the southeast to Mingora and Bajaur Hills in the northwest. The entire frontier region of West and North-west Pakistan, marked by Sulaiman and Kirthar Mountains, forming the western limit of Punjab and Sindh respectively, presents a mountainous terrain. It may be added that the northern mountainous region of Pakistan presents some of the highest peaks associated with the Karakoram Range and Nanga Parbat. Four of the peaks associated with the Karakoram Range rise to a height of over 8,000 m ASL, and 70 peaks scattered all across the Karakoram and Hindukush Ranges attain a height of over 7,000 m ASL. Added to this mountainous zone is the entire plateau of Balochistan, formed by mountain ranges, plateaus and a few small river valleys.

The plains of Pakistan, formed by the Indus and its tributaries, are represented by the featureless Punjab Plain and Sindh. Besides, there are small riverine plains formed by the tributaries of the Indus, as in the case the river Kabul forming the Peshawar Basin and the river Swat forming the lower Swat Plain.

The south-eastern part of Pakistan, falling in Punjab and Sindh, is largely dry with desert landscape. And, though the canals from the principal rivers, Indus and Sutlej, have led to the reclamation of part of the Cholistan and Thar Deserts, much of this dry part still awaits development.

The drainage of the country determines the slope characteristics. The principal and the most important river of Pakistan is undoubtedly the Indus, the western most of the five rivers (Sutlej, Ravi, Chenab, Jhelum and the Indus) after which the province of Punjab (literally means five waters) is named. The river Indus originating from the Mount Kailash in the Trans-Himalayan region, first flows northwest following the Indus suture, traverses the Baltistan–Gilgit region in the highest zone of Pakistan, south of the Karakoram Ranges, and finally turns southwards completely changing the direction of its course. During its course in the northern mountainous region, it collects the waters of the region from north as well as from south. On its course southward, the river travels almost 200 km before it is dammed at Tarbela. From Tarbela, the river flows south in its flood plain, with multiple braided channels and river islands. In Punjab, it receives the waters of Sutlej, Ravi, Chenab and Jhelum, all merged into one Panjnad, on its left bank and flows further down, furrowing the plain of Sindh to join the Arabian Sea. Besides its tributaries in Punjab, the Indus receives other rivers like Swat from the north in Khyber Pakhtunkhwa, river Kabul from Afghanistan and a number of small rivers descending down the western mountain zone, flowing eastward, to join the Indus.

Thus, while the Indus remains the lifeline of Pakistan, its tributaries in Punjab are not less important, gifting the country with a fertile alluvial plain, a perennial source of water, and significant amount of hydroelectricity, not to mention the vital source of irrigation without which the agriculture in Pakistan would not survive.

Notes

1 This difficulty is specific to Jammu and Kashmir, part of which is under the control and administration of Pakistan, while the remaining part is under the administration of India.

2 The section on geology and structural framework of Pakistan including the Himalayas are based on the following sources:

Geological Survey of Pakistan (1964) Geological Map of Pakistan (1:2,000,000)
Valdiya, K. S. (1998) *Dynamic Himalaya*, Universities Press, Hyderabad.
Kazmi, A. H. and M. Qasimjan (1997) *Geology and Tectonics of Pakistan*, Graphic Publishers, Karachi.
Searle, M. P. and B. R. Hacker (2018) 1 Structural and Metamorphic of the Karakoram and Pamir following India-Karakoram-Asia collision from P. J. Trelor and M. P. Searle (eds-2019) *Himalayan Tectonics-A Modern Synthesis*, Geological Society, London, spl publication No. 483.
Sorkhabi, Sorkhabi Rasoul (2010) Geologic Formation of the Himalayas, *Himalayan Jour*, vol. 66, pp. 87–102.

3 Marcus, Nüsser and Sussane Schmidt (2017) Nanga Parbat Revisited: Evolution and Dynamics of Socio-hydrological interactions in the Northwestern Himalaya, *Annals of the American Association of Geographers*, vol. 107 (2), pp 403–415.

4 Hewitt, K., C.P. Wake, G.J. Young and C. David (1989) Hydrological investigations at Biafo Glacier, Karakoram Himalaya: An important source of water for the Indus River, *Annals of Glaciology*, vol. 13, 103–108.
Wissman, H. von. (1959) *Die heutige Vergletscherung und Schneegrenze in Hochasien mit Hinweisen auf die Vergletscherung der letzren Eiszeit*, Abh. Math.-Naturwiss. Kl ..Akad. Wiss . Lit , Mainz, Jahrg, vol. 14, pp. 1101–1407.

3

RIVERS OF PAKISTAN

The Drainage Network

Rivers are a vital link in the hydrological cycle and form the constituent units in a natural drainage system. A drainage network is a natural device to dispose of the surplus water received from rains or snowfall, left over after evaporation and infiltration. Yet, there are regions and countries, which, despite having scant rainfall, are endowed with a well-developed drainage network and mighty rivers, not warranted by the amount of rainfall they receive. The rivers in such a drainage system are termed allochthonous, originating beyond the region. The rivers of Pakistan are largely allochthonous, having their origins, hundreds or even thousands of kilometres beyond the international border, in the Himalayan or the trans-Himalayan region. They are perpetually fed by rainfall or snowmelt in their source regions or the higher regions of their river basins.

For Pakistan, like other monsoon countries, the rivers as a source of water are vitally important for the country's agriculture. It has to be emphasised that in monsoon countries, rainfall is largely confined to four summer months from June to September, and the soil moisture, subjected to a very high rate of evapotranspiration, gets depleted by the end of October. In such a situation, all the monsoon countries need high doses of irrigation. This underlines the significance of rivers and canal irrigation for the agriculture of Pakistan.

The importance of rivers in the life of the Indian subcontinent is so important that many regional units including India have an eponymous character. The two constituent units of Pakistan, viz. Punjab and Sindh, are named after rivers, the former after the collective nomenclature of the five rivers of Punjab (Sutlej, Beas, Ravi, Chenab and Jhelum) and the latter after the Vedic name of Indus, the Sindhu.

DOI: 10.4324/9781003510147-4

3.1 The Rivers of Pakistan

Almost the entire Pakistan, with the exception of two-thirds of the area of Balochistan, is drained by the river Indus and its tributaries. These may be enumerated as follows:

1. Indus – the principal river
2. Main left-bank tributaries from west to east
    a. The Jhelum – a sub-tributary of the Indus and a tributary of the Chenab.
    b. The Chenab – a tributary of the Indus that joins the Indus as *Panjnad* (five rivers), representing the combination of all the five sub-tributaries, (from east to west – Sutlej, Beas, Ravi, Chenab and Jhelum). The stretch of the Chenab, known as *Panjnad,* is a short stretch of 60 km, between Bakhri, the point of confluence of the Sutlej with the Chenab and Mithankot, the place where *Panjnad* joins the Indus.
    c. The Ravi – a tributary of the Chenab joining the latter at a point, around 60 km north of Multan at the Chenab.
    d. The Beas – a right tributary of the Sutlej, joins the latter at a point 50 km south of Amritsar (India). The river doesn't enter Pakistan.
    e. The Sutlej – The easternmost tributary of the Indus: for much of its course, it flows in India and flows for over 350 km in Pakistan.
3. Main right-bank tributary
    a. The Kabul River
    b. The Zhob River
    c. Numerous short streams descending down the Sulaiman Range and joining the Indus

Other tributaries of the Indus – The tributaries mentioned above are those that join the Indus in Punjab and give the territory the name *Punjab.* There are several other tributaries, some beyond Pakistan and some in the High Himalayas and the Karakoram snowfield. These are enumerated below, from the source of the river.

Starting from the source, the following tributaries join the river Indus (Table 3.1 and Table 3.2).

Besides these tributaries, there are not-so-important left bank tributaries of the Indus in the piedmont zone. Starting from the west of Nanga Parbat, there are three left bank tributaries. These are, from east west, Astore, Bunar and Thak Garh joining the Indus, after it takes a westward turn at 35°30' N and 74°40' E, after flowing north–south for over 60 km. Further down, as the river turns south, after an east-west course of 150 km, it is joined by a number of small tributaries. These include Pasal and Allai, both joining the left bank from the east, and another river Siran. At Tarbela, the Dor flows into the Indus from the east.

TABLE 3.1 Major Tributaries of the Indus

Name of the Tributary	*Location of Its Confluence with the Indus*	*Remarks*
Tributaries in the Himalayan region		
Zanskar	Left bank tributary 220 km from the source of the Indus about 25 km north of Leh (3,200 m ASL).	Joins the Indus on the left bank.
Shyok	Right bank tributary 480 km from source, 240 km north of Leh.	Joins the Indus at a height of about 2500 m ASL.
Shigal	Right bank tributary, joins the Indus from the right bank at Skardu.	The river flows in a reverse direction from NW to SE.
Hunza	Right bank tributary, joins the Indus from north in Hunza district, 1300 km from the source of the Indus.	The river Hunza flows from a glacial field, first south, then west, then again south before it joins Gilgit River at Daimar; this is a sub-tributary of the Indus.
Gilgit	Right bank tributary, joins the Indus near Jaglot about 50 km south of Gilgit town.	Gilgit is the principal river of Gilgit, flowing NW-SE before joining the Indus. It is a 225 km long river.
Tributaries joining at levels below 650 m ASL – Swat River	The river is formed by joining of the two head-streams, the Gabral and the Ushu, joining at Kolan to flow south as the Swat River.	The river commences its course in the Kalam Valley of Swat Kohistan with the confluence of two main tributaries Ushu and Utrar.
Swat	A left bank tributary, joins the river Kabul before the latter joins the Indus at Attock.	The most important hilly river originating from Falaksar (5,918 m ASL), close to the KP–Gilgit borders, the river passes through the important town of Mingora in the Swat Valley.
Panjkora	A right bank tributary of Swat.	
Kabul	Right bank tributary, joins Indus at Attock (630 m ASL).	The river originating from near Kabul collects the waters of its left tributaries notably Panjsher, Alising, Alinagar and Kohar, all originating in the Hindu Kush region.

TABLE 3.2 Minor Tributaries of Indus

Minor Tributaries of the Indus	*These include the tributaries of the Indus that originate in the highlands of frontier region, between Pakistan, more specifically FATA (Federally Administered Tribal Area) and Afghanistan. These tributaries coming from the western mountain ranges and as enumerated from north to south are as follows*
Kurram	A right bank tributary of the Indus joining the river near Mianwali
Tochi	A tributary of Kurram, joining it from southwest
Takwara	A right bank tributary just north of Dera Ismail Khan
Gomal and Zhob	Gomal and its southwest tributary Zhob joining the Indus on right bank close to Dera Ismail Khan
Sawana	Joining the Indus on its right bank, south of Dera Ismail Khan
Vehowa	Right bank tributary
Sangarh	Right bank tributary joins the Indus near Taunsa
Bolan and Nari	Originating from the hills close to Quetta, flow NW-SE in the Kachhi Plain and in their lower course are channelised to join the Indus on its right bank
Gaj	An important right bank tributary of the Indus, joining the later at a point south of Dadu

Down the Tarbela Dam, the Indus is joined on its left bank first by Haro, around 20 km south of Attock, and by the river Soan, which, flowing from NE to SW, bisects the Potwar Plateau. Further south, the Indus is bounded on its left by Sind Sagar Doab and the Thal Desert .

All these tributaries of the Indus form the network of drainage in Pakistan.

The Rivers of Pakistan – Here is a brief account of the five rivers of Punjab, and a couple of them flowing in Khyber Pakhtunkhwa is given.

3.2 Indus – the Principal River of Pakistan

Indus is the 12th largest river in the world, with a length of 3200 km and a basin area of 1.12 million km^2, spread across four countries, Pakistan, India, China and Afghanistan (Figure 3.1). The country-wise distribution of basin area, as estimated by FAO of the United Nations Organisation is as follows (Table 3.3):

3.2.1 The Course of the Indus

The Indus originates in the trans-Himalayan region at a height of 5,200 m ASL at 31°15' N and 81°40'E, northeast of Mount Kailash. The river is

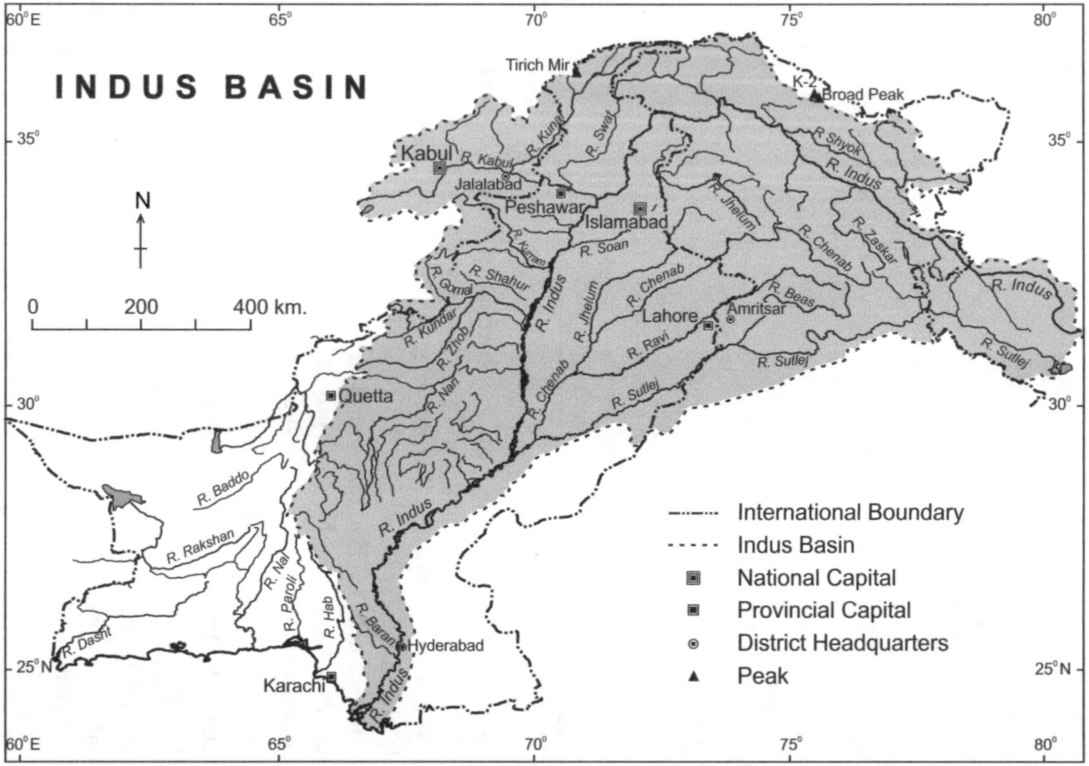

FIGURE 3.1 The Indus Basin Including the Basin Head with Tributaries Extending into Afghanistan, India and China besides Pakistan

TABLE 3.3 Country-wise Distribution of the Basin Area of the Indus

Basin Area of the Indus (km²)	*Countries Included in the Basin*	*Area of the Country in the Basin (km²)*	*As % of the Total Area of the Basin*	*As % of the Total Area of the Country*
1,120,000	Pakistan	520,000	47	65
	India	440,000	39	14
	China	88,000	8	1
	Afghanistan	72,000	6	11

Source: AQUASAT – FAO (UNO) http://www.fao.org/nr/water/aquasat/basins (accessed on 6/8/2016)

known as Sindh–Ka-bab, for about 260 km from its source, before it receives its first tributary Ghar. Flowing further northwest, it skirts Leh at 3,200 m ASL. A little down Leh, the Indus is joined by a left bank tributary, Zanskar River, from southwest. The river descends to 2,500 m ASL in Baltistan, before it receives its right bank tributary Shyok, which flows sub-parallel to the Indus in the northwest direction following an inter-montane valley between the Zanskar Range in the southwest and the Ladakh Range in the northeast of the river. Shyok has a substantial discharge as it is fed by its tributary Nubra which emerges from the Siachin Glacier to the northeast of the Satpara Range .

At Skardu (35°18' N and 74°41' E, at 2,317 m ASL), the Indus receives another right bank tributary, the Shigar which flows from northwest to southeast, bringing enough discharge, received from the snowmelt of the Harmosh Massif. Skardu is a small, little plain, a flat that is the site of some subsistence farming and a moderate concentration of population, giving Skardu the status of the headquarters of Baltistan. The Indus at Skardu is 150 m wide and 10 m deep in winter.

Further down, the river flows in a very narrow valley, for 120 km, before it makes an acute angle turn to the south at 35°50' N and 74°45' E between Sassi in the north and Telichi in the south. Flowing further down, the river receives Hunza and Gilgit, the former a tributary of the latter, on its right bank, coming from the northwest glaciated central part of the Gilgit Mountain. Another river that joins the Indus on its left bank is Astor, flowing SE-NW between Nanga Parbat on the west and Deosai Plateau on the east. It joins the Indus near Telichi, south of Bunji. At Gunar, where the river turns westward, it flows under the shadow of Nanga Parbat (8,126 m), hardly 50 km to the southeast. From Gunar (35°40' N and 74°45' E), the river takes a complete turn to the west, flowing east-west for 140 km before turning south, entering the Khyber Pakhtunkhwa province of Pakistan. South of 35° N, the river enters the sub-montane and piedmont zone of the Himalayas. In the sub-montane zone, flowing north–south, the Indus makes the boundary between the mountainous Kohistan region in the east and Swat and Chitral region in the west, both part of Khyber Pakhtunkhwa.

From its source till it enters the piedmont zone, the Indus flows for over 1,200 km, about 40 percent of its entire course, in the mountainous terrain, with deep gorges and acute angle bends, following lines of structural weakness. It negotiates a course that is umbrella-like, arcuate, passing through gorges abutted by glaciated mountains.

The most significant aspect of the journey of the Indus, through the Himalayan heights, is the glaciated terrain that holds the second largest storehouse of glaciers in the world, next only to the Polar Regions. The perennially glaciated area of this mountainous area is estimated to be of the order of 22,000 km^2.[1] The snow and the glacial melt water is a major contribution to the flow of the river. It is estimated that annually, the Upper Indus caries about 116 km^3 of water, that is roughly half of the total supply of water in the Indus system. It is this discharge from the glacial fields, marked by mighty glaciers like Siachin, Hispar, Baltoro, Gasherbrum and Chokolungma, (refer map 9.1) that open the prospects of several major and minor hydroelectric projects like Daimer Bhasa, in the course of the Indus. Kazmi and Qasumjan (1997) have identified 30 glaciers (7 in the Himalayan, 19 in the Karakoram region and 4 in the Hindu Kush Mountains), ranging in length from a few km to over 70 km, all contributing to the discharge of the river Indus. The contribution of snowmelt is so significant that it is estimated that about 80 percent of inflows to Tarbela reservoir are from ice and snowmelt (Hashmi Daniel, 2014).

The piedmont zone of the Indus course extends from 35° N coming down to 34° N, the latter almost coinciding with the latitude of Attock (33°53' N and 72°17' E). This is the region, where suitability of relief, a relatively higher rainfall and lower evaporation have combined to create a proper site for the building of a dam on the Indus at Tarbela. Attock (634 m ASL), almost midway in the course of the Indus, is the last point where the river, flowing between the Kohat Hills on the west and the Potwar Plateau in the east, leaves the higher terrain, though still contained by hard rocks as at Kalabagh.

Between Attock and Kalabagh, further down, a distance of around 150 km, the river flows in a relatively well-defined channel. The terrain is plateau-like. At Kalabagh, it cuts through the hills, between the offshoots of the Sulaiman Range in the west and the outliers of the Salt Range in the east. South of Kalabagh, the river flows in its own flood plain, swinging widely in the meander belt, often laced like a braided channel, with multiple channels and cut-offs in the transverse cross section. For almost 500 km, from Kalabagh downstream, the Indus flows marking the western limit of the Punjab Plain and parallel to the Sulaiman Range on the west. This is the dry tract bordered by the Sind Sagar Doab on the east. Named after the three *deras*, Dera Ismail Khan, Dera Fateh Khan and Dera Ghazi Khan, the strip of land, between the Indus on the east and the Sulaiman Range on the west, is known regionally as *Derajat*. The river, forming the western border of Punjab, is marked by river islands, sandbars, cut-offs and bits of deserted course. Crossing the river, at some points, requires several streams to be crossed, all parts of a braided Indus. An evidence of the change of course of the river, in its lower course, is in Muzaffargarh district, in south-western Punjab, in a village is named "Bet Basera". It is suggestive of a change in the course of the Indus, bringing the village between the two streams of the Indus, a deserted one and the present one.[2] The stretch of Indus, between Kalabagh and the river's confluence with Panjnad, the combined trunk stream of five rivers (Sutlej, Beas, Ravi, Chenab and Jhelum) at Mithankot (79 m ASL), is marked by a number of barrages that have been built for irrigation. The six barrages (Jinnah barrage, Chasma barrage and Taunsa barrage in Punjab and Guddu, Sukkur and Kotri barrage) represent the irrigation works on the Indus, beside the Tarbela Dam, the largest reservoir.

The Indus enters Sindh at Kashmore and is so wide that locally it is also called "*Darya*. During its course of 700 km in Sindh, it assumes a width ranging from 500 m during winter to 1,500 m during the monsoon months. Before the barrages were built, embankments were constructed to contain the floodwaters. Ever since the three barrages were built at Sukkur, Guddu and Kotri, the flood fury of the river is partly contained. In an unprecedented flood, however, like the flood of August; 2022; these barrages hardly matter; as on an imperceptible slope. The entire area gets submerged.

The stretch of the Indus, further down in Sindh, is marked by inundation during floods, utilised by inundation canals: In Dera Ghazi Khan, there were 16 inundation canals, built during the 17th and 18th centuries and improved during the Sikh regime in the 19th century. It has to be noted that, before the building of the Sukkur barrage in 1932, large areas in Punjab and Sindh were irrigated by inundation canals, fed by the outflow of the flood waters. Some of these inundation canals are still active.

Traversing a distance of 700 km in Sindh, the Indus joins the Arabian Sea, forming a delta, having roughly an area of 7,000 km^2 and a sea frontage of 200 km. The delta is characterised by a number of distributaries of the Indus, radiating from it at the apex of the delta that is close to Thatta, a city around 100 km east of Karachi. The main branch of the Indus recognised as Indus Proper joins the sea at Keti Bundar (24°8' N, 67°27' E).

3.2.2 Amount and Seasonal Variations in Discharge in the Rivers of Pakistan

The discharge data for the rivers of Pakistan are available, for a period of almost hundred years. The number of gauging stations, no doubt, has increased after 1947. The discharge in the Indus as well as its tributaries varies from month to month, with a maximum discharge in August and lowest in the winter month of January. There is a fair degree of correspondence between the figures quoted by AQUASTAT (UNO), the Irrigation Department of Pakistan and the actual measured discharge of the rivers in Pakistan.

The discharge of the Indus is based on the records of the river gauging station at Attock (33°9' N and 72°25'E). This represents the total flow of the Indus and its principal tributaries in the mountainous region and that of Kabul and its tributaries. The discharge figure for the other rivers, Jhelum, Chenab, Ravi and Sutlej, is taken from the gauging station on Panjnad at 29°35' N and 71°3' N longitude). After the confluence of Panjnad with the Indus at Mithankot, the Indus enters the arid regime without receiving any significant tributaries. The discharge data of the Indus at Kotri, a twin city of Hyderabad (Sindh), is significant only to the extent that it gives an idea of the volume of annual flow of water into the Arabian Sea (Table 3.4 and Table 3.5).

It is to be noted that the discharge measurement of the Indus is at Attock, downstream from the Tarbela reservoir. Also the irrigation by the tributaries of Kabul like Swat and the storage at the Warsak Dam at Kabul, divert considerable volume from the river Kabul, causing a reduced discharge of the river Kabul into the Indus. Equally important is the fact that the annual discharge of 63.6 km^3 at Panjnad, represents the residual flow of the tributaries of the Indus, after they are tapped for irrigation, upstream. To that extent, the discharge of Panjnad does not represent the total discharge of the tributaries of the Indus. At best, it is an approximation. The total flow of water obtained by the addition of annual discharge of the Indus and Panjnad at some critical points is only indicative and does not offer a complete picture.

TABLE 3.4 The Discharge Data of Pakistan Rivers from NASA Archives[3]

River	*Gauging Station*	*Exact Location of the Gauging Station*	*Observation Period*	*Mean Annual Discharge Rate m^3/sec*	*Mean Annual Discharge Rate km^3/yr*
Indus	Attock	33°9' N and 72°25' E	1973–79	2,544	80.3
Chenab Panjnad	Panjnad	29°35' N and 71°3' E	1973–79	2,016	63.6
Indus	Kotri	25°35' N and 68°35' E	1973–79	2,644	71.3
Total of the Indus and Panjnad					143.9

TABLE 3.5 Capacity of the Major Reservoirs in Pakistan[4]

Name of the Reservoir	*Nearest City*	*River*	*Year of Completion*	*Height (m)*	*Capacity (km³)*
Mangla	Mangla	Jhelum	1968	118	10.15
Tarbela	Ghazi	Indus	1976	137	11.96
Chashma (barrage)	Mianwali	Indus	1971	–	0.87
Total					22.98

The total flow of the rivers of Pakistan is roughly 167 km³, which includes 23 km³ storage and 144 km³ of flow measured at two critical points, one at Attock on the Indus and the other at Panjnad near Mithankot; the former gives the approximate flow of river Indus and the latter the combined flow of the tributaries of the Indus.

3.2.3 River Flow Estimate of Pakistan by WAPDA (Water and Power Development Authority)

The estimate by WAPDA is more authoritative. They have given the measured flow of the rivers. The only difference is that the discharge of the Indus is measured at the rim station at Tarbela or at Kalabagh. Since the discharge at Kalabagh also includes the contribution of Kabul and its tributaries, the flow at Kalabagh is taken to represent the Indus flow. The measured flow of the rivers of Pakistan under the caption "River Flow Availability" by WAPDA, presents the flow data, in its Compendium on Environment, for 19 years from 1996–97 to 2014–15. These give an idea of the actual flow in these principal rivers Indus, Jhelum and Chenab (Table 3.6).

The annual flow, including both, *kharif* and *rabi* season, calculated as the mean of 19 years (1996–97 to 2014–15) is around 130 million acre/ft, and when converted into cubic metre, that comes to 160 billion m³. A higher figure of 160 billion m³, as opposed to 144 billion m³ (the figure obtained by adding the discharge average of only six years) obtained from NASA, based on two stations is understandable. For all practical purposes, the annual discharge in the three principal rivers, Indus, Jhelum and Chenab, adds up to 130 million acre ft or 160 billion m³ ($160*10^9 m^3$). Out of the total flow of these rivers, the Indus accounts for roughly two-thirds, i.e., 65.8 percent, Chenab 18.25 percent and Jhelum 16 percent. It must be mentioned that the storage in the reservoir in the Indus at Tarbela and Jhelum at Mangla are not taken into account.

The location of monitoring stations, up or downstream on a river, determines the volume of discharge. Normally, the downstream stations show a

TABLE 3.6 River Flow Availability – During Years of Maximum and Minimum Flow. (It is observed that 1996–97 has been the year of maximum flow while 2001–2002 is the year of minimum flow.)

Years	*Water Flow Availability in Different Seasons (million acre/ft)*									
	Jhelum River at Mangla			*Chenab River at Marala*			*Indus River at Kalabagh*			*Grand Total*
	Kharif	*Rabi*	*Total*	*Kharif*	*Rabi*	*Total*	*Kharif*	*Rabi*	*Total*	
1996–97 the year of maximum flow	24.93	4.11	29.04	27.48	4.41	31.89	85.08	15.23	100.31	161.25
2001–2002 the year of minimum flow	8.23	3.68	11.89	16.00	2.9	18.9	55.65	10.72	66.37	97.16
Mean flow for 1996–97 to 2014–15	15.56	4.94	20.51	19.24	4.5	23.74	71.19	14.30	85.49	129.84

Source: WAPDA Compendium on Environment, p. 192.

higher discharge than the stations upstream. In case of the Indus, the discharge recorded at Tarbela and at Kalabagh, around 240 km downstream, varies greatly, as at latter point the contribution of the river Kabul and its tributaries is added. The Indus flow at Tarbela is 15 percent less than at Kalabagh.

3.2.4 The Discharge of the Eastern Tributaries, Ravi and Sutlej, of the Indus

The flow of the eastern tributaries is limited as they are tapped upstream in India. In fact, both these rivers Sutlej and Ravi that flow in Pakistan have been assigned to India in the Indus Water Treaty (1960) between India and Pakistan. They have been rejuvenated in Pakistan, by linking them with Chenab, through a triple project linking first Chenab with Ravi, and then Ravi with Sutlej. The flow of these two rivers Ravi and Sutlej are given for points above the link with Chenab (Table 3.7).

The total flow in the streams of Pakistan, at rim stations (Kalabagh on the Indus, Mangla on the Jhelum, Marala on the Chenab, Baloki on the Ravi and Suleimanki on the Sutlej) adds up to an average of 132.5 million acre ft or 163 billion m^3.

Thus, in whatever way one calculates, the total annual stream flow in Pakistan is roughly 160 billion m^3.

3.3 An Overview of the Tributaries of the Indus

The Jhelum – Referred to as Vitasta in Rigveda, the Jhelum originates in a deep pool at Vernag, about 60 km east of Anantnag at a height of 1900 m ASL and has a length of about 800 km and a catchment of 39,200 km^2. The river flows northwest between the two parallel ranges of Pirpanjal on the left and Zanskar on the right. Passing by Srinagar and through the Dal Lake and 30 km further down through the Wular Lake, the river reaches Sopore and Baramula from where the river turns westward making a concave loop, and reaching Muzaffarabad flowing over a distance of about 100 km. Near Muzaffarabad, the Jhelum is joined at its right bank tributary Kishenganga/ Neelam. From Muzaffarabad, the river turns south, making an acute angle bend. From Muzaffarabad, the Jhelum assumes a north–south course, in the direction opposite to its flow in the valley of Kashmir. Down Muzaffarabad, right up to Kohala, a distance of 25 km, the river flows in a gorge, hemmed in by the offshoots of Pirpanjal on the left and the Kajnag Mountain on the right. From Muzafferabad to Mirpur, the site of the Mangla Dam, the river flows north–south for 150 km, where one of Pakistan's major reservoir is developed on the river.

Leaving the Mangla Dam, the river enters the plain of Punjab at Jhelum town, about 25 km south of the Mangla Dam. From Jhelum the river takes a south-westerly course, flowing leisurely, south of the Salt Range till Khushab,

TABLE 3.7 Eastern Rivers (Ravi and Chenab) at Rim Stations in Million Acre ft

Year	*Ravi at Baloki* [(a)]			*Sutlej at Sulemanki* [(b)]			*Total*		
	Kharif	*Rabi*	*Total*	*Kharif*	*Rabi*	*Total*	*Kharif*	*Rabi*	*Total*
2011–12	1.48	1.08	2.56	3.57	0.22	3.79	3.79	1.30	6.35
2012–13	0.97	0.64	1.61	0.15	0.02	0.17	0.17	0.66	1.78
2013–14	3.79	0.98	4.77	2.10	0.05	2.15	2.15	1.03	6.92

Source: Water Resource Management Directorate WAPDA

(a) Ravi at Baloki, above the link with Chenab and (b) Sutlej at Suleimanki above with the link with Ravi where it again turns south and joins the Chenab of the Punjab Plain, near a place called Dhiraj, about 20 km southwest of Jhang.

Though considered a tributary of the Chenab, the Jhelum is the second most important river of Pakistan in terms of its water resources. The Mangla Dam on the Jhelum impounds a huge reservoir with a capacity of 10.15 million m^3 of water.

The Chenab – The river originates in Lahul near the Baralacha Pass in the snow-covered area of northeast Himachal Pradesh (India) and is formed by the union of two lower order streams, Chandra and Bhaga. It has a length of over 1,200 km and commands a catchment of 41,750 km^2. The Chenab has several hundred kilometres of southeast–northwest course in Himachal Pradesh and another 200 km of east–west course in Kashmir in Jammu–Kashmir, before it enters Pakistan. The river flows through Pangi in Himachal, and turning westward cuts across the outer ranges of the Himalayas near Kishtwar. Passing by Doda, it further negotiates the Pir Panjal heights through Ramban, where it is bridged for the passage of the Jammu–Srinagar Highway. Further down, the river enters the Pakistan territory near Gigrial, 40 km north of Sialkot. The river enters Pakistan in a plain territory, and 60 km downstream, it is crossed over by the Grand Trunk Road between Wazirabad and Gujrat. About 20 km west of Wazirabad is the head-works of the Lower Chenab Project at Khanki, in Gujranwala district of Punjab.

It has to be emphasised that the Lower Chenab Canal is one of the earliest irrigation works in Punjab. Completed in 1898, the irrigation project brought about three million acres of land under irrigation.

Chenab, flowing southwest ward drains the central part of Punjab and passing through Chenabnagar, finally joins the Jhelum south of Jhang.

The Ravi – The river Ravi rises in the Himalayas near the Rohtang Pass in Kangra district of Himachal Pradesh. The river is 880 km long and has a catchment of 24,960 km^2. Under the Indus Water Treaty (1960), the Ravi is assigned to India. The river enters Pakistan at a point about 40 km southwest of Pathankot (India). For some distance, it runs along the Indo–Pakistan border. Leaving the international border, about 50 km northeast of Lahore, the river flows by Lahore, the capital of Punjab. Lahore, situated on the left bank of the river is undoubtedly the most important town on the Ravi. Flowing for about 500 km southwest of Lahore, the river joins the Chenab near a place called "*Khalpur Sind*" about 60 km northwest of Multan.

The river is dammed for irrigation at Madhopur in Gurdaspur district of Punjab (India), to construct the head-works of Bari-Doab Canal, which irrigates the northern district of Indian Punjab. After that, it remains an emaciated stream, until it receives an infusion of water at Baloki from the Chenab.

The world famous Harappa ruins, a part of Indus valley civilisation are on the left bank of the Ravi in Sahiwal district, the former Montgomery district of the British colonial era.

The Sutlej – Near only to the Indus, the Sutlej is the largest river of Punjab, though only a section of it flows in Pakistan, through Bahawalpur district of Southeast Punjab. It is 1,536 km long and has a catchment area of 25000 km^2. The river joins the Chenab at Bakhri, close to the head-works of Panjnad barrage in Pakistan.

The Sutlej emerges from close to the Mansorowar Lake of Mt Kailas and the northern glaciers of Mt Kamat. Flowing northwest ward at a height of roughly 4,500 m ASL, the Sutlej cuts a deep gorge, about 1,000 m deep. According to Krishnan, in its gorge section, the Sutlej is comparable to the Grand Canyon of Colorado. At Shipki la, (approx 3,000 m ASL), the river enters the Indian Territory. At Shipki, the river takes a right angle turn flowing southwest for about 50 km, before it takes a westward turn. The straight flows and right angle turns suggest a structural control of the river. Taking a series of turns, the river crosses the Himalayan Range, traverses the hilly stretch of Himachal Pradesh, and reaches a point about 20 km north of Shimla. Further down, the river is dammed at Nangal, impounding a northwest–southeast reservoir, known in India as Govind Sagar, after the last Guru of the Sikh Guru Govind Singh. Popularly known as Bhakra Nangal Project with a dam height of 226 m, one of the highest in the subcontinent, Govind Sagar has a capacity of holding 5,620 million m^3 of water.

Traversing the plain of Indian Punjab, the Sutlej touches Pakistan about 15 km east of Kasur and flows parallel to the Indo–Pakistan international boundary, before it turns southwest in Bahawalnagar district of Punjab, near Fazilka in Indian Territory.

The Sutlej, flowing through Bahawalnagar and Bahawalpur districts of Punjab – both parts of the old Bahawalpur princely state – joins the Chenab at Alipur, located at the northern extremity of Panjnad. In Pakistan, the Sutlej flows for 250 km, with only limited potential for irrigation. As it is assigned to India under the 1960 Indo–Pakistan Treaty, the river is fully exploited in India, leaving only limited water for irrigation in Pakistan. To rejuvenate the river, it has to be fed by other rivers, like the Chenab and the Jhelum. A barrage at Suleimanki is meant to irrigate the area of Bahawalpur, bringing the water through, Jhelum–Chenab–Sutlej triple project with Qadirabad–Baloki and Baloki–Suleimanki links.

3.3.1 Drainage Characteristics of Pakistan

The drainage of Pakistan in the three provinces of Pakistan, Punjab, Sindh and Khyber Pakhtunkhwa is well integrated. In the high mountainous region of Gilgit and Baltistan and even Khyber Pakhtunkhwa, the drainage lines follow lines of structural weakness guided by a fault or a fracture line or some synclinal depression. The course of the Indus is like an arch with a convexity to the north and an inflexion point around 35°50' N and 76°40'

E where it is joined by Gilgit. Often referred to as an antecedent river, the Indus flows presumably following the Indus suture line defining the contact of the Indian and the Central Asian plate. The great central thrust of the Himalayas also follows an arcuate outline starting from the eastern end of the Himalaya to Chaman, a point 130 km northwest of Quetta. The course of the Indus marks roughly the northern subterranean outline of the Indian plate. From its origin, right up to 74° E longitude, the Indus flows in furrow like valleys bordered by mountains on either side. These valleys are structurally guided. The north-south course of the river in Khyber Pakhtunkhwa is also a response and adaptation to the alignment and structural arrangement of the mountain ranges. Once in the plain, the mighty Indus adjusts its course, to side-step the Salt Range, with a stratigraphic range, extending as far in the past as the pre-Cambrian. Further south, the river turns east avoiding the subterranean ridge of Jurassics at Jacobabad, and flows into Sindh basin in a deep syncline 6,000 m deep, carrying sediments from Jurassic to Cretaceous to a complete sequence of Tertiary sediments.

The tributaries of the Indus, well-spaced make a dentritic pattern with the Indus as a trunk in Sindh and four tributaries Sutlej, Ravi, Chenab and Jhelum as branches.

Like the Indus, all its tributaries follow an arcuate course, first flowing northwest from the source region following inter-montane valleys, then suddenly turning south or southwest like the Indus and flowing in Punjab Plain before joining the Chenab, a major tributary of the Indus. Thus, a sharp turn is noticed in all the tributaries of the Indus, especially seen in the Jhelum and the Chenab. The course of these tributaries, the near straight courses, often connected with a transverse link that follows a fracture line, suggests structural control. The flow of the Jhelum from Sopore on Wular Lake to Uri in a straight NE–SW course and then from Uri to Muzaffarabad, before the river takes a complete southward turn is certainly the course of the Jhelum following the geological contact zone, fracture lines. The Uri–Muzaffarabad stretch of the Jhelum follows the boundary fault.

3.3.2 The Western Trans-Mountain Rivers

The rivers emerging from the central highlands of Afghanistan and joining the Indus almost at right angle, represent a different set, draining smaller basin. These rivers cutting across the highlands of Afghanistan–Pakistan border descend east in a torrent to fall into the Indus. The most important is the Kabul River originating from highlands west of Kabul and collecting a large number of tributaries, especially from the Hindu Kush Mountain, it cuts the frontier mountains, makes an inter-montane basin of Peshawar, before joining the Indus near Attock. Another river, though much smaller, is Kunar River, which originates close to Tirich Mir Mountain (7,690 m) in Chitral, and flowing in northeast–southwest Chitral Valley crosses the offshoots of Hindu Kush into Afghanistan before joining the river Kabul. The Kabul and its tributaries, especially those originating in the glaciated terrain of the northern region, are very significant for Pakistan, as they bring substantial volume of snowmelt from the glaciers of Hindu Kush, and add to the discharge of the river Kabul and the Indus.

The other two tributaries Kurram and Gomal , both coming from across the western mountains, make narrow valleys and some intermontane narrow plain. Kurram joins the Indus on the right bank, opposite Mianwali, which is located on the east bank of Indus. Another river Gomal originating in the highlands of Paktika province of Afghanistan crosses the Pak–Afghanistan border at around 32°N, flows through Waziristan tribal area, receives the water of Zhob on its way and joins the Indus near Dera Ismail Khan.

3.3.3 Some Special Features of the Pakistan Rivers

Unlike most South and Southeast Asian countries, where major rivers have their lower courses and deltas in a relatively high rainfall area, Pakistan's rivers flow from snowfield glaciers or relatively high rainfall source region to their confluence with the Indus in a relatively dry area. The Indus delta receives hardly 125 mm of annual rainfall. This makes the tributaries of the Indus and the contribution from the upper reaches far more vital. In fact, rivers in Pakistan are as important as monsoon in India. In India as in other South Asian countries, monsoon rainfall is the major source of water. The annual flow of the stream and the ground water recharge to store water depend on rainfall. In contrast in Pakistan, with two-thirds of the area receiving less than 250 mm of annual rainfall, there is heavy dependence on stream flow fed by the glacial melt in the mountainous region and the inflow of water from beyond the border of the country. Of the total annually renewable water resource of the country amounting to 246 (10^9m^3/year), 191.81 (10^9m^3/year) or 77 percent comes from beyond the border of Pakistan through its rivers.

The rivers not only transport water from far off regions and flow through the plains of Punjab and Sindh, but these are also instrumental in recharging the aquifers, especially in Punjab, which are tapped through several hundred thousand tube wells for irrigation.

The variation in the amount of discharge in the principal rivers; at rim stations; Tarbela on the Indus, Mangla on the Jhelum and Marala on the Chenab, is minimal regardless of rainfall. These are, no doubt, lean years as in 1999–2000 and 2000–2001, in the discharge of rivers, but deviation from a mean is not very significant, and not reflected in the acreage irrigated in those specific years. To quote an example, the mean annual flow in the Indus at Tarbela is 86 million acre ft, with a maximum flow of 105 million acre ft in 2010–11 and a minimum of 70 million acre ft. In 2000–2001, yet, the total area irrigated in 2000–01 is 17.82 million hectares, not much less than 18 or occasional 20 million hectares irrigated in a specific year. The shortfall in stream flow resulting in lower flow in the canals is made up by tube wells that pump out water from the aquifers storing a considerable volume of water, as a result of the seepage from the canals.

The effect of monsoons on agricultural economy and surface water resources of Pakistan is minimal, though it may affect life in a variety of ways.

3.4 Rivers, the Life Line of Pakistan

Needless to say, rivers are the life line of Pakistan, vital for a healthy socio-economic existence. The beautifully organised natural network of rivers, especially in Punjab, is like an arterial system in a human body, through which the rivers and the canals perpetually supply the much needed life blood for a healthy body economic of Pakistan. What is denied to the country by way of the scanty rains is compensated by the rivers and canals. The agricultural economy of the two provinces, Punjab and Sindh, and their water supply systems are so deeply integrated with the perennially flowing rivers and the canals that the latter have become a part of their daily life and entered in the folklores of the villages and towns in their basins. It is worth noting that both Punjab and Sindh are eponymous, the former to *Punj ab,* the five waters, and the latter directly to the Sindhu, the original name of the Indus.

It may be mentioned that, unlike India, there is little religious-cultural significance attached to the rivers in the monotheistic Islamic society of Pakistan except in Sindh, where a section of the population appears greatly attached to the river Indus for its bounty. The Indus, the only perennial river, is important to the people of Sindh, as it traverses from north to south and nourishes the economy and culture of the region. A source of perennial water, with a wide network of canals, the Indus is the life blood of Sindh, parallel to the case of Punjab where the five rivers of the province infuse the proverbial life blood in the socio-economic life of the province.

To Punjab, the Indus remained, for long, marginal to its existence, physically and economically. It formed the western boundary of the province except in the southwest corner, where in Dera Ghazi Khan where the provincial boundary transgresses the Indus. While the canals from Ravi, Chenab and Jhelum follow a natural slope to irrigate the entire stretch of the agricultural land in Punjab, the alignment of the Indus is not favourable to lay canals to irrigate the plains of Punjab. The thalweg of the Indus, representing the lowest point on the transverse profile of Pakistan, occupies the western margin of the province, while much of the province and its agricultural land occur east of the Indus. The canals from the Indus could not easily transport water against the natural slope. Thus, despite an abundant source of water, the Indus remained untapped for irrigation in Punjab.

In Sindh, the irrigation potential of the Indus was utilised earlier through Sukkur barrage. Built in 1932, the barrage not only extended the limits of agricultural land in the province but also boosted its agricultural productivity. It is only lately that a number of barrages on the Indus have been built that benefit largely the narrow strip of land between the Indus and the Sulaiman Range, and to a limited extent the river`s left bank flood plain. The waters of the Indus could not be easily and effectively brought to reclaim and irrigate the land of Punjab. And to the farmers of Punjab, Ravi , Chenab and Jhelum are far more relevant than the Indus.

The situation is gradually changing as the power generation potential of the river is being realised. Unlike the utilisation of Indus-water for irrigation, confined largely to Sindh, the power generation will benefit the country as a whole. It is this energy potential of the Indus which has made it nationally important.

Indus has a vast potential as a source of energy, and the river is being harnessed for generating hydroelectric power and storage of water for emergency situations. The Tarbela Dam on the Indus, at a location about 100 km from Islamabad, is a hydroelectric power project with an installed capacity of 4,888 MW. The reservoir, with a surface area of 260 km^2, can hold 106 million cubic metre of water. This is truly a national project, as the hydroelectricity generated through the project and fed into the national grid, can be transmitted to all parts of Pakistan. Similarly, the Daimer-Bhasha hydroelectric project on the Indus, about 300 km north of Tarbela, impounding 10 billion cubic metre of water and an installed capacity of 4,800 MW, likely to be completed by 2028, is another project of national importance.

For Khyber Pakhtunkhwa, the Indus is a barrier and benefits the region marginally by the inundation canals that irrigate the land adjacent to it. But, undeniably, the main sources of water for agricultural growth and productivity in Khyber Pakhtunkhwa are the three principal rivers, the Kabul, the Swat and the river Kurram – all tributaries or sub-tributaries of Indus – and their tributaries. Thus Kabul, Swat and other rivers of Khyber Pakhtunkhwa are as important to the economy and culture of the province as Ravi, Chenab and Jhelum are to Punjab.

That the potential of the Jhelum and the Indus was not fully exploited, was realised only in the late 1950s of the last century, but it took a long time to build the two most important reservoirs, the Tarbela Dam on the Indus, completed in mid-seventies, and Mangla on the Jhelum, completed in the late sixties. These have added significantly to the water resource of Pakistan. In fact, the two reservoirs together store around 22 billion m^3 of water, more than the combined storage of all the reservoirs on the tributaries of the Indus in India.

The network of canals, the annually recharged aquifers, the storage of water in some larger reservoirs, a high degree of irrigation, especially in Punjab and Sindh, are the most reassuring aspects of Pakistan`s water economy. Any disruption in the flow of these rivers, or even a perception, that the discharge of any of these rivers is going to be affected adversely, is reason enough to cause some anxiety to the people of Pakistan. As most of rivers flow in their upper courses through India, any river project in India involving these rivers sends alarm signals, real or false, that the flow of the specific river is going to be affected. Such situations are remedied by periodic meetings of experts on both sides. The engineers and the other technical personnel in both the countries are, as expected, quite sensitive to each other`s position on the issues involved, and differences are amicably resolved. Only in rare cases, arbitration or a court is needed to adjudicate.

Rivers are the most precious gift of nature to Pakistan, and it must be said to the credit of the people of Pakistan that they have one of the best planned river water utilisation systems in the world.

Notes

1 Ojeh, E. (2006) Hydrology of the Indus Basin, Pakistan, quoted from AQUASAT, FAO (UBO) 2016.
2 Imperial Gazetteer, vol. 13, p. 361.
3 NASA Earth Observation System, Oak Ridge National Laboratory Distributed Active Archive Centre (ORNLDAAC), http://daac.ornl.gov/data/rivdis/Stations (accessed on 13–8–2016).
4 FAO/UNO -AQUASTAT (2011) Indus Basin, http//www.fao.org./nr/water/aquastat/Basins (accessed on 6–8–2016).

4

BIO-CLIMATIC BACKGROUND

Climate, Soil, Vegetation and Natural Hazards

4.1 The Climate

Among the many elements of weather that constitute the climate, the two most significant elements are the temperature and rainfall and their seasonal distribution. Several other elements, like atmospheric pressure, wind direction and velocity, humidity and the number of rainy days, duration of sunshine and the number of frost-free days, storms and variability in weather, are important in giving a special effect to the climate of a place. Yet, temperature conditions and rainfall during the year, with their seasonal variations, remain the most basic element of a region. To a large extent, the temperature of a region is primarily determined by its relative location on the globe, between the equator and the poles, as the lower latitudes having the advantage of receiving more concentrated solar radiation, are warmer than the higher latitudes. Besides the latitude, the altitude of a region or a place above sea level shows its effect through the decrease in pressure and rarefied atmosphere that fails to create a blanket, thick enough to retain the heat, developed during the day through insolation. Thus, a higher latitude and a higher altitude affect the temperature equally; the difference is that latitudinal effect is global while the altitudinal impact is highly localised. Besides, the proximity to oceans with a large body of water, the vapour content and the daily movement of the sea and land breezes affect the temperatures of the coastal areas, sometimes as deep as 100 km inland. As for the rainfall, there are a few wind systems that bring rainfall. These may be part of a planetary system like the trade winds, regionally active monsoon winds or the ever blowing westerlies and the cyclonic storms. These systems are often regionally modified, but their underlying existence is ever present.

DOI: 10.4324/9781003510147-5

Pakistan occupies a position in the warm temperate zone of the globe, in the northern hemisphere, extending between 24° N latitude, just north of the Tropic of Cancer, and 37° N latitude, a distance of roughly 1,400 km from south to north, and an west–east distance of around 1,100 km along 28° N latitude. With an area of 796,096 km^2, Pakistan has a long coastline, a mosaic of altimetric zones, plains, plateaus and high mountains, as well as river valleys, alluvial plains and deserts, all reacting differently to atmospheric elements and thus causing some degree of modification of climate. Besides the above factors that are constantly at play, there is also an increasing role of human activity, specifically the burning of fossil fuels and destruction of forests, a growing concentration of CO_2 in the atmosphere with a blanketing effect to hinder the escape of terrestrial radiation and accumulation of heat leading to warming of the atmosphere. The latter being fluid, the risk of global warming and consequently a change in climate is real. In this chapter, it is thought prudent to adopt the temperature and rainfall normals for understanding the climate of Pakistan, leaving the wider question of global warming and climate change to the specialists in the area.

4.1.1 Temperature Conditions

The location of Pakistan, away from the equator, has introduced an element of seasonality and seasonal variations in temperature, and given rise to a well-delineated summer and winter season. Some countries in South Asia, notably Bangladesh and India, also have a rainy season as a third component in the seasonal rhythm of the year. These countries receive prolonged copious rains and have introduced rainy season as a separate season in their climatic analysis. By contrast, Pakistan does not experience heavy and prolonged rains. The rainy season in Pakistan, except in a few areas, is not very pronounced. Pakistan has two main seasons – summer and winter. One may add autumn and spring and any other for the transitional period.

4.1.1.1 Summer Temperatures

Summer temperatures are universally high varying with latitude, altitude and continentality. June is by far the hottest month in Pakistan and July is equally hot (Figure 4.1). There are two broad temperature zones in Pakistan: (1) The lowlands of Sindh, Balochistan and Punjab in the south and south-eastern part of the country and (2) The northern and northwest mountainous regions, coinciding with Khyber Pakhtunkhwa and the mountainous part of Balochistan. Mountainous regions are, to a larger extent, in the higher latitudes as in Khyber Pakhtunkhwa or in mid-latitudes as in Balochistan. The lowlands of Pakistan, confined between 24° N and 32° N latitude, have indeed very hot summers, moderated only by the influence of

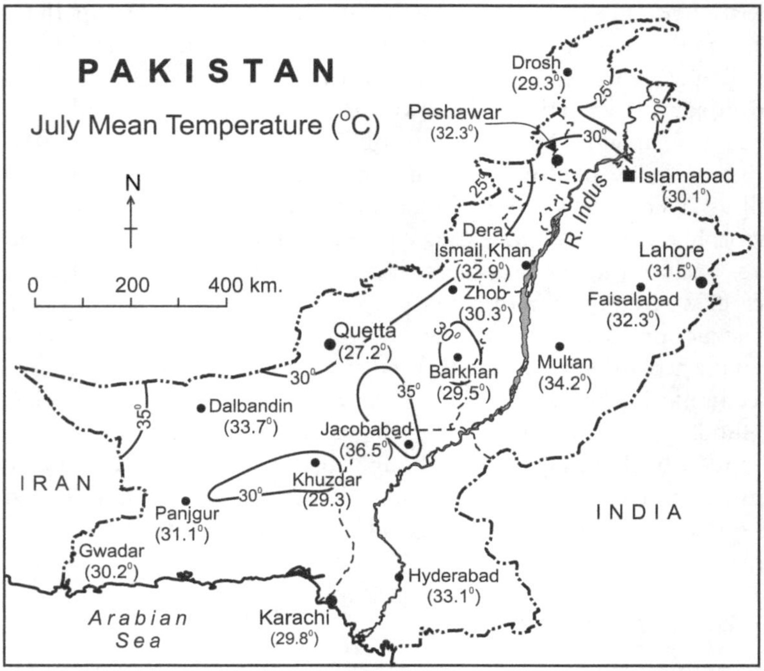

FIGURE 4.1 Mean Temperature Condition in July in Pakistan

sea breezes in the coastal areas, or occasionally by a localised mountainous terrain.

4.1.1.2 Temperatures in Lowlands during Summer

The mean temperatures in the plains of Punjab and Sindh in the month of June vary between 30.5° C on the coast, as in the case of Karachi and Gwadar, to 35.5° C at Multan (123 m), deep inside Punjab. Besides, places with extreme continentality and lower altitude record exceptionally high temperatures. For example, the mean temperatures for the month of June for Sibi (Balochistan), Jacobabad (Sindh) and Rahim Yar Khan in Punjab are 38.5° C, 34.9° C and 36° C, respectively. The June maxima of temperature of all these places are still higher, with lower values for the coastal areas.

4.1.1.3 The Impact of Coastal Location

As can be observed from the above table, despite being located at a lower latitude, the coastal stations show a much lower mean maximum temperature, and an overall mean temperature in June than the inland stations. The impact of coastal location, primarily through daily sea breezes is obvious. Before the

TABLE 4.1 Summer (June) Temperatures for some Stations in the Lowlands of Pakistan

Stations	*Coordinates*	*Altitude (m. ASL)*	*Mean Temperatures for June*			*Diurnal Range of Temperature*
			Mean Maximum	*Mean Minimum*	*Mean Temperature Recorded*	
Coastal Stations						
1. Karachi	24° 54’ N 66° 56’ E	22	33.3	27.9	30.6	5.4
2. Gwadar	25° 08’ N 62° 20’ E	30	34.0	26.9	30.5	7.0
Inland Stations						
1. Lahore	31° 33’ N 74° 20’ E	214	40.4	27.4	33.9	13.0
2. Multan	30° 20’ N 71° 26’ E	122	42.2	29.0	35.5	13.2
3. Dera Ismail Khan	31° 49’ E 70° 55’ E	171	41.8	27.0	34.4	14.8
4. Faisalabad	31° 26’ N 73° 08’ E	185	41.0	27.6	34.3	13.4
5. Jacobabad	28° 18’N 68° 28’E	56	44.3	29.4	34.9	14,9
6. Sibi	29° 33’N 67° 53’N	134	46.0	31.1	38.5	14.9
7. Sukkur	27° 42’ N 68° 50 `E	67	43.2	28.3	35.7	14.9
8. Peshawar	34° 02’N 71° 56’E	237	40.4	25.7	33.1	14.7

afternoon temperatures shoot up, the sea breezes following the pressure gradient created by the sudden rise of temperature on land flow in lowering the temperatures in the coastal areas. A coastal zone of 60–80 km is usually influenced by sea breezes, lowering the temperatures in summer and moderating the cold conditions during winter by injecting humidity in the atmosphere that prevents rapid terrestrial radiation and creating a less severe winter.

In contrast, in the inland locations, the rise of temperatures during the day and radiation during the night remain unimpeded and show a very high diurnal range of temperature. Thus, while the diurnal range of temperatures in coastal areas lingers between 5° and 7°C, the continental locations show a much higher diurnal range of temperatures during summer.

Thus, the lowlands of Pakistan have very hot summers, the mean day temperatures rising above 40°C and the night temperatures not dipping to any point below 25°C, the diurnal range of temperature varying between 13° and 15°C. The coastal locations, however, have a lower temperature, the mean day temperatures often below 35°C and the minimum temperatures in the night varying between 25° and 27°C, with a much lower diurnal range of the temperature.

As in summer so in winter, the coastal lowlands show a higher temperature in January than the stations inland (Figure 4.2). The day temperatures in inland locations are slightly lower than the temperatures in the coastal

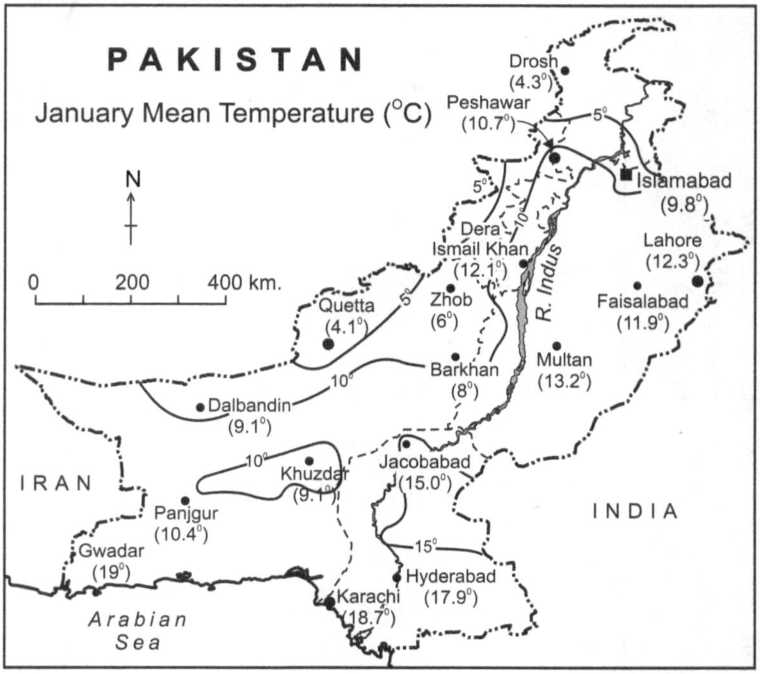

FIGURE 4.2 Mean Temperature Condition in January in Pakistan

areas, but night temperatures of the coastal locations are unmistakenly higher, sometimes leading by 10°C. Similarly, the diurnal range of temperatures in the inland locations is significant higher than the stations on the coast. The coastal locations have the advantage of sea breezes that moderate the temperatures during the day, and the build-up of humidity during the day keeps the night relatively warm.

The continental locations like Lahore, Faisalabad and Multan in Punjab, Jacobabad in Sindh and Sibi in Balochistan have extreme conditions, having 20 to 22°C temperatures during the day and 4° to 6°C during the night, showing a diurnal range of temperature as high as 15 to 17°C.

The lowlands of Pakistan, leaving the coastal areas have severe winters with mean temperatures not rising above 16°C and even during the warmest hours of the day the temperatures linger around 20°C. Lahore and Faisalabad have a mean maximum of less than 20°C. Figure 4.2

4.1.1.4 Summer Temperature Conditions of North-western and Western Highlands

The highlands of Khyber Pakhtunkhwa and Balochistan fall in this category. These include areas in the extreme north of Pakistan around 36° N latitude, represented by stations like Chitral (1,494 m), Dir (1,420 m), Abbottabad (1,234 m) in the north and Miram Shah (923 m) and Parachinar (1,728 m) in the south of Khyber Pakhtunkhwa. Similarly, a large part of Balochistan has highlands, represented by the temperatures of Khuzdar and Kalat, Quetta and Zhob. The temperatures of these stations are given in the table below.

Most of the above stations are located either at higher latitude or altitude or both. Chitral and Dir, the northern stations in this group have very different temperatures from the stations in lowlands. Most of the stations have mean maximum temperature in June, between 30° and 35°C. The contrast is very striking from the plains where the mean maximum temperature in June ranges between 40 and 45°C. Similarly, the mean maximum temperature in the mountainous terrain ranges from 15° to 20°C, and at higher altitudes, as at Kalat (2,015 m), the mean minimum dips as low as 13°C.

While the summers in the highlands present a pleasant weather, in the plains these are severe and very uncomfortable. In the mountainous region while the maximum day temperatures linger between 30°C and 35°C, it is the mean minimum which shows greater variations. The mean minimum declines with an increase in height, and so also the diurnal range of temperature. Kalat (29.2°C) at a height of 2,015 m shows the lowest mean monthly minimum for the month of June at 13.5°C. Contrast this with Khuzdar, which, though at a lower latitude, shows a higher mean minimum because of the lower altitude.

TABLE 4.2 Summer (June) Temperatures of some Stations in North-West Highland

Stations	*Coordinates*	*Height in m*	*Mean Maximum Temperature*	*Mean Minimum Temperature*	*Average Temperature*	*Diurnal Range*
Chitral	35° 51' N 71° 50° E	1,494	31.8	18.0	24.9	13.8
Dir	35° 12' N 71° 51' E	1,420	34.6	20.3	27.4	14.3
Abbottabad	34° 09' N 73° 13" E	1,234	33.7	21	27.3	12.7
Miram Shah	33° 57' N 57° 67' E	923	35.8	24.2	30.8	13.3
Parachinar	33° 37' N 70° 05' E	1,728	30.9	18	24.4	12.9
Quetta	30° 15' N 66° 53' E	1,589	35.3	15–9	25.6	19.4
Zhob	31° 21' N 69° 28' E	1,407	37.2	22.6	29.9	14.6
Kalat	29° 02' N 66° 35' E	2,015	31.9	13.5	22.7	18.4
Khuzdar	27° 50' N 66° 36' E	1,232	37.6	20.5	29.0	17.1

4.1.1.5 Summary of Summer Temperatures

Pakistan has two distinct temperature domains: the temperatures in the plains and temperature in the highlands. In the plains, like the Punjab Plains, the average daily maximum temperature in June is between 40°C and 45°C, whereas in the mountainous areas of Khyber Pakhtunkhwa and Balochistan, the mean daily maximum for June is between 30°C and 35°C, a difference of 10°C. Similarly, the average daily mean minimum temperatures in June for various stations varies between 27°C and 30°C, whereas in the mountainous areas, it varies between 18°C and 22°C. The diurnal ranges in both areas, plains as well as mountains, remain more or less similar lingering around 15°C. The coastal areas, though a part of the lowlands present a contrast with summer temperatures moderated by sea breezes.

4.1.2 Winter Temperature Conditions

As in the case of summer, the two principal regions – plains and highlands – show similar contrasting conditions in winter with coastal areas showing some moderating effect. In January, the temperature contrasts between the lowlands and the highlands can be confined as very cold winters in the plains and severely cold winters in the highland region. In all regions, January is the coldest month and January temperatures represent the winter conditions. Except the moderating effect of the sea in coastal locations, the temperatures decline with higher latitude and higher altitude. Unlike the impact of the sea on the temperatures of the coastal region, the impact of the latent heat released by western disturbances is not clearly seen. Most likely, it is offset by higher altitude in the northern region.

4.1.2.1 Winter Temperature Conditions in Lowlands of Punjab and Sindh

As in summer so in winter, the coastal lowlands show a higher temperature in January than the stations inland. The day temperatures in inland locations are slightly lower than the temperatures in the coastal areas, but night temperatures of the coastal locations are unmistakenly higher, sometimes leading by 10°C. Similarly, the diurnal range of temperatures in the inland locations is significant higher than the stations on the coast. The coastal locations have the advantage of sea breezes that moderate the temperatures during the day, and the build-up of humidity during the day keeps the nights relatively warm.

The lowlands of Pakistan, leaving the coastal areas have severe winters with mean temperatures not rising above 16°C and even during the warmest hours of the day the temperatures linger around 20°C. Lahore and Faisalabad have a mean maximum of less than 20°C.

TABLE 4.3 Winter (January) Temperatures for some Stations in the Lowlands of Pakistan

Province	*Stations*	*Coordinates*	*Altitude (m ASL)*	*Mean Temperatures for January*			*Diurnal range of temperature*
				Mean Maximum	*Mean Minimum*	*Mean temperature recorded*	
Coastal Stations							
Sindh	1. Karachi	24°48’ N 66°59’ E	22	25.6	14.1	19.85	11.5
Balochistan	2. Gwadar	25°08’ N 62°20’ E	30	24.1	13.8	19	10.3
Inland Locations							
Punjab	1. Lahore	31°33’ N 74°20’ E	215	19.8	5.9	12.8	13.9
	2. Faisalabad	31°-26’ N 73°-08’ E	186	19.4	4.8	12.6	14.6
	3. Multan	30°20’ N 71°26’ E	122	21.0	5.5	13.25	15.5
	4. Mianwali			19.6	3.9	11.75	15.7
	5. Rahim Yar Khan	28°26’N 70°20’E	83	21.8	4.9	13.35	16.9
Sindh	6. Jacobabad	28°18’N 68°28’E	55	22.6	7.7	15.1	14.9
	7. Sukkur	27°40’N		24.8	8.3	16.5	16.5
Balochistan	8. Sibi	29°33’N 67°53’N	134	22.3	5.7	14	16.6

Source: The data are taken from sources or alternatively from NOAA (National Oceanic and Atmospheric Administration), Climate Data Processing Centre (CDPC) and Pakistan Meteorological Department, Karachi.

The continental locations like Lahore, Faisalabad and Multan in Punjab, Jacobabad in Sindh and Sibi in Balochistan have extreme conditions, having 20°C to 22°C temperatures during the day and 4°C to 6°C during the night, showing a diurnal range of temperature as high as 15°C to 17°C.

4.1.2.2 Winter Temperatures of North-western Mountainous Region and Mountainous Balochistan

The winters of the Northwest mountainous region as well as the mountainous Balochistan show a marked contrast from the winter temperatures of the plains of Punjab and Sindh. The day temperatures of most of the stations north of 34° N latitude vary between 8 and 13°C, depending on the latitude while the minimum temperatures linger around 0°C or even dip below 0 to -2 or -3°C. The mean temperatures vary between 4 and 7.5°C and the daily range of temperature varies between 4°C and 7°C, a condition far removed from the plains of Punjab where the mean temperatures are around 12°C to 13°C.

The mountainous regions of north-western and western Pakistan, coinciding with the mountainous regions of Khyber Pakhtunkhwa and Balochistan have very severe winters. Temperatures below freezing point in January are common. The mean temperatures for January, however, vary between 4.1°C in Chitral (1,494 m) to 7.5°C for Abbottabad (1,234 m), in South Khyber Pakhtunkhwa where Bannu (381 m) and Kohat have mean temperatures of 11.7°C and 11.1°C in January. The lowest temperature in the region is recorded in Parachinar, at the head of the Kurram Valley, which records mean temperatures of less than 4°C in January. The mean of the minimum, however, does not fall below 4°C at most stations and varies between 0°C and 2.5°C in Chitral–Dir region. The mean maximum ranges from 6°C in Chitral region in the north to 17–18°C for Kohat and Bannu in the southern part of Khyber Pakhtunkhwa.

The situation in winter is similar in the mountainous Balochistan where the mean January temperatures range from 3.5°C for Quetta (1,589 m) and Kalat (2,015 m) to 6°C for Zhob (1,407 m), 9°C for Khuzdar (1,232 m) and 19°C at Gwadar on the Arabian Sea coast. The minimum in most places falls below zero at most the mountainous places in Central and North Balochistan. Quetta and Kalat, both on the central highland, have a mean minimum of 3.4 and 3.5°C respectively. The annual range of temperatures in these highland places is rather high, often over 21°C. It may be pointed out that though the mean minimum is -3.4°C in Quetta, the city experienced very low temperatures in some winters. The record low temperature for Quetta is -18.3°C. Even the mean low temperatures for winters linger at below zero degrees centigrade for at least three months – December, January and February – during the year.

TABLE 4.4 Winter Temperatures of the Northwest and Western Mountainous Region

Province	*Stations*	*Height in m*	*Mean Maximum Temperature*	*Mean Minimum Temperature*	*Average Temperature*	*Diurnal Range*
K. P.	Chitral	1,494	8.3	0.0	4.1	8.3
	Dir	1,420	8,4	0.0	4.2	8.4
	Abbottabad	1,234	12.7	2.4	7.5	10.3
	Mingora	984	13.0	2.1	7.6	10.9
	Miram Shah	923	15.2	1.9	8.5	13.3
	Parachinar	1,728	9.7	-1.9	3.9	11.6
Balochistan	Quetta	1,589	10.8	-3.4	3.7	14.2
	Zhob	1,407	12.7	-0.6	6.0	13.3
	Kalat	2,015	10.3	-3.5	3.4	3.4
	Khuzdar	1,232	17.0	1.3	9.1	15.7

Source: The data are taken from sources from NOAA (National Oceanic and Atmospheric Administration), Climate Data Processing Centre (CDPC) and Pakistan Meteorological Department, Karachi.

As a broad generalisation, one may observe that the mean temperatures for these highland regions range from around 4°C in January to 25°C in June, a range of over 20°C. The severity of winter increases especially with altitude. A combined effect of altitude can be seen in the northern weather stations of Chitral, Dir and Drosh. The most severe winter is experienced at Kalaun (2,001 m) which is located at a height of over 2,000 m.

Most of the places in the mountains of North or Western Balochistan experience snowfall in January and February. Winter months are the rainiest months in these parts, because of western disturbances. A combination of freezing point temperature coupled with moisture imported by western disturbances leads to rainfall as well as snowfall. Quetta receives 10 to 20 cm of snowfall, though there are years when there is no snowfall at all. In contrast, the Chitral–Dir region is a region of heavy annual snowfall because of combined effect of winter rains and very low temperatures in January.

4.1.2.3 Some Recent Trend Analyses of Temperature

There are many analyses on temperature changes in Pakistan, but only two of these studies are summarised as they add to the existing knowledge of temperature variations in Pakistan. According to Hussain (2005)

> the trend analyses of the historical data for the period 1970–2000 show that winter season temperature have increased in both sub-mountain and high mountain regions during the past thirty years. Relative higher increase in maximum winter temperatures was observed, wherein minimum temperature during winter season showed a slight decline.

Similar to the analysis of climate change and variability in the mountain regions of Pakistan is another study by Iftikar Ahmad[1] et al. (2010) that analyses the January temperature over entire Pakistan for a 45-year period (1961–2006) with a geographical perspective. According to these authors, "change in (January) temperature varies from region to region and decade to decade in Pakistan". To evaluate the regional trends they divided Pakistan into five regions (1) Northern parts, (2) South-western parts, comprising Balochistan plateau, (3) the Upper and Lower Indus Plains, (4) North-western hill areas and 5. Western and western rugged territories. While the northern parts, the Balochistan Plateau and the Indus Plain have a warming trend, the north-western hilly areas show a cooling trend in January. The northern and western rugged territory shows a larger inter-annual temperature variability in January than the plains of the country. A cool phase in the 1970s and warm phases in 1980, 1990s onwards is obvious. While these trend analyses have a great value and add to the existing knowledge of temperature variations in Pakistan, these may, however, be a part of a natural cycle of temperature variations. A negative and a positive change in each decade is quite marginal and suggests only a modest annual variation in temperature. This does not fit into a general pattern of global warming or certainty of change.

4.1.3 Rainfall in Pakistan: Summer Monsoons and Western Disturbances – The Two Principal Sources of Rains

There is copious literature on the climate of Pakistan, more specifically on rainfall variability, its intensity and numerical modelling in the pages of "Pakistan Journal of Meteorology" and some in the files of "Pakistan Geographical Review". Rainfall in Pakistan is brought by two principal streams the southeast monsoon (southwest monsoon of India) in summer, and the western disturbances that bring rainfall principally to the western half of the country in winter. Though, both the streams cover a large area of Pakistan, their intensity and the amount of rainfall they bring, vary considerably. On the whole, summer monsoons, though of shorter duration, bring greater amount of rainfall than the western disturbances, which have a sustained presence over several months (December to April). The contribution of the western disturbances, though widespread in area and duration, is limited in volume. Indeed, as suggested by Ghazala Naheed and Ghulam Rasul (2011),[2] monsoon precipitation is the lifeline of Pakistan, which not only caters to the national power supply and the standing crops' water demands but helps gather the reserves to meet the requirements of low-flow in the next four to five months.

4.1.3.1 The Summer Monsoons

The origin of the summer monsoons is linked to the larger picture of southwest monsoons in South Asia. Writing in "Pakistan Geographical Review" way back in 1962 Sibte Nabi Naqvi and M Rahamatullah noted that that during May and June "*The Southeast Trades of the Southern Hemisphere cross the equator towards the end of May and under the combined influence of Coriolis force and the particular geography of East Africa assume a south-westerly direction*".[3] The two branches of summer monsoon – the Arabian Sea branch and the Bay of Bengal branch – meet the Inter-Tropical Zone of Convergence, though the position of the ITZC fluctuates considerably. The tropical depression that forms at the head of the Bay of Bengal move in a north-westerly direction; some of them recurve in the north and northeastward and cause heavy rainfall in the catchment area of the Indus basin. This explanation of Naqvi and Rahamatullah is rather tenuous. The northwest monsoons and the depressions associated with them follow the median line of low pressure trough that develops in the northern plains of India and Pakistan. The monsoon winds filling this trough move northwestward to

Delhi, thence to Lahore and finally in a very feeble stream even to Peshawar, following the Kabul Valley. One may even hazard a statement that both streams of rainfall – summer monsoons and the western disturbances – have a long sweep. While the summer monsoon sweeps the entire sub-continent from Bangladesh to Khyber Pakhthunkhwa, i.e., from Dhaka to Peshawar, the western disturbances that bring rains during winter also sweep the entire sub-continent. These cover not only Pakistan but also more than half the plains of North India, running down as far down the Ganga Valley as Patna. It is only a matter of intensity and the amount of rainfall that these two systems vary, but both bring rainfall – one in summer and the other in winter – to different areas of South Asia.

The summer monsoons have a well-defined track and a belt which they sweep moving from southeast to northwest from Lahore to Peshawar, spreading laterally and penetrating deep into the valleys, especially in Khyber Pakhtunkhwa, bringing heavy rainfall in July and August (Figure 4.3). There is no part of Pakistan except the extreme northwest of Khyber Pakhtunkhwa, like Chitral and Drosh and the extreme western part of Balochistan,

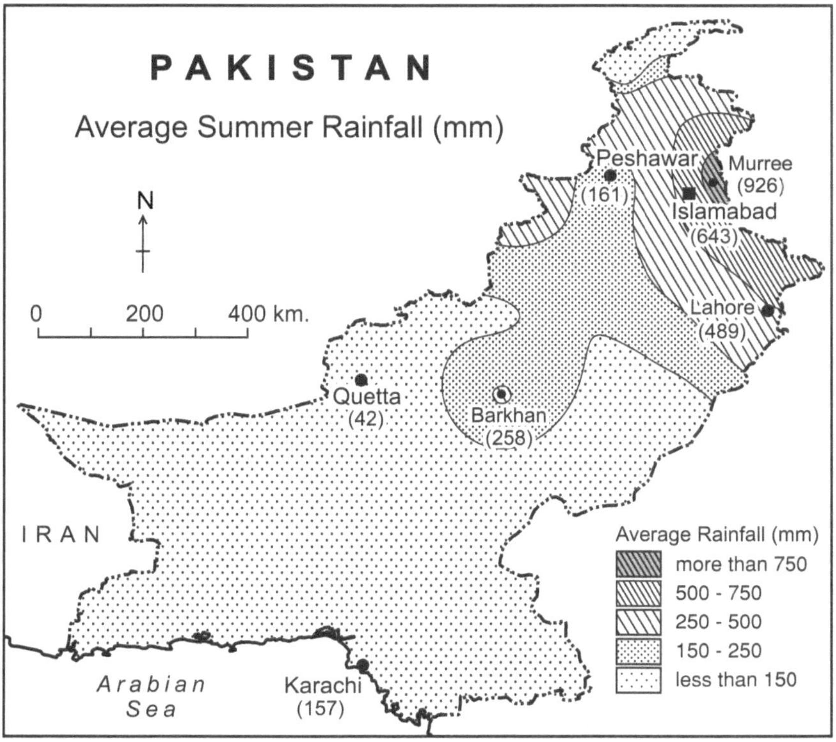

FIGURE 4.3 Summer Rainfall Distribution in Pakistan

represented by weather stations like Quetta, Kalat, Norkundi or the port city of Gwadar, where the summer monsoons don't make an impact.

The western disturbances bring the heaviest rainfall in the north-western part of Khyber Pakhtunkhwa, with its focus at Dir (1,420 m), which receives 768 mm of rainfall during winter. Other stations like Abbottabad (520 mm), Mingora (488 mm), Balakot (619 mm), Kohat (529 mm) and Parachinar (442 mm) at the head of the Kurram Valley, also receive substantial rains from the western disturbances during winters. Islamabad and Rawalpindi, two of the rainiest places in Pakistan, receive 337 mm and 369 mm of rainfall from the western disturbances during winter. The rainfall from the western disturbances declines as one moves further east with Jhelum receiving 243 mm and Lahore 148 mm of rain from the western disturbances during winter (Figure 4.4). The region depending almost exclusively on winter rains brought by the westerlies is the high plateau of Balochistan. Quetta (1,719 m), the capital of Balochistan, having an annual rainfall of 261 mm, receives 236 mm of rain in winter that is roughly 90 percent of its total annual rainfall. Similarly Kalat, receiving a total amount of 163 mm of annual rains gets 128 mm during winter that is 78 percent of the annual rainfall. Zhob (earlier known as Fort Sandeman) in the extreme northern part of Balochistan, close to the southern border of Khyber Pakhtunkhwa,

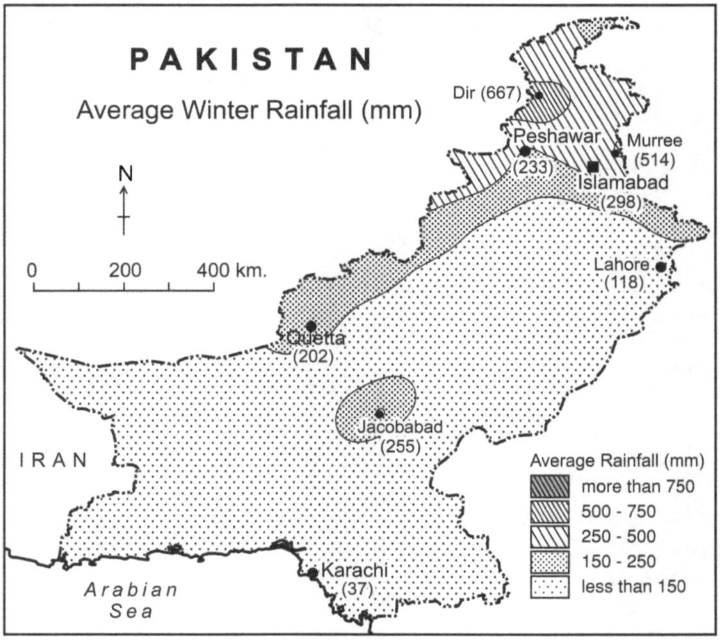

FIGURE 4.4 Winter Rainfall Distribution in Pakistan

though within the zone of influence of summer monsoons, receives more than 50 percent of its rain during winter. The coastal area of Balochistan, though receiving only 100 mm of rains annually, receives almost 90 percent of it during winters. For instance, Gwadar with an annual rain of 105 mm receives 93 mm of rains during winter. Even Sindh is not untouched by the westerlies. Though the province as a whole, like Balochistan, is deprived of much rainfall, the 100 to 200 mm rainfall that it receives at different stations also has a share of rains by western disturbances. Thus, Sukkur, Jacobabad, Hyderabad, Karachi and Badin receive 43 percent, 25 percent, 12 percent, 12 percent and 5 percent of their annual rainfall in winter from the western disturbances.

The southeast monsoons usually arrive in late June or early July, and the summer monsoon is active in Pakistan during July and August. July is often the rainiest month accounting for roughly 30 percent of the annual rainfall; and July and August together usually bring more than 50 percent of the total rainfall at a station. As one moves westward, the late arrival of summer monsoon results in August becoming the month of heaviest rainfall during summer. The rainfall brought by summer months has a greater intensity, is often copious, and in most parts of Pakistan, accounts for more than 50 percent of the annual rainfall. The divide line, between the dominance of western disturbances and summer monsoons, is blurred – yet it is possible to draw an irregular north–south line that separates the regions where the summer monsoon brings over 50 percent of rainfall. The following list is illustrative.

TABLE 4.5 Distribution of Rainfall between Summer and Winter in mm and Percentage

Rainfall Station	*Summer Rains*		*Winter Rains*		*Total Rainfall*	
	(mm)	+ %	*(mm)*	+ %	*(mm)*	+ %
Lahore	480	76	148	24	628	100
Faisalabad	269	72.5	102	27.5	371	100
Rawalpindi	958	71	389	29	1347	100
Islamabad	886	71	357	29	1243	100
Multan	122	65	65	36	187	100
Bahawalpur	104	73	39	27	143	100
Jhelum	610	71.5	243	28.5	853	100
Karachi	140	83	28	17	168	100
Sukkur	50.1	57	37.5	43	87.6	100
Peshawar	158	39	246	61	400	100
Parachinar	343	44	442	56	785	100

Note: The author has included a six-month period from June to November as summer months and December to May as winter months for the purpose of rainfall.

The westernmost points, where the summer monsoons reach, are Peshawar and Parachinar. Though these stations receive more rains in winter than in summer, the share of summer rainfall is substantial.

4.1.3.2 The Orographic Impact on Monsoon Rainfall

As a rule, the amount of rainfall brought by the southeast monsoon should decrease as one moves westward. The trend, though generally observed, is disturbed wherever the monsoon winds encounter greater heights. The monsoon winds with a specific load of moisture, when ascending vertically to overcome the heights run into lower temperatures, with a lower moisture holding capacity, resulting into condensation and immediate release of moisture as rainfall. Thus Rawalpindi and Islamabad, falling right in the path of the normal course of southeast monsoon, receive much higher rainfall than the adjacent areas. Lahore, much to the east, at a height of 216 m, receives an annual rainfall of 628 mm. In contrast, the twin city of Islamabad–Rawalpindi, located at a distance of over 300 km to the northwest, but at a far greater height, receives much higher rainfall. As seen in the above table, Rawalpindi at 50 m receives an annual rainfall of 1,347 mm while Islamabad almost at the same height receives slightly lower rainfall of 1,245 mm. Other places like Murree (2,127 m), located at the Himalayan foothills, and developed during the colonial period as a hill resort, receives 1,440 mm rainfall annually, of which 656 mm, 63 percent of the summer rains is recorded in the two months of July and August. Other towns located at higher altitude and receiving higher r ainfall are Abbottabad, on the Karakoram Highway, not more than 60 km north of Islamabad, and Balakot, a small town overlooking the Kagan Valley. While Abbottabad at 1,234 m, receives an annual rainfall of 1,262 mm, of which 60 percent is from summer monsoon, Balakot at 996 m gets an annual rainfall of 1,429 mm with a greater portion of the rains received from the southeast summer monsoon.

Thus, Murree, Balakot, Abbottabad, Islamabad and Rawalpindi, all within a radius of 30 to 40 km, except Balakot that is further north, are the stations located on the Himalayan spurs, enjoy heavy rainfall and a pleasant weather. Murree is by far the rainiest place in Pakistan.

By contrast, the low-lying areas of the Indus Plain receive lower rainfall as one moves from the Himalayan foothills to the south. In a north–south alignment, while Sargodha has an annual rainfall of 532 mm, Faisalabad receives 375 mm, and further south, Multan receives only 187 mm of rain. One may observe that the southeast monsoons follow a narrow SE–NW corridor, not exceeding 250 km in width. The higher areas closer to the Himalayas receive higher rainfall, the latter thinning out as one moves south. Bahawalpur, further south, receives 143 mm, and the towns in Sindh receive less than 200 mm of rains. The highest annual rainfall in Sindh occurs at Badin (222 mm) in the Indus Delta, followed by Karachi 168 mm, Hyderabad (171 mm) and Sukkur with the lowest annual rainfall of 87.6 mm.

To summarise, while Khyber Pakhtunkhwa, especially its north-western part, and much of Balochistan receive about 60 percent of the annual rains from the western disturbances during winter, Punjab and Sindh receive much of their rains from summer monsoon. The rainiest places in Khyber Pakhtunkhwa, like Dir (1,183 mm), receive about a third of their total rainfall in summer. The entire Punjab and Sindh receive much of their rains from southeast summer monsoons, during which July and August are the rainiest months. An important aspect of rainfall in Punjab and Khyber Pakhtunkhw is that there is no station which does not receive rainfall from both the streams of rainfall, the winter western disturbances and the summer southeast monsoon.

Sindh receives very little rains during winter, though Karachi, Hyderabad and Jacobabad receive 12 percent, 12 percent and 25 percent, respectively, of their total rainfall in winter.

4.1.3.3 Distribution of Rainfall

In terms of distribution of rainfall (Figure 4.5) the highest Rainfall occurs in the Himalayan foothill region of Islamabad, Rawalpindi, Murree and Abbottabad, the amount of rainfall ranging from 1,200 to 1,500 mm. Nowhere in Pakistan, the annual rainfall exceeds 1,500 mm, the highest being 1,440 mm at Murree. The rainfall declines south and south-westward. Balochistan, Sindh and South Punjab are relatively dry, and Northern Punjab and Central and northern Khyber Pakhtunkhwa receive a moderate rainfall. It is the Himalayan foothills that receive the highest amount of rainfall.

4.1.3.4 Impact of Planetary Oscillations of Temperature and Pressure on the Rainfall of Pakistan

There are two well-defined oscillations of temperature and pressure on a global scale. The southern oscillation giving rise to El Niño, and generally termed as ENSA, and the Northern Atlantic Oscillation of temperature condition (NAO), coinciding with La Niña. Both these phenomena have their impact on the precipitation situation of Pakistan. While El Niño is marked by a rise in the sea-surface temperature, at least by one standard deviation above the monthly mean for at least four consecutive months in the South Pacific,[4] La Niña is an extreme cooling of the central and eastern Equatorial Pacific over a period of several months.[5] Concluding his findings, Rashid observes "*in short it can be said that in an El Niño year, if there is warming in AMJ (April, May, June), or if even it is normal and it is likely to continue in JAS (July, August, September), then there it is a likelihood that*

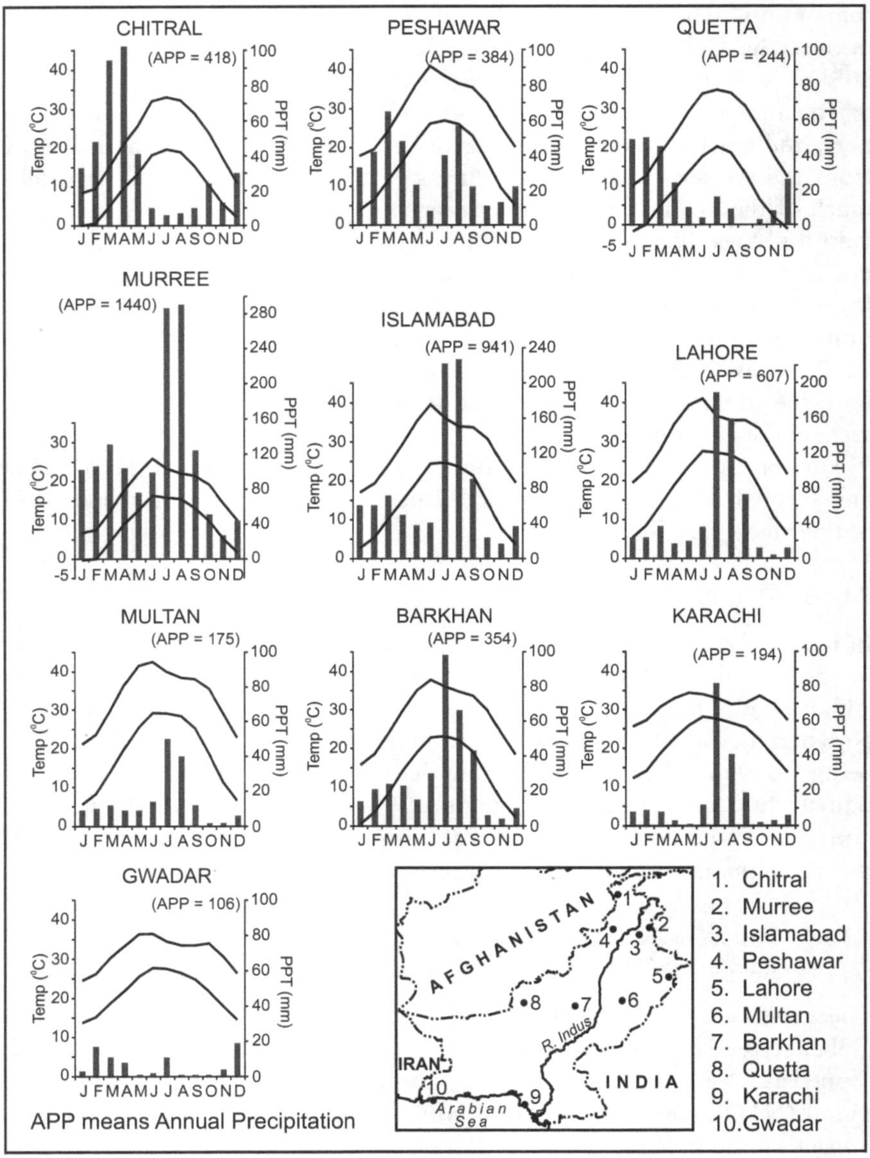

FIGURE 4.5 Mean Monthly Temperature and Rainfall at Some Selected Stations

monsoon rainfall over Pakistan will be deficient".[6] In other words, the situation becomes worse in La Niña years; while the El Niño years produce a deficit, La Niña causes severe droughts. Similarly, writing about "The influence of La Niña phenomena on Pakistan's precipitation, and the 1998 drought", Khan has observed that the period of 1998–2001, "was the period of cold episode in Pacific Ocean, termed La Niña phenomena, and it was believed that history's worst drought condition over South Asia were triggered by La Niña phenomena."[7] While explaining the relationship between El Niño and Pakistan monsoon rainfall, Abdul Rashid concludes "from the analysis of data it is evident that El Niño has a negative effect on summer monsoon (July-September) rainfall in Pakistan."[8] Besides these two phenomena of El Niño and La Niña, "the North Atlantic oscillation also affects the precipitation. During a positive NAO, which consists of stronger Icelandic low and Azores high, a positive precipitation anomaly is found in response to a positive NAO phase over central Southwest Asia, including northern half of Pakistan".[9] This also shows that strong westerly and El Niño conditions are the opposite for better snowfall over Hindukush, Karakoram and the Himalaya. To summarise the epoch of NAO, El Niño and La Niña, the El Niño has a negative effect on summer monsoon rainfall, a strong La Niña produces droughts as in 1998–2001, and a positive NAO coupled with a strong El Niño create conditions for better snowfall in the northern mountains of Pakistan.

4.1.3.5 Variability of Rainfall in Pakistan

Variability is the change in the amount of precipitation at a station over a long period of time. It is a relationship between the standard deviation and the mean, statistically expressed as

$$CV = \frac{\sigma}{\bar{x}} \times 100$$

CV is the coefficient of variability

σ is the Standard deviation and is the mean value of precipitation at a station, "x" number of years

As a general rule, weather stations (or regions) with low rainfall have a higher rainfall variability and those with heavier precipitation have a lower variability. In other words, the coefficient of variability is higher in drier areas than in areas receiving higher rains. Rainfall variability in Pakistan is investigated by Ghazla Naheed and Ghulam Rasul (2010).[10] Analysing the rainfall over a period of 50 years (1960–2009), they found the coefficient of rainfall variability of 251 percent in Balochistan, 247 percent in Sindh and 208 percent in Punjab, an increasing trend from north to south as the rainfall decreases. Also "the variability was highest in post-monsoon and pre-monsoon seasons as compared to the winter monsoon seasons".[11] There, however, appears a positive change as reported by IPCC. According to them, during the 20th century (1900–2000), there has been an average increase in rainfall of +20 percent in Northern Pakistan, +10 percent in central part and +40 percent in southeast.[12] A 40 percent increase in rainfall in Southeast Pakistan, perhaps caused by the strengthening of summer monsoons, would definitely be an ameliorating feature for Pakistan's climate and economy. This observation of IPCC presents a feature very different from the findings of S Salma, S Rehman and M A Shah, who have demonstrated a decreasing trend of rainfall of –1.18 mm/decade all over the country.[13]

4.1.4 Weather Elements that disturb the Rhythm of Human Activity

Besides temperature and rainfall, the two principal components of climate, there are other weather elements that appear occasionally and disturb regular human activities. These are:

A
1. Fog
2. Storms of different kinds and intensity

B Weather conditions that result in natural disasters
3. Severe cyclonic storms
4. Prolonged and intense rains resulting in floods
5. Prolonged deficiency of rainfall or no rainfall resulting in droughts

4.1.4.1 Fog

Fog is defined as "cloud of small water droplets near the ground level, sufficiently thick to reduce horizontal visibility to less than 1000 m (3281')".[14] The word also refers to clouds of smoke and ice particles. If the visibility is greater than 1 km, it is referred to as mist or haze. The most common situation of fog occurrence is created by the cooling of the ground level air with warm air above, a kind of inversion of temperature. Cooling of the surface air is often effected by adiabatic expansion, mixing of masses of air with different temperature or by excessive radiation from the ground. Such conditions where the ground level temperatures remain very low during January occur in northern and central Punjab, northern Sindh and parts of Khyber Pakhtunkhwa. The month of January is especially known as foggy winter when low temperatures are accompanied by low visibility due to fog. Islamabad, Rawalpindi, Lahore and many other towns occasionally experience very low visibility during the nights and mornings because of fog. Fogs have some positive effect of bringing moisture to the soil as part of the fog droplets settle on the surface, charging the soil in the cultivated fields with moisture.

The harm and disturbance caused by fog, however, far outweighs its utility as a lean supplier of moisture to cultivated fields. A prolonged fall even harms the crops, by depriving the crops of sunshine. The most affected activity is the air traffic at airports. Air traffic is severely disrupted in January due to fog at airports like Lahore, Islamabad and Peshawar. Karachi, the capital of Sindh, is not affected by fog because of the impact of warm sea breezes that prevent the surface air from excessive cooling.

4.1.4.2 Thunderstorms

There are many types of storms, like dust storms, thunderstorms or severe cyclonic storms. Dust storms are known to occur on the margins of Bahawalpur or Thar desert and the eastern part of Sindh. Dust storms locally known as *Andhis* occur in summer, where pressure gradients in adjacent regions create gusty winds laden with dust. The frequency and severity of dust storms have declined because of increase in land under irrigation and much larger area of land under crops than what the situation was before the large scale canal irrigation was introduced in Punjab and Sindh.

A thunderstorm is a violent short-lived atmospheric disturbance that is produced by clouds of great vertical extent and that is always accompanied by lightning and thunder. The instability of the atmosphere caused by an ascending mass of air is often promoted by orographic obstruction, often a mountain ridge. The storms created by rising warm air could be short-lived or may last for several days.

Thunderstorms in Pakistan have been studied by many scientists in Pakistan. The frequency of thunderstorms over Pakistan is studied by Zubair Ahmed Siddiqui and Abdul Rashid (2008).[15] In their study of thunderstorms of Pakistan for a period of 30 years (1961–1990), based on 48 weather stations, the authors have found that Khyber Pakhtunkhwa with their hilly terrain and steep slopes have the largest number of thunderstorms. July and August are the months of maximum frequency of storms at most of the stations, except the southern part of Balochistan where Jiwani, a coastal city on an island tip, has the maximum frequency of storms in winter.

Summer thunderstorms follow a definite belt extending from Lahore to Parachinar at the head of the Kurram Valley on Pakistan–Afghanistan international border. Most weather stations in Khyber Pakhtunkhwa have more thunderstorms than elsewhere. Parachinar and Kakul (NE Khyber Pakhtunkhwa) have the distinction of having the largest number of thunderstorms during the year. Islamabad, Murree, Jhelum and Sialkot, lying in the northern summer monsoon belt, receive more than 70 to 80 thunderstorms a year. July and August, the rainy monsoon months, have maximum frequency of thunderstorms in all the months. The country experiences on an average over 150 thunderstorms every year, with maximum frequency in July and August. These two months account for roughly 40 percent of the total thunderstorms in the country. The preceding and the following months of July and August, i.e., May and June and September also receive a fair number of thunderstorms. January and February are the months least disturbed by thunderstorms.

In terms of regional distribution, the entire northern belt of Pakistan including Khyber Pakhtunkhwa and Northern Punjab have the highest frequency of thunderstorms, occurring largely in summer. It is difficult to estimate the damage caused by these thunderstorms. There is certainly a loss of lives caused by thunder. Associated hail storms cause huge damage. Standing crops also suffer great damage. These storms bring volumes of water which may charge the soil with moisture and enhance storage in tanks, but on balance these are quite damaging.

4.1.4.3 Summary of Climate

Lying beyond the tropics between 24° N and 37° N latitude, Pakistan experiences a variety of climates from arid climate of Thar and Cholistan Deserts (BW) type of Koeppen to rainy cold winter, with no dry season and cold summer, of northern highlands of Khyber Pakhtunkhwa, the Cfc of Koeppen's classification. Extreme climate with a degree of continentality, with hot summers and cold winters, receiving monsoon rains supplemented by western disturbances, would summarise the climate of northern Punjab. Southern Pakistan is dry as in Sindh, dry summer and high temperatures reaching 50°C in extreme case and a rainfall average 150 mm annually is the reality. North Balochistan, a highland plateau with heights ranging from 1,500 to 2,500 m is cool and dry in summer, and cold and rainy in winter. Western Balochistan bordering with Iran is hot and dry during summer and cold in winter. The highland area like Quetta has cool summers and was considered the Paris of the East by the British. It may be remembered that it was here in Ziarat, about 75 km from Quetta, where Muhammad Ali Jinnah spent his last days before his death in Karachi, as the place in the midst of pine trees was considered health giving.

4.2 Soils

Soil is a complex natural product consisting of minerals, organic matter, air and water. The formation of soils takes a long time, involving weathering of rocks, development of mature soil profile with distinguishable horizons ranging from O and A at the top to D at the base denoting the parent rock. The process starts with the weathering of rocks, with variable resistance. The weathering is defined as the disintegration and decomposition of rocks by atmospheric agents. While the mechanical weathering produces clasts or rock fragments and takes a long time to produce soils, the process is rather rapid in climates with higher temperature and rainfall, as in the tropics or subtropical regions. With the weathering of rocks, different horizons are formed from the surface downward indicating not only the extent of weathering but even the process of leaching from the surface downward, thus depriving upper horizons of certain soluble minerals and enriching or accumulation of the dissolved minerals in the lower horizons.

A mature soil profile has well developed horizons. This top horizon is rich in organic matter and clayey particles proving suitable environment for the growth of plants. Such soils, develop *in situ* are also known as residual soils. In contrast, soils developed in one region are transported by water and wind, giving rise to alluvial soils or loess. Being a relatively dry region, much of Pakistan is devoid of deep weathering profiles or mature soil profiles, and in most parts of the country "*the transported soils*" dominate. The provinces of Punjab, Sindh and Khyber Pakhtunkhwa are largely covered by alluvial soils, transported and deposited in their flood plains. There are regions like Potwar Plateau filled with loess and glacial deposits or the Cholistan and Thar Deserts where the aeolian deposits predominate and develop a mixture of clay and sand in depressions where water accumulates during occasional rains.

4.2.1 Role of Climate

For long the soils have been classified on the basis of climate and organic matter, ranging from Polar regions to Taiga and the *chernozem* of Russian steppes or the *podzols,* the ash-coloured soil, of the temperate regions. While the chernozems are rich in organic matter, podzols are produced following the leaching of the bases and their replacement by an acidic horizon. In the Tropics, in areas with a relatively heavy rainfall where chemical weathering predominates, the parent rock usually determines the type of soils. In climate, excessive leaching of bases leads to the development of sols rich in aluminium and iron, commonly known as "pedalfers" and in relatively warm and dry climates where the bases (calcium, magnesium and sodium) remain concentrated in top layers, the soils are called "pedocals". One would normally expect calcareous or alkaline soils in most of the rainfall deficient areas of Pakistan. The soils in Punjab and Sindh or in large parts of Khyber Pakhtunkhwa are transported and bring a new kind of soil with better granulometric composition. The extreme summer temperatures and the higher degree of evaporation do pose a problem and lead to the concentration of salts in the upper horizon, a fact highly detrimental to agriculture as crops don't thrive in extreme cases of high alkalinity or acidity. The people and the Government of Pakistan are quite alive to the problem of salt accumulation in the soils and have devised ways to drain it out.

4.2.2 A Word about Systems of Classification

The simplest classification is based on texture, a granulometric composition, that classifies soils from gravelly to clayey, all in various proportions. These classes are expressed as gravelly, sandy, sandy loam, loam, clayey loam and clayey. But a more scientific classification based on environment and climatic zones, held the field for several decades and is still en vogue. Known as the zonal classification of soils, the system is based on soil variation from the poles to the equator depending on the climate and vegetal cover. These include soils like tundra soils, podzols, prairie soil and chernozem of the steppes and the lateritic and red soils of the tropics, with many regional variations. The transported alluvial soils are the transported soils, unlike the locally developed soil profiles of a region and could occur anywhere.

The pre-existing soil classifications were discussed by soil scientists in an attempt to evolve a more elaborate system of soil taxonomy, to be acceptable universally. What came out is the "Seventh Approximation" Soil Classification published in 1967. The Seventh Approximation is largely based on soil morphology and the character soil profile with different horizons. The entire system can be seen in a publication of the United States, Science Education, Administration, 1975.[16]

4.2.3 Soil Surveys, Classification and Mapping in Pakistan

In Pakistan, the survey, classification and mapping of soils, is largely done by the Soil Survey of Pakistan. An inventory of soils in the country was initiated in 1962 through a Soil Survey Project undertaken by the Government of Pakistan with the assistance of the United Nations Development Programme Organisation. This Soil Survey Project (2012), meant to assist the Government of Pakistan, became a permanent institution, *The Soil Survey of Pakistan* in 1973.

The Soil Survey of Pakistan (SSOP) has surveyed approximately 715,000 km^2 of land covering over 80 percent of the total area of the country. An up-to-date inventory till 1993 consists of 55 reconnaissance soil surveys, consisting of five integrated soil and vegetation reports, along with maps of landforms "Soils, Land capability, and present Land Use / Natural Vegetation at 1:125,000 or 250,000 scale, and of a Generalized Agricultural Development Potential at 1:750,000 scale".[17] The Surveys identified 750 different kinds of soils, grouped into eight land capability classes.[18]

4.2.3.1 Classification of Soils According to Pakistan Soil Survey

Class I to IV – Arable Land
- Class I Soils with Very Good Agricultural Land
- Class II Soils with Good Agricultural Land
- Class III Soils with Moderate Agricultural Land
- Class IV Soils with Poor or Marginal Agricultural Land

Class V to VIII Soils for Forestry or Rangeland
- Class V Soils with Good Forests or Rangeland
- Class VI Soils with Moderate Forest or Rangeland

Class VII Soils with Poor Forest of Rangeland
Class VIII Soils with No Potential for any type of Agriculture, Forest or Rangeland

The Survey's reports are meant to typify the soils and to recommend their agricultural potential. The classification system adopted is derived here largely from a soil map of Punjab[19] on a scale of 1:100,000 prepared by SSOP in 1978 and a soil map of Pakistan prepared in 1993.[20] Both these maps are in public domain. There is yet another map of Pakistan based on satellite images that define 14 types of soils.[21] Considerable information is also available from the report titled "Soil Survey of Pakistan: History, Achievement and Impact on Agriculture".[22]

4.2.3.2 Types of Soils in Pakistan

The following account of the soils and soil types in Pakistan is based on the reports and publication of the soil maps of different provinces of Pakistan by the Soil Survey of Pakistan. Consistent with the classification of soils (7th Approximation) referred to earlier, the Soil Survey of Pakistan has identified the following orders and sub-orders of soils in Pakistan;

Hierarchical classification of soils identified in Pakistan and their numbers:

Soil orders	06	1. Alfisols 2. Aridisols 3. Entisols 4. Inceptisols 5. Mollisols 6. Vertisols
Sub-orders	21	
Great Groups	38	
Sub groups	112	
Families	353	
Soil Series	894	

The Soil Survey of Pakistan has added another three (1. Type of terrain, 2. Texture, 3. Alkalinity, i.e., calcareous or non-calcareous) very useful elements, in the soil classification of Pakistan, to the one defined by 7th Approximation. The first element is the landform, like (1) River plain or flood plain, (2) Piedmont plain, (3) Loess and weathered rock plain and (4) Mountains and hills. This broad grouping of soils, according to terrain type, is extremely useful in understanding the genesis and nature of the specific soil series, as well as, in determining the land capability. In the Generalized Map of Pakistan according to 7th Approximation, the 25 types of soils that are mentioned carry the typology, terrain, texture and the level of alkalinity.

Most of the classes of the 7th Approximation are expressed by using a prefix, usually a Greek or Latin syllable that denotes a climatic condition or any geomorphic agent or simply the appearance and texture of the soil profile horizon. These prefixes are referred to by the 7th Approximation as formative elements. To give an example a prefix "alb", a Latin syllable, means white or "arg", a Latin syllable meaning clay, or "cry" a Greek syllable implying cold, or "ud", a Latin syllable means humid. Thus, Udifluvents soils of recent river plains or cryorthents are soils of cold regions.

The most commonly understood and used typology, as given in Punjab soil map is given here. A detailed classification as appearing in the Pakistan Generalized Soil map 1993 (1:2,000,000).

The classification used in Punjab soil map (1:1,000,000) 1978:

A. River plains and terraces
  1. Non calcareous and loamy soils
  2. Non calcareous clayey soils
  3. Calcareous sandy soils and dunes
  4. Calcareous loamy soils
  5. Calcareous clayey soils
  6. Salt affected soils
  7. Seasonally flooded soils
  8. Sand dunes and sandy soils
B. Piedmont Plains
  9. Non-calcareous loamy soils
  10. Non-calcareous clayey soils
  11. Calcareous sandy soils
  12. Calcareous loamy soils
  13. Calcareous clayey soils
  14. Salt affected soils
  15. Sand dunes and sandy soils
C. Loess and weathered rock plains
  16. Non-calcareous silty soils
  17. Non-calcareous loamy soils
  18. Calcareous loamy and clayey soils
  19. Calcareous silty soils – gully land complex
  20. Gullied land and badlands
  21. Rough broken land
D. Mountainous and Hills
  22. Mountainous land with nearly continuous soil cover
  23. Mountainous land with patchy soil cover

In describing the soil types of other provinces, specific landforms are added. For example, in the case of Sindh, besides river plains and piedmont plains,

Deltaic Plains and Aeolian sand plains are added as landforms. Similar additions and adjustments have been effected in mapping soil types in Balochistan. In the Generalized Map of Pakistan, allowance is made to enter the landforms indicating the terrain. This additional information is similar to the basic method adopted by the Soil Survey of Pakistan.

In the 7th Approximation classification, there is a formative element expressed by a Latin or Greek syllable (most commonly Latin), but the soil map of Pakistan elaborates on it and makes it simpler and clearer. In the Generalized Soil Map of Pakistan, two more categories are added, i.e., (1) Clayey loam and (2) Severely saline soil (Salorthids) or Salt-laden soils.

4.2.3.3 Soils in Different Regions of Pakistan

> The bulk of soils in the Indus Plain belong to scierozem or brown soils of the old USDA (1938) classification, or to the camborthids of the soil classification of the 7th Approximation, as revised in March 1967, by Soil Survey Staff (USDA) 1967.[23]

The following account gives a sketch of soils of different regions (Figure 4.6)

Soils of Punjab – More than 95 perc ent of the area of Punjab is covered with one or the other of the following types of soil.

1. Soils of the Sub-recent Plains – These are mainly calcareous loamy and clayey soils, occupying much of the Ravi–Sutlej doab, including some areas under Vehari, Bahawalpur and Rahimyar Khan districts. Low rainfall and high summer temperatures, in the south, have led to a concentration of calcium carbonate in the soils. In the classification of the 7th Approximation, these soils are termed as Camborthids and Torrifluents.
2. Non-calcareous Loamy and Clayey Soil of Old River Terraces – The Ravi–Chenab Doab, centred around Faisalabad and Chenab–Jhelum Doab have non-calcareous, loamy and clayey soils occupying the old river terraces. The famous Chenab Colony has such soils. According to 7th Approximation, these soils are identified as Camborthids with some Natrargids.
3. Soils of Alluvial and Loess Plains – These are non-calcareous loamy soils of loess plain. These are the soils of the Potwar Plateau. Friable, but often gravelly, these are gullied soils. In the 7th Approximation classification identifies as Ustochrepts and Haplustraifs.
4. The Soils of the Piedmont Plain – These soils occur in the piedmont areas east of the Sulaiman Range, covering much of Dera Ghazi Khan and extending further north into Dera Ismail Khan. These soils are clayey

on the flood plain of rivers and loamy on the narrow terraces, merging with sandy loam in higher areas. According to the 7th Approximation, these are identified as Camborthids and torrifluents.

5. Sandy Soils of the Aeolian Desert – These soils occur in the Sind–Sagar Doab, Thal and Cholistan Deserts, identified by 7th Approximation as Torripsamments (note psamm means sand in Greek and torr (Latin) means hot and dry).
6. Loamy and Gravelly Soils of Mountains – The northern part of Punjab around Murree and other Siwalik foothills have a kind of soil, classed as Hapludalfs.

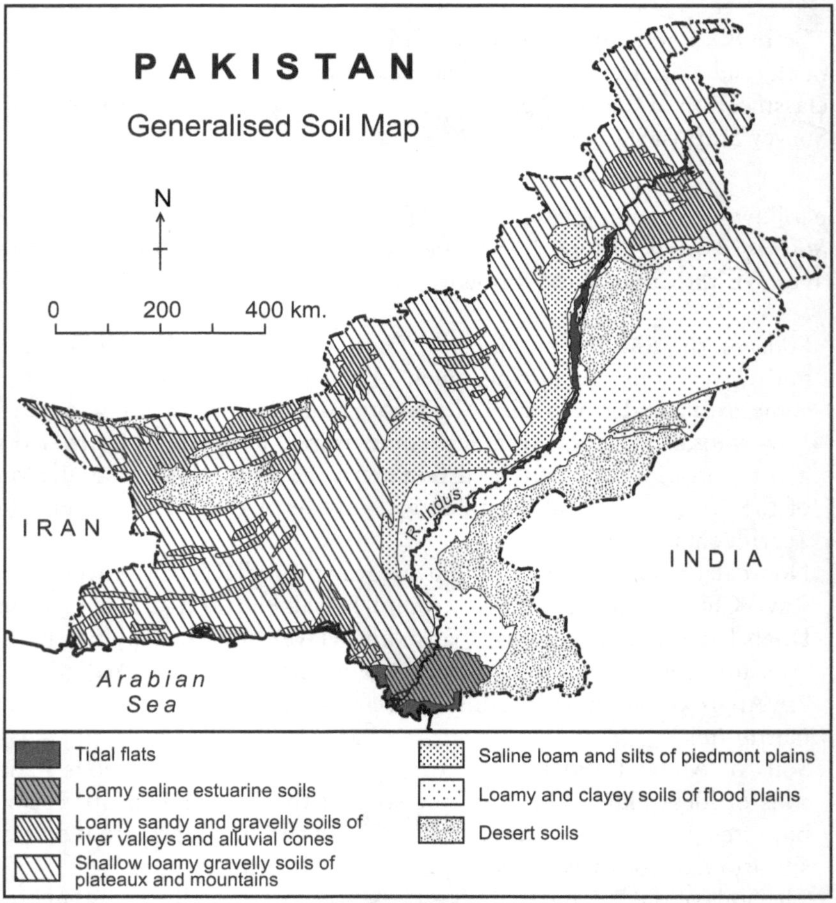

Source: Generlised Soil Map, Soil Survey of Pakistan, Lahore 1993

FIGURE 4.6 Generalised Soil Map of Pakistan

Soils of Sindh: These are seven types of soils, including one of the deltaic region and another of the desert area.

1. Loamy and clayey soils of sub-recent river plains, covering much of the alluvial plain on both sides of the Indus, these are the older alluvium of the central longitudinal plain of Sindh. These older alluvial plains, with loamy soils, cover completely the north-western part of Sindh, along the Kandhot – Shikarpur – Larkana axis, including Dadu on the right bank of the Indus, and along Ghotki – Khairpur – Naushahro axis, on the left bank of the Indus.
2. Loamy stratified soils of the recent river plain, including the flood plains, confined to a narrow 25 km wide zone, on either side of the Indus (Torrifluents and Torripsamments).
3. Rocky outcrops with very patchy cover of heterogeneous soil materials of Kirthar Mountains and the Piedmont: The valleys in the region having mainly loamy, part gravelly soils (Camborthids, Calciorthids and Torriorthents).
4. Soils of Piedmont in a narrow zone on the eastern margin of the Kirthar Mountain. Loamy, clayey in part saline soils (Salarthids and Camborthids with some Torrifluvents).
5. Soil of Estuarine Plain in the Indus Delta – mainly loamy saline soils (Salarthids and Camborthids with some Torrifluvents).
6. The silty and clayey wet saline soils of the Tidal Plains (Haploquents and Salorthids). Figure 4.6
7. Rolling sandy soils of the Aeolian Desert. These sandy soils cover one-third of the entire Sindh province, extending from south to north, including part of the Thar Desert (Torripsamments and Camborthids with some Salorthids).

Soils of Balochistan – The most common soils are the mountainous soils, categorised according to height and wetness into three categories.

The Mountain Soils – Rock outcrops with very patchy cover of heterogeneous soil materials. The valleys in these mountainous regions are covered with loamy and gravelly soils (Camborthids, Calciorthids and Torriorthents). These heterogeneous mountain soils cover much of western Balochistan, with the exception of some serpentine patches marked by wadis, often associated with non-perennial streams. In these serpentine valleys, the soils are mainly loam, supporting some cultivation. The rest of the mountain region is covered with rock outcrops with very patchy cover of heterogeneous materials (xerothents and xerochrepts). The north-western part of Balochistan, on Afghanistan border, presents a dry appearance and carries very shallow loamy soils (xerothents and xerofluvents).

Soils of Khyber Pakhtunkhwa – This north-western province of Pakistan has four principal types of soils.

1. The rock outcrops with some loamy and very shallow soils of the cold northern region of Khyber Pakhhtunkhwa (cryorthens, cryochrept with some cryboroles. Note cry (greek) means cold). These very shallow soils cover much of the northern region of Khyber Pakhtunkhwa, including Chitral and Drosh region.
2. The mountain and valley soils of eastern Khyber Pakhtunkhwa that includes Abbottabad, and Mansehra on the east, and Swabi and Swat districts on the west of the Indus River. The valleys have non-calcareous soils. The soils in the mountains are identified as Hapludolls and Eutrochrepts with some rock outcrops. The soils in the valleys in the region are identified as Eutrochepts and Hapludalfs.
3. Loamy and clayey soils of the Peshawar basin. These consist of loess and river alluvium.
4. Loamy shallow soils of the mountainous region of Safed Koh and Kurram Valley (ustorthents, Ustochrepts).
5. 5Glacier and snow caps in the extreme north of Khyber Pakhtunkhwa – Close to Wakhan Corridor are some snow caps and glaciers, devoid of any soils.

4.2.3.4 Quality of Soils, Soil Potential and Land Capability

Usually, the quality of soils is determined by their richness in all the nutrients required for plant growth. An example of a good quality soil is a clayey loam with a good colloidal fraction and a granular texture amenable to irrigation. Excess of alkalinity or acidity, caused by the concentration of salts or ferro-aluminous minerals is harmful to plant growth.

4.2.3.5 The Land Capability Classification of the Soil Survey of Pakistan (SSP)

The Soil Survey of Pakistan (SSP), having surveyed 715,000 km^2 of land, covering 81 percent of the area of Pakistan, has worked out a Generalized Agricultural Development Potential, based on landforms, hydrology, land capability and the present land use, together grouped into eight land capability classes. The land capability classification is primarily based on the relative severity of limitations of a particular piece of land for agricultural use.[24] The SSP came out with eight land capability classes. It takes into account besides irrigated and dry land, ten other factors. (1) Erosion hazard, (2) Relief, (3) Excessive wetness, (4) Permeability of soils (sandy), (5) Slow impermeability (clayey), (6) Salinity, alkalinity and sodicity hazards (non-gypsous), (7) Salinity, alkalinity and sodicity (gypsous), (8) Surface stability due to river

TABLE 4.6 Land Capability Classification and Distribution in Different Provinces of Each Class of Soil

No.	*Land Capability Class*	*Distribution among Provinces (in 000 ha)*					
		Punjab	*Sindh*	*Khyber Pakhtunkhwa*	*Balochistan*	*N. A.*	*Pakistan*
1.	Very good agricultural land	3486.4 (17 %)	1097.8 (11.9 %)	187.3 (2.64 %)	463.2 (2.41 %)	2.4 (0.07 %)	5237.1 (8.47 %)
2.	Good agricultural land	3679.2 (17.83%)	2326.9 (25.23 %)	524.4 (5.15 %)	443.9 (1.28 %)	9.6 (0.13 %)	6986.0 (11.3 %)
3.	Moderate agricultural land	2395.1 (11.6 %)	1496.9 (16.23 %)	665.6 (6.54 %)	196.5 (1.02 %)	21.8 (0.57 %)	4776.1 (7.72 %)
4.	Marginal agricultural land	1439.9 (6.98%)	219.4 (2.38 %)	581.6 (6.36 %)	699.8 (3.65 %)	46.5 (1.26 %)	2987.2 (4.83 %)
5.	Good Forest or Range land	–	–	70.1 (0.76 %)	–	101.1 (2.74 %)	171.1 0.27 %)
6.	Moderate Forest or Range land	261.8 (1.26%)	8.3 (0.1 %)	827.0 (9.04 %)	84.6 (0.44 %)	88.6 (2.4 %)	1270.3 (2.06 %)
7.	Poor Forest or Range land	4610.6 (22.35%)	993.4 (10.7 %)	2603.8 (28.5 %)	6551.2 (34.2 %)	651.6 (17.68 %)	15410.5 (24.93 %)
8.	Agriculturally unproductive land	4159.7 (20.16%)	2714.7 (29.4 %)	2974.0 (32.5 %)	10,599.1 (55.3 %)	2749.8 (74.6 %)	23,197.3 (37.5 %)
9.	Unclassified	592.3 (2.87 %)	364.9 (3.95 %)	704.9 (7.7 %)	102.3 (0.53 %)	14.0 (0.38 %)	1778.4 (2.87 %)
10.	Grand total	20,625.0	9,222.3	9,138.9	19,140.6	3,685.2	61,812.0

cutting or burial by fresh sediments, (9) Climatic limitations, extremes of temperature and rainfall and (10) Snow or ice cover.

4.2.3.6 Soil Degradation and Soil Conservation in Pakistan

Soil degradation in Pakistan, like anywhere else, is caused by excessive use, erosion by wind and water, salinisation or even deforestation. There are several estimates of the soil degradation by different authors.

Water Erosion – This is severe in areas with steep slopes, as mountainous areas with intense rainfall. Khyber Pakhtunkhwa is the most severely affected province because of water erosion, which accounts for roughly 40 percent of the total degradation of soils. Loessic soils, as on Potwar Plateau, suffer from water erosion observed in gullying and ravines. Land denudation also occurs as a result of excessive grazing and degradation of pasture lands.

Wind Erosion – This is typical of sandy and arid areas with loose sandy soils. The Cholistan desert area and Bahawalpur district of Punjab, as well as the Thal area in Bhakkar and Leah districts of Punjab suffer from wind erosion. The Thar region in eastern Sindh, especially in Khairpur district, is subjected to severe wind erosion. The south-western Balochistan, especially Kharan Desert, show the impact of wind erosion.

Salinisation – Salinity and sodicity are the two main problems of soil that affect most the fertility and suitability of soils for agriculture. "The soil materials in the plains of West Pakistan are almost low in salts when first

TABLE 4.7 Soil Degradation in Pakistan According to Estimates by Different Authors

Type of Degradation	*FAO (1994) 1*		*SSP (1965–88) 2*		*Khan et al. (2012) 3*	
	Million ha	*% of Land Area*	*Million ha*	*% of Land Area*	*Million ha*	*% of Land Area*
Water erosion	7.2	9.0	11.17	14.0	11.0	13.8
Wind erosion	10.7	13.4	4.76	5.97	3.5	4.4
Water logging	1.0	1.3	1.55	1.94	11.0	13.8
Salinisation	4.2	5.3	5.32	6.67	5,3	6.7
Soil fertility decline	3.2	6.5	–	–	–	–

Source: 1. Lal, Rattan (2018) Managing Agricultural Soils of Pakistan for Food and Climate, Soil Environ 32 (1)pp.1–10 (FAO, figures are taken from here)

2. Mian, Alim & Mirza, Yasin Javed (1993) Pakistan soil Resources, prepared by Govt. of Pakistan (Environment & Urban Affairs Division and World Conservation Union, pp. 13–19.

3. Khan, M. A., M. Ahmad & H.S. Hashmi (2012) Review of Available Knowledge on Land Degradation in Pakistan, OASIS Country Report 3, ICARDA, quoted from Lal Rattan (2018).

deposited. In most of the area calcium carbonate is the most soluble salt present in solid form".[25] In southern parts of Punjab, from Multan southward, the saline soils contain gypsum. In the coastal belt, large amounts of sodium chloride and some calcium and magnesium chloride and sulphate are present in the soil, attributed to deposition from sea-water. The soil in South Sindh, in Karachi – Badin belt is loaded with sodium chlorides.

Areas with rainfall exceeding the amount of evaporation – In areas where the rainfall exceeds the amount of evaporation, as along the north-western border of Punjab, the soils have been calcified. In the mid-northern part of Khyber Pakhtunkhwa, with a lower degree of evaporation, the non-calcareous soils have developed from non-calcareous rocks and sediments. This is the case with Swat Valley and other parts of Khyber Pakhtunkhwa.

Areas with rainfall and irrigation water less than evaporation – A contrasting situation emerges if the rainfall and irrigation supply is less than the potential evapotranspiration[26] and the groundwater is shallow. In such situations, there is an upward mobility of water through plant roots or evaporation. This leads to accumulation of salt in the upper horizons of soil, the nature of salt depending on the salt content of groundwater. The resulting salts are carbonates, or chlorides or even sulphate.

Soils with clay and the properties of swelling and shrinking – Such soils as the black cotton soils of the Deccan Plateau are rare in Pakistan. The "*Jhalpat*" series soils of Pat area (the flat alluvial plain between Sibi and Jacobabad) have a presence of swelling soil, with the properties of highly clayey soils. It is concluded that they may have been derived from the parent material with a high percentage of clay.

Specific causes of salinity and sodicity in Pakistan – Salinity and sodicity are considered the most adverse to soil factors that have proved detrimental to agricultural development. Generally, this salinity in the soil is commonly attributed to canal irrigation. Canal irrigation, however, may not be the sole reason of increased salinity in soils. Other reasons could be (1) Rising water table, (2) Inadequate rainfall, (3) Excessive irrigation and several other minor factors.

It may be noted that more than 5 million hectares of land (5,327,700 ha) are affected in Pakistan by alkalinity or sodicity, with more than 80 percent the affected land laying in Punjab. Of the affected land lying in Punjab and Sindh, each having over two million hectares with soils affected by salinity and sodicity.

Soil Nutrient Depletion – Besides the soil salinity, the most serious problem of soil degradation is the depletion of soil nutrients, generally associated with loamy sands and sandy loam soils.

Restoring the health of soils – Restoration of soil health is possible by adopting certain agricultural practices, such as a balanced use of farm inputs like fertilisers and pesticides, and above all a judicious approach to irrigation, avoiding especially over-irrigation.[27] Besides irrigation and drainage, excessive use of fertilisers, especially nitrogen poses a threat to soil quality.

There is no doubt, Pakistan has chalked out strategies for the restoration of soil health by not only draining the salts from irrigated fields, especially on the margins of the desert, but also by fixing the upper limits of irrigation water, guiding actively the farmers using tube well water, and even lowering of the surface of the cultivated soil for leaching off the salts. Restoration of soil health is the uppermost among the measures adopted by Ministry of Agriculture, Government of Pakistan, and vigorously promoted by Pakistan Soil Survey and different Departments of the Federal and Provincial Governments. Soils in Pakistan have received all possible attention from the agricultural scientists, as reflected in the state of agriculture, and ever increasing agricultural productivity in the country.

4.3 Natural Vegetation: Forests and Grasslands

The term natural vegetation includes all varieties of plants that grow naturally in an ecosystem, as determined by the regional environmental conditions, especially climate (temperature and rainfall) and soils. Vegetation has different life-forms, like trees, shrubs, perennial herbs, annuals, epiphytes (growing on plants), water plants, marsh plants and lichens-moss, etc. These plants live in large or small communities, like forests and grasslands which again vary with climatic conditions, largely temperature and rainfall.

In case of Pakistan, the natural vegetation includes forests, shrubs and mangrove of the coastal region. In addition, a large part of agriculturally unproductive land is officially designated as "*rangelands*", meaning grazing areas. While in the "Land utilisation data" of Pakistan,[28] there is no mention of "range lands", the latter figure prominently in the FAO assisted Land Use Survey of Pakistan, as well as the Soil Survey of Pakistan. It appears that the "land not available for cultivation" accounting for 29.48 percent of the total geographical area of Pakistan is what appears as a "*rangeland* in the "Land Use Categories of Pakistan" given by the Soil Survey of Pakistan. The term "rangeland" has gotten currency in Pakistan as soil scientists use it more frequently. Arshad (2015),[29] a soil scientist of great repute and Tamgha-i-Imtaz, uses "range lands" as one of the soil utilisation categories in his lecture, where he quotes a figure of 45.2 million ha or 60 percent of the total area of Pakistan as "rangelands".

Presently, the forests, scrublands and tidal mangroves are discussed as natural vegetation.

4.3.1 The Forests of Pakistan

The total forested area of Pakistan is 4.24 million hectares (Statistical Year Book of Pakistan 2018). This is 5.32 percent of the total geographical area

TABLE 4.8 Forest Area of Pakistan in Million Hectares

Source	*Forest area of Pakistan in Million Hectares*	*Percentage of the Geographical Area of Pakistan*
Pakistan Statistical Year Book 2018	3.91	4.91
Forestrypedia	4.247	5.33
Forestry Sector Master Plan (1997)	4.224	5.3
FAO assisted Land cover and Land use of Pakistan	4,523	5.4

Source: 1. Pakistan Statistical Year Book 2018, table 1

2. *Forestrypedia – Classification of Forests and Forest Types in Pakistan (detailed note) https://* forestpedia.com/classification of forests and – fores/# Forest-Types of Pakistan – accessed on 5-9–2020
3. Taken from working paper by Dr. K. M. Siddiqui, Director General, Pakistan Forest Institute, Peshawar on "Forests of Pakistan' – Working paper no. 11, Asia Pacific Forestry Sector Outlook Study Working Paper series, "Asia-Pacific Forestry towards 2010,' FAO, Rome, 1997
4. Land use – Land cover Atlas of Pakistan's Provinces, based on remotely sensed data GCP/PAK/125/USA, the Govt. of Pakistan and FAO of the UNO, and the US Dept. of Agriculture Table 4.9

of the country. Other organisations follow closely the land utilisation figures of Pakistan Year Book 2018.

One may say, Pakistan has 4 to 4.5 million hectares of land under forests. Of the various estimates given above, the following table reproduces the figures given in "Forestry Sector Master Plan of Pakistan", as stated in a working paper of Siddiqui (1997, together with a broad two-fold classification of forests).

Commenting on this, Siddiqui has postulated that *farmland trees* (466,000 ha) *linear planting* and *canal and road plantations* (16,000 ha) don't constitute forests. So is the case with miscellaneous forests (155,000 ha). The actual area under forests in Pakistan is 3.88 million hectares; that is 2.1 percent of the area of Pakistan.

4.3.2 Typology and Regional Distribution of Forests

The distribution of forests in percentage terms is grossly disturbed by the addition of forest categories such as "farmland tree" and linear and miscellaneous planting or even irrigation plantations. These are not natural forests and could not be justifiably included in the category of forests. The Pakistan Year Book mentions the following distribution of forests.

TABLE 4.9 Estimates of Forest Area in Pakistan based on Satellite Imagery Interpretation, (,000 ha)

Types of forests	*Provinces*						*% of Each Category*
	Balochistan	*K P K*	*Punjab*	*Sindh*	*Others*	*Total*	
Conifers	42	940	30	–	901	1913	45.3
Scrub	504	539	132	–	16	1191	28.2
Riverain	20	13	27	112	1	173	4.1
Mangrove	2	–	–	205	–	207	4.9
Irrigation Plantation	1	–	79	23	–	103	2.43
Farmland trees	23	70	306	54	13	466	11.1
Linear Planting	–	2	14	–	–	16	0.3
Miscellaneous Planting	–	120	20	–	10	155	3.66
Total	592	1684	608	399	941	422.4	
In percent	14.01	39.86	14.4	9,44	22.27	100	

Source: Land use – Land cover Atlas of Pakistan's Provinces, based on remotely sensed data GCP/PAK/125/USA, the Government of Pakistan and FAO of the UNO, and the US Dept of Agriculture

Note: Others include Azad Kashmir and Northern Areas (GB)

TABLE 4.10 Regional Distribution of Forests in Pakistan

No.	*Province*	*Forest Area in Million ha*	*% of Total Forest Area of Pakistan*
1	Punjab	0.48	12.27
2	Sindh	1.03	26.34
3	Khyber Pakhtunkhwa	1.24	31.71
4	Balochistan	1.10	29.66
	Total	3.91	99.98

Source: Pakistan Year Book 2018 (Provisional), Government of Pakistan, p.1

An elaborate note on the "Forests and forest types of Pakistan" appears in *Forestrypedia* which gives the following types of forests in Pakistan:

1. Tropical Littoral and Swamp Forests (Mangroves)
2. Tropical Thorn Forests
3. Tropical Deciduous Forests
4. Sub-Tropical Broad Leaved Evergreen Forests
5. Sub-Tropical Chir-Pine Forests
6. Himalayan Moist Temperate Forests
7. Himalayan Dry Temperate Forests
8. Sub-Alpine Forests
9. Riverain Forests
10. Irrigated Plantations
11. Linear Plantations
    a. Roadside Plantations
    b. Railway track Plantations
    c. Canal side plantations

This classification, like the previous one, includes natural as well as man-made plantations. The natural forests types that are found in Pakistan are the coniferous forests, shrub forests and the littoral mangroves. The elaborate typology follows a climatic pattern, but the types mentioned may not exist in reality. The most important types and their regional distribution is discussed below. For convenience, the present writer divides the forests into two broad categories: primary and secondary forests. The primary forests are the natural forests regardless of their state of preservation and the secondary forests are the man-made forests.

4.3.3 Primary Forests

Coniferous Forests – These are the most significant and genuine forests in Pakistan and are largely confined to Khyber Pakhtunkhwa province. The province has over a million hectares (1,105,000 ha) under coniferous forests.

These generally occur in areas above 1,000 m in higher latitudes. The northern part of Hazara division with Abbottabad and Mansehra, and the extreme and mid-northern districts of Chitral, Drosh, Dir and Malakand, with higher reaches of Swat, carry conifers, most often with exclusive stands. The conifers here include species such as blue pine (*Pinus wallichiana*), *Chir* pine (*Pinus roxburghii*), besides deodar (*Cedrus deodara*) and some varieties of spruce and firs like *Morinda* spruce (*Picea smithiana*) and fir (*Abies pindrow*). The higher north western part of Balochistan has some coniferous forests, covering an area of 130,000 ha of land. The most common pine known locally as *Chilghoza* (*Pinus gerardiana*), an edible variety, and juniper are the principal coniferous trees in the province.

Scrub land – (scrub forests), climatically, the sub-tropical dry forests; scrub forests, the second important category in terms of area covered, are typified as sub-tropical dry forests. These forests are highly degraded and occur in culturable waste lands or in areas unsuitable for farming. Much of the scrub forest is confined to the two western provinces of Khyber Pakhtunkhwa and Balochistan. Each of these two has 500,000 ha of land under scrub forests. These scrub forests are always found at lower altitude often below 1,000 m. Peshawar, Mardan, Kohat and a significant part of what was earlier known as Federally Administered Tribal Area, carry the scrub forests. In Balochistaan, a large region in the shadow of the Sulaiman Range and the non-desertic part of southern Balochistan carry scrub forests. The principal floral species are *Acacia modesta* locally known as *phulai* and *Kau* (*Olea cuspidata*). The scrub forests also occur in over 100,000 ha of land, in areas not suitable for cultivation, like the gullied piedmont slopes, exemplified by the northern ravine land of Jhelum, Gujrat, Rawalpindi and Attock.

Thorny Forests (The Tropical Thorn Forests) – These patches of scrub land are dominated by xerophytic species. Southern Punjab, the Cholistan Desert of Bahawalpur and Khairpur area of Sindh, on the margin of the desert, are the principal tracts of thorny xerophytic bushes. A large part of western Balochistan also carries xerophytic thorny bushes.

The Littoral Forests: the Mangroves – These are the tidal forests found in the Indus delta and along the western Balochistan coast, wherever the land is flat. Mangroves are a salt loving species and remain tied to the coasts, especially muddy coasts and the deltas. There are many species of mangrove, but the principal species in Indus delta is *Avicennia marina*. The mangrove forests cover over 200,000 ha (207,000 ha) of coastal area in Pakistan, principally confined to Indus delta in Sindh.

4.3.4 Secondary Forests

The word secondary is not used in the sense of "succession" but to identify the man-made forests and plantations. These are forests nurtured by irrigation in stand-alone patches, or along the roads, railways and canals, attaining a linear form. Sometimes, these forests are natural as in Sindh, along the river Indus and its tributaries. These are preserved by the government as well as the local population as green patches in otherwise dry area. The most common species is *shisham* (*Dalbergia sissoo*), the most valued wood for furniture. *Tamarix dioica* is another species, which is often found in dry areas. *Khejri* (*Prosopis cineraria*) a highly valued species, known for surviving in dry climate, is a revered tree in some parts of Rajasthan (India).

Plantation Forests – Also known as irrigated forests, these were promoted during the British colonial period. The Changa Manga Forest, in Lahore district, is a patch of plantation forest covering 37 sq miles of area. It was preserved to grow *shisham* (Dalbergia sissoo), a timber producing species known to yield quality wood for furniture. This patch of forest was irrigated by Bari-doab canal. The Changa Manga Forests in Lahore district are still preserved though these have shrunk in area. Similar plantation forests exist elsewhere as Chinchwatni plantation in Sahiwal district or Khipro reserved forests in Sanghar district. Besides, a large area in Thal Desert is also put under plantation forest. These plantations are grown by canal irrigation.

4.3.5 Preservation of Forests

The Government of Pakistan adopted a National Conservation Strategy (1993) to preserve natural forests and wild life. Under the plan, some important patches of forests are earmarked as protected forests. The conifers of Chitral are being protected as Deodar-Chilghoza-Oak forests. Similar is the forest area in the northern heights of Sulaiman range, abounding in pine forests, and these are protected as "Sulaiman-Chilghoza"[30] pine forests. Similarly, the Ziarat Juniper forests are also being preserved. Ziarat, a small town not too far from Quetta, was the summer residence of the British Governor during the British colonial period. The residence was also the place of convalescence of Quaid-i-Azam Muhammad Ali Jinnah, during the last days of his life. The juniper forests here are a protected species. As a matter of government policy, wherever there are luxurious natural forests, these are being protected and preserved by the forest departments of the respective provinces

4.3.6 Forestry and Forest Products in Pakistan

Climate does not favour a luxuriant growth of forests as in the equatorial or other tropical areas with heavy rains. Every year, Pakistan derives over 300,000 m^3 of forest products from its forests. These products include timber and fire wood. The value of timber is, however, very low amounting to just 20 percent, and 80 percent of the total volume of forest products is constituted of firewood, normally derived from the scrub land. More alarming

TABLE 4.11 Forest Products and Their Value (in million PKR) from 2008–09 to 2017–18

Year	*Quantity (000 m³)*			*Value (million PKR)*		
	Timber	*Firewood*	*Total*	*Timber*	*Firewood*	*Total*
2008-09	63.92	244.73	308.65	837.11	417.91	1255.02
2011–12	63.31	257.55	315.85	943.96	25.94	969.90
2015–16	53.25	336.89	390.14	2041.06	422.02	2463.08
2017–18	34.63	221.05	255.68	1068.48	582.15	1650.63

Source: Pakistan Statistical Year Book 2018 (Provisional) Government of Pakistan, Pakistan Bureau of Statistics, p. 71

is the fact that the volume of timber has gradually declined over the last decade, though in value terms, the timber product has increased. In contrast, the volume of firewood has not declined much.

4.3.7 The Future Prospects

Gone are the days of the colonial rule when ruthless exploitation of timber and wanton destruction of the subcontinent's valuable forests took place. Today, forests are no longer seen only as a valuable source of timber and fuel wood as well as a source of revenue. The forests, today, are viewed as an environment resource, functioning as the lungs of the earth. Increasing pollution of atmosphere, a very high concentration of CO_2 in the atmosphere as a result of the burning of fossil fuels and consequent danger of climate change, have enhanced the value of forests manifold. These are called the primary sink of CO_2 and an indispensable tool in preserving the purity of the atmosphere.

4.3.8 The Grasslands and Rangelands of Pakistan

"*Rangelands are vast natural landscape of grasslands, shrub lands and woodlands.*[31] *Areas characterised by natural and semi-natural woody and herbaceous vegetation with aerial stems, generally less than 6 m tall with individuals and clumps not touching to interlocking*". The rangelands, as a class of land use, do not figure in the "Land Utilization Statistics of Pakistan". In most cases, in the opinion of this author, the rangeland figures are derived from the "unculturable waste" and "land not available for cultivation" categories of the usual land use categories of the Government of Pakistan. In any case, the so-called rangeland may be the land available for foraging and grazing, but these are not the grasslands like the Russian Steppes.

Yet, there are a few grassland patches, the high altitude summer pasturage of the nomadic people, practicing transhumance. In Pakistan, largely confined to northern Khyber Pakhtunkhwa, there are a few nomadic communities which move to higher altitude during the summer, with their flock of sheep and goats to graze on the grasses that sprout and grow in summer. In Swat and Hazara regions, the nomadic communities, after their winter hibernation, move in search of green pastures at higher altitudes. The Dana Meadows in Kagan (Kaghan Valley), the Khanan Meadows in Mansehra district or further up Jangth Meadow near the base camp of Nanga Parbat, are some of the noteworthy pastures. Mention of these pastures is often found in the study of transhumance in Khyber Pakhtunkhwa province. One such study is made by Khan Sanaullah and Ahmad Mukhtar[32] of the pastoral systems and transhumance in Malakand division. They refer to alpine pastures, visited by grazing communities especially the Ajars and the Gujars who move with their flock of sheep and goats. The pastures were generally regarded as community grazing grounds. Besides, the landowners are also reported to rent out pastures to nomads for grazing. The pastures were often located on some fixed transhumance routes. Of late, these pastures are fast disappearing and being brought under afforestation. The Swat, Buner and Kohistan are experiencing afforestation with losses of pastures. Beyond these, there are hardly any large expanse of grasslands.

4.3.9 National Parks

To preserve environment, conserve bio-diversity, and protect wildlife, Pakistan has developed a number of National Parks, all concentrated in the sub-Himalayan region of northeast Khyber Pakhtunkhwa and northern Punjab. These are the regions with higher rainfall and lower evapotranspiration, suitable for most plant species to grow, develop greenery, conserve bio-diversity and protect natural environment. There are 28 national parks in Pakistan, most of them established after the passage of "Modern Protected Area Legislation", passed in 1983. Some were established as the Conservation Scheme of the International Union for the Conservation of Nature. Further development of parks was undertaken under the National Conservation Strategy of 1992, and many more national parks sprang up after 1992. Together, these national parks cover an area of over 30,000 km^2 that is over 3.5 percent of the total area of Pakistan. Some of the largest national parks are in Khyber Pakhtunkhwa, notably in Chitral, where Broghil Valley National Park has an area of 134,744 ha. There are several other national parks in the same province like Lalusar–Dudipatsar in Mansehra district with an area of 30,375 ha and Chitral Gol (7,750 ha). Closer to the Federal Capital Islamabad, there are a few national parks. Margala Hills National Park in the backdrop of Islamabad having an area of 17,386 ha and Murree – Kotli Sattiam – Kahuta National Park, spread over an area of 57,581 ha, are important parks in the vicinity of the federal capital. Parks in drier areas include Hingol in South Balochistan having an area of 165,000 ha and Hazarganji–Chiltan park near Quetta with a smaller area of just 15,000 ha.

One of the oldest parks in the dry area of Bahawalpur is the "Lal Suhanra" National Park. Started in 1972, it has an area of 87,426 ha. A similar park located in a dry area in the piedmont zone of Kirthar is Kirthar Park with a large area of 308,733 ha, by far the largest designated park in Pakistan.

4.4 Natural Hazards in Pakistan

Besides an established rhythm of climate that forms the basis of near-predictable state of temperature and rain or snowfall in an annual cycle, or the possibility of a moderate flood, there are unpredictable episodic events in nature that cause unprecedented damage to life and property, bringing untold miseries. Such episodic events are known as natural hazards. On a global level, natural hazards include, high intensity excessive rainfall causing floods, or scant rains resulting in droughts; cyclonic storms, especially the tropical cyclones; earthquakes, tsunamis, and such other events that cause huge damage to life and property, disrupting the rhythm of life in a region.

Unlike some other South Asian countries, especially India and Bangladesh, Pakistan is only marginally affected by severe cyclonic storms, though high intensity prolonged rains in specific localities produces some adverse effects. The main natural hazards in Pakistan as seen in the last several decades are: (1) Earthquakes and (2) Floods.

4.4.1 Earthquakes

Earthquakes are caused by an accumulated stress in the crust of the earth in certain areas. The release of such stress with the movement of the rock mass, overcoming the friction, results in an earthquake. Most earthquakes are associated the movement of tectonic plates or reactivation of a fault. In case of South Asia, the northern part of the sub-continent is very susceptible to earthquakes. This region lies on the northern plate boundary, where the Indian plate collides with the Central Asian plate with an under-thrust, creating a sub-duction zone. This plate boundary runs all along the Himalayan Frontiers. The entire northern and north-western and western part of Pakistan is, thus, susceptible to earthquakes. This seismic zone starting from Balochistan with the Chaman Fault, extends further north into Khyber Pakhtunkhwa, and Gilgit–Baltistan all lying in the northern mountainous region. This explains why the epicentres of most severe earthquakes lie either in Balochistan or in Khyber Pakhtunkhwa, or the Kashmir region of Pakistan.

Earthquakes are known to have occurred all through the history. Their records are available for over 200 years. These are compiled by the geologists of the Geological Survey of India in the 19th century.[33] One of the severest earthquakes that occurred in the beginning of the 19th century (16-6–1819) was in Kutch, on the border of Sindh. It transformed the local topography with the sudden appearance of a dyke- or bund-like feature, locally known as Allah's bund (the bund created by God). Many more earthquakes occurred in Pakistan, confined largely to Balochistan, except one that shook Lahore in 1827. The first, well recorded earthquake of great severity with 7.7 on the Richter Scale was the Quetta earthquake in 1935 (May 31, 1935). An estimated 30 to 60,000 people were killed and the entire town appeared flattened. The Quetta Earthquake of 1935 has lingered in the memory of the region as the most severe earthquake in history.

The 1945 earthquake with 8.1 magnitude on Richter Scale had its epicentre on the Makran coast and generated even a tsunami, but despite the severity of the earthquake the damage was minimal, largely because of sparse population of the region. In contrast, the 1974 earthquake (28-12–1874) with a magnitude of 6.2 on Richter Scale wrought a lot of damage. It affected the area of Pattan, Hazara region, Swat Valley and the northern parts of Pakistan. It resulted in the death of 5,300 people and injury to 17,000 people.

In Pakistan, after 1947, there have been more than three dozen earthquakes with variable intensity excluding frequent tremors in earthquake prone regions of Balochistan, Kyber Pakhtunkhwa, Gilgit and parts of Kashmir. More than 50 percent of these earthquakes occurred in Balochistan. The rest occurred in Khyber Pakhtunkhwa or Kashmir. In Khyber Pakhtunkhwa, the Hazara syntaxial region of Haripur – including Balakot Batagram – Mansehra districts witnessed greater frequency of earthquakes. In Kashmir region, including Muzaffarabad, Uri has been of late, a very active seismic region. Of the recent earthquakes that wrought unprecedented destruction was the Muzaffarabad earthquake of 1905, with a 4 km deep focus and an intensity of 7.6 on the Richter Scale. The earthquake resulted in the death of 80 to 90,000 inhabitants and about 100,000 people were injured. The loss of property was incalculable. This earthquake left more than four million people homeless, and was followed by a wide occurrence of diseases that claimed many lives. Among the more recent earthquakes, the one that occurred on 24-9–2013 at Kech in Awaran district (Balochistan) having a magnitude of 7.7 on the Richter Scale. With a 37 km deep focus, the earthquake also caused a tsunami on the Makran coast. Despite its severity, the loss of lives and property was not comparable to the destruction caused by Muzaffarabad Earthquake that occurred eight years earlier on August 10, 2005. More than 800 people died in the Awaran Earthquake. The limited destruction is attributed to a very sparsely peopled area with a declining population unlike other parts of Pakistan. The Awaran Earthquake is

TABLE 4.12 Earthquakes in Pakistan during the Decade 2010–20

	Date	*Region*	*Depth of Focus*	*Intensity on Richter Scale*	*No of Lives Lost*
1.	20-02-2009	Kashmir	48 km	5.4	0
2.	10-10-2010	Khanpur, Haripur	25 km	5.2	1
3.	12-11-2010	Quetta	19 km	4.7	0
4.	18-01-2011	Quetta (Gorhi Khairo)	23 km	7.2	3
5.	24-09-2013	Awaran, Kech	37 km	7.7	825
6.	28-09-2013	Awaran, Kech	45 km	6.8	22
7.	08-05-2014	Nawabshah (Sindh)	23 km	4.5	2
8.	01-10-2016	Mansehra (Naran) K. P.	33 km	5.4	1
9.	07-02-2017	Pasni Makran Coast	39 km	6.3	0
10.	24-09-2019	Mirpur (Kashmir)	55 km	5.8	39

Source: Earthquakes in Pakistan since 1950 WorldData Info (http://www.worlddata.info/asia/pakistan/earthquakes.php Retrieved on 10-3-2021

specially known for its recurrence after four days. The two successive earthquakes of more than 7 on the Richter Scale were experienced in a span of four days, first on September 24, 2013 and the second on September 28, 2013. During the last decade 2010 to 2020, the following earthquakes were experienced in Pakistan.

Since 1950, more than 83,100 people died as a direct consequence of earthquakes. A couple of these earthquakes also caused tsunami which brought about additional damage.

It is also a fact that Pakistan, as much as other countries of South Asia, experiences earthquakes which have their epicentres outside Pakistan. For example, the tremors a severe earthquake (21-1–2018), having its epicentre in South Afghanistan, were experienced all over Pakistan. Similarly, the earthquake of April 16, 2013, having a magnitude of 7.9 on the Richter Scale, with its epicentre in South Iran, was widely felt in Pakistan. The provinces of Balochistan and Khyber Pakhtunkhwa are seismically the most susceptible regions of Pakistan.

Earthquakes, either in Pakistan or elsewhere, cannot be predicted much in advance, nor can they be prevented, but their impact can be minimised by building earthquake-proof structures and buildings in the seismically active zones, and educating the people about measures to be adopted during the earthquakes.

4.4.2 Floods

Floods are the most destructive natural hazards. These represent a high water stage in the flow of a river when the volume of water to be drained exceeds the bank-full discharge of a river and overflows on to the adjacent flood plain and sometimes even beyond. Such a situation develops following a prolonged high intensity rainfall in the catchment of the rivers. In South Asia, floods are associated with high intensity monsoon rains for a prolonged duration, or following copious rains brought by tropical storms. Floods caused by cyclonic storms are quite common in South Asia, especially India and Bangladesh.

In Pakistan, floods are caused during summer monsoons, when heavy and intense rainfall causes the rivers to swell. Cyclonic storms also cause floods, though these are not as frequent as in India and Bangladesh. Occasionally, high water flow is seen in the Indus in summer, caused by glacial melting in the upper reaches of the river basin. Most floods in Pakistan occur during July–September period, coinciding with summer monsoons. It may be noted that the barrages and canal network hold considerable volume of water, lessening the possibility of frequent floods. Only the exceptional rains, lingering over a week or so, cause flood in the rivers of Pakistan.

4.4.3 Recent Floods in Pakistan

Some of the severe floods in Pakistan occurred in 1995, 2003, 2007, 2009, 2010, 2011, 2012, 2013, 2018, and 2019. Of all the floods, the one in 2010 was the most destructive. The severity of most of the floods, recorded above, was confined to specific regions.

It is also a noteworthy fact that Sindh, being in the lower part of the Indus basin, is more susceptible to flood than other areas, despite the fact that it has a low rainfall. To take an instance, Sindh had an exceptionally high rainfall, experiencing 285 mm of rains in a space of two days, in 2003. Similarly, Sindh had the highest ever recorded rainfall between August 11 and September 14, 2011. This caused severe flooding in the Indus. Roughly 500 people died and millions of acres of land were rendered temporarily unproductive. As recent as in 2020, Karachi had the highest single day rainfall of 231 mm in 12 hrs. In August 2020, Karachi received 484 mm rainfall, the highest in the last ninety years. Similar stormy weather with intense rainfall had affected Karachi in December 1965, and then in August, 1973, when a very severe flood submerged a large part of the province. Sindh, in fact, is more prone to experience cyclonic rains than other regions of Pakistan. With a featureless flat plain and minimum gradient towards sea, the province, though a relatively dry part of Pakistan, experiences more frequent floods than any other region of Pakistan.

By contrast, Punjab is least susceptible to floods, though Indus-flood of 2010 quite adversely affected parts of Punjab. The dry region of Balochistan occasionally experiences flash floods, as it happened in 2007, when more than 800 people died in flash floods.

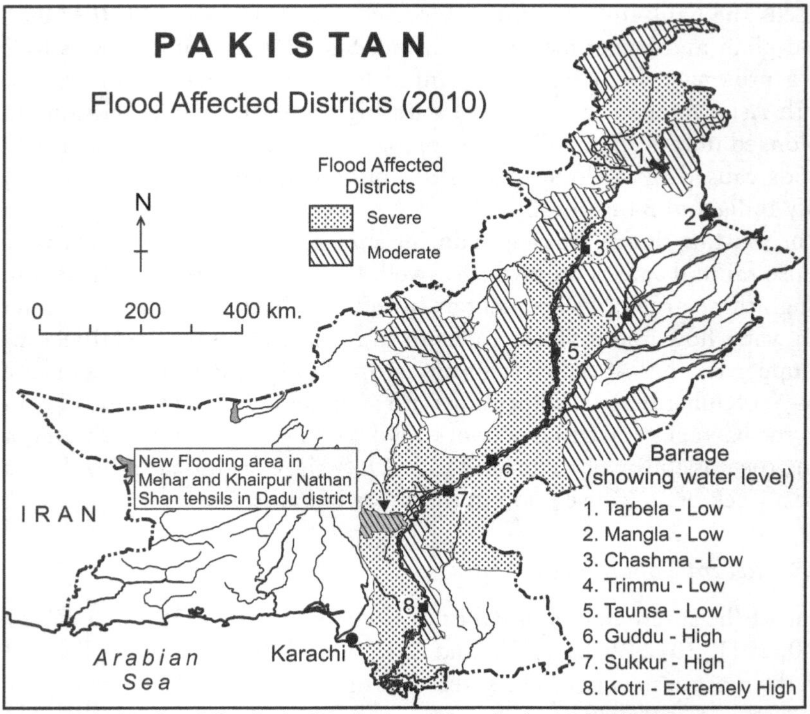

FIGURE 4.7 Flood Affected Districts of Pakistan during 2010 Flood

Source: United Nations Office for the Coordination of Humanitarian Affairs – UNOCHA

4.4.4 The Great Indus Flood of 2010

The great Indus Flood of 2010, often referred to as the flood of the century, was largely caused by heavy monsoon rains during July till late August 2010. "Rainfall began on July 22, 2010, and led to flash floods in the Northwest and East Pakistan and downpour continued till end of August".[34] At the end of July 2010, Khyber Pakhtunkhwa was hit by a five-day period of exceptional precipitation, including 60 hours of continuous rainfall, which exceeded the usual total for three months of monsoon season in this region by 30 percent. The National Weather Forecasting Centre in Islamabad described these as "once in a century" levels of rainfall. A highly destructive tsunami-like wave of water tore through the maintain valleys that feed into the main channel of Indus creating floods that travelled southwards to affect the entire river course in the centre of Pakistan. The result was excessive flow, several times the discharge in normal years. Exceeding the bank-full discharge, the flood spread all over Indus Plain. It is reported that at times the discharge was

40 times the normal discharge.[35] As the rains continued, the flooding in the river moved southward engulfing in its sweep, parts of western Punjab, a small part of Balochistan, specifically the Kacchi Plain and much of Sindh province. Mid and late August rains in the upper reaches of the Indus further worsened the situation. The cumulative impact of combined discharge of the Indus and its tributaries resulted in an unprecedented flood in Sindh. At Sukkur barrage, the discharge in the Indus was 32,000 m^3/ per second.

4.4.5 The Impact of 2010 Flood

The worst affected parts of Pakistan were Khyber Pakhtunkhwa and Sindh. The Indus flows through both these provinces, though it also forms the western boundary of Punjab in some sections of its course. The Punjab districts of Dera Ghazi Khan, Rajanpur and Rahim Yar Khan were as badly hit as some parts of Khyber Pakhtunkhwa. Northern part of Sindh and the land west of Indus were worst affected. Many towns were threatened with flood water and large number of people had to be evacuated to safe places. Very severe flood was restricted to a 60 to 120 km wide zone, spreading asymmetrically on both sides of Indus, submerging many villages and entering many towns (Figure 4.5).

The damage caused to people and their settlements, suggested by UNO "Office for Co-ordination of the Humanitarian Affairs" is as follows. Table 4.13

According to the Congressional Research Service, about 2,000 people were believed to have been killed, about a fifth of the country was submerged and around eight million Pakistanis displaced. In all, about 20 million people of Afghanistan were affected. The total damage as estimated by the Wold Bank and the Asian Development Bank was to the tune of US $ 9.7 billion.

The After-effects of the Flood – There was an enormous damage to agriculture and infrastructure. Many school buildings were damaged, transport was disrupted following breaches in roadways, poor house colonies were ruined and crops destroyed. It is to the credit of Pakistan that reconstruction was effected rather speedily with aids pouring in from all international funding agencies and other organisations.

4.4.6 Other Natural Hazards

Other natural hazards in Pakistan include droughts, though not with enormous impact primarily because of the continuous flow of the Indus, sustained by glacial and snow melt water, and in equal measure because of a large canal network that saves Pakistan from any severe impact of drought. Occasional attacks of locusts, and once in a long period, the arrival of a tsunami are the other hazards.

TABLE 4.13 Impact of 2010 Flood on Population and Settlements of Pakistan

Province	*Deaths*	*Injured*	*Houses Damaged*	*Population Affected*	*Damaged Schools*	*Schools Used as Shelters*
Balochistan	48	98	75,261	1,060,162	997	434
K. P. K.	1,121	1,165	192,605	4,365,909	923	785
Punjab	103	350	500,000	8,200,000	4,497	3,404
Sindh	141	840	470,910	4,158,160	4,029	1,473
A.j & K	71	87	7,069	245,000	389	N.A
Gilgit and Baltistan	183	60	2,820	8,561	81	N.A.
Total	1,667	2,600	1,248,665	18,037,792	10,916	6,097

Source: Map Action: OCHA (Office for the Coordination of Humanitarian Affairs)

The country is, however, well prepared to face and ameliorate the impact of minor natural hazards, with its accumulated experience and technical expertise.

Notes

1 Iftikar, Ahmad, Sun Zhaobo, Deng Weitao, and Romana Ambreen (2010) Trend Analysis of January Temperature in Pakistan over the Period of 1961–2006: Geographical Perpective, *Pakistan Journal of Meteorology*, vol. 7 (13), pp. 1–22.
2 Ghazala, Naheed and Ghulam Rasul (2011) Investigation of Rainfall Variability in Pakistan, *Pakistan Journal of Meteorology*, vol. 4 (14), p. 25.
3 Sibte, Nabi Naqvi and M. Rahamatullah (1962) Weather and Climate of Pakistan, *Pakistan Geographical Review*, vol. 17 (1), pp. 1–18.
4 Abdul, Rashid (2004) Impact of El-Nino on Summer Monsoon Rainfall of Pakistan, *Pakistan Journal of Meteorology*, vol. 1 (2), p. 36.
5 Azmat, Hayat Khan (2004) The Influence of La-Nina Phenomena on Pakistan Precipitation, *Pakistan Journal of Meteoroloy*, vol. 1 (2), p. 23.
6 Abdul, Rashid (2004) Impact of El-Nino on Summer Monsoon Rainfall of Pakistan, *Pakistan Journal of Meteorology*, vol. 1 (2), p. 36.
7 Azmat, Hayat Khan (2004) The Influence of La-Nina Phenomena on Pakistan Precipitation, *Pakistan Journal of Meteoroloy*, vol. 1 (2), p. 23.
8 Abdul, Rashid (2004) Impact of El-Nino on Summer Monsoon Rainfall of Pakistan, *Pakistan Journal of Meteorology*, vol. 1 (2), p. 35.
9 Afzal, M., M. A Haroon, A. S. Rana and A. Imran (2013) Influence of North Atlantic Oscillation and Southern Oscillations on the Winter Precipitation of Northern Pakistan, Pakistan *Journal of* Meteorology, vol. 9 (18), p. 1.
10 Ghazla, Naheed and Ghulam Rasul (2010) Investigation of Rainfall Variability for Pakistan, Pakistan *Journal* of *Meteorology*, vol. 7 (14), p. 15.
11 Ghazla, Naheed and Ghulam Rasul (2010) Investigation of Rainfall Variability for Pakistan, Pakistan *Journal* of *Meteorology*, vol. 7 (14), p. 25.
12 IPCC Technical Paper V (2002) quoted from Ghazla Naheed and G. Rasul (2010) as quoted in (1).
13 Salma S., S. Rehman and M. A. Shah (2012) Rainfall Trends in Different Climatic Zones of Pakistan, *Pakistan Journal of Meteorology*, vol. 9. p. 37.
14 Encyclopaedia Britannica (1985) vol. 4, p. 857.
15 Zuabair, Ahmed Siddiqui and Abdul Rashid (2008) Thunderstorm Frequency Over Pakistan (1961–1990), *Pakistan Journal of Meteorology*, vol. 5 (9), pp. 39–63.
16 United States, Science Education Administration (1975) *Soil Taxonomy: A Basic System of Soil Classification for making and interpreting Soil Survey or Soil Conservation Service*, United States Department of Agriculture, Washington, DC, Agricultural Handbook 436.
17 Mian, Amin, M. Mirza and Yasin Javad (1993) *Pakistan Soil Resources: A Pakistan National Conservation Strategy Sector Paper 4*, published by World Conservation Union, Pakistan, Karachi.
18 Khan, Muhammad, Ahmad Munir and Hashmi Hassnain Shah (2012) *Review of Available knowledge on Land Degradation in Pakistan: Oasis Country Report 3*, ICARDA International Centre for Agricultural Research in Dry Areas, Aleppo, Syria, p. 4.
19 Soil Map of Punjab (1978) 1:1,000,000 prepared under the direction of Dr. M. Bashir Choudhry, Director Soil Survey of Pakistan, Lahore.

20 Soil Survey of Pakistan, Lahore (1993) Pakistan: Generalized Soil Map, 7th Draft June 1993, scale 1:2,000,000.
21 Khan, Imran Ahmad and Muddassar Hassan Arsalan (2018) Satellite Based Assessment of Soil Moisture and associated Factors for Vegetation Cover. A Case Study of Pakistan and Adjoining Region, *Pakistan Journal of Botany,* vol 50 (2), pp. 699–709 (map on page 701).
22 Khan, Javanar Attar (2012) Soil Survey of Pakistan: History, Achievement and Impact on Agriculture.
23 Brinkman, R. (1971) *Soil Genesis in West Pakistan*, Central Soil Research Institute, Lahore, Pakistan Soils Bulletin, No. 4, p. 2.
24 Mian, Alim and Yasin Javed Mirza (1993) *Pakistan Soil Resources. A Pakistan National Conservation Strategy*, paper 4, prepared by the Goverment of Pakistan (Environmental & Urban Affairs Division) in collaboration with the International Union for Conservation of Nature (IUCN), pp. 2–3.
25 Brinkman, A. (1971) *Soil Genesis in West Pakistan*, Central Soil Research Institute, Lahore, p. 5.
26 Brinkman, A. (1971) *Soil Genesis in West Pakistan*, Central Soil Research Institute, Lahore, p. 6.
27 Taken from the slides presented by Prof. Muhammad Arshad, Distinguished National Professor, based on a workshop on "Asian Soil Partnership Consultation Workshop on Sustainable Management and Protection of Soil Resources', Bangkok, 13–15 March 2015.
28 Pakistan Statistical Yearbook (2018) Government of Pakistan, table 1
29 Arshad, Muhammad (2015) Pakistan, Soil Resources, Issues, Threats, On-going Activities and their Sustainable Management – Lecture delivered on 13–15 May 2015, in "Asian Soil Partnership Consultation Workshop on "Sustainable Management and Protection of Soil Resources' Bangkok.
30 Chilghoza is an edible pine.
31 This is the definition given in the introduction to "The Land Cover Atlas of Punjab' as part of the project "Agricultural Information System in Building Provincial Capacity in Pakistan for Crop Estimation, Forecasting and Reporting based on integral use of Remotely Sensed Data": GCP/PAK/125/USA. the Government of Pakistan, FAO of the UNO and the United States Department of Agriculture, 2014.
32 Khan, Sanaullah and Mukhtar Ahmad (2003) High Altitude Pastoral Systems in Malakand Division, Pakistan, in J. M. Shuttle and S. G. Reynold (eds.) *Transhumant Grazing Systems in Temperate Asia, Chapter XII – Pakistan Case Study* 2, Plant Production and Protection Series no. 31, FAO, Rome, pp. 239–248.
33 Oldham, T. (1883) Catalogue of Indian Earthquakes from the Earliest Times to the End of AD 1869, Mem, *GSI*, vol. 19, p. 3.
34 Michel, Morrish (2011) Geofile on Line, September 2011 (Pakistan Floods, 2010 – Causes, Impacts and Aid)
35 Kronstadt, K. Alan, A. Sheikh Parvez and Vaughn Bruce (2010) *Flooding in Pakistan: Overview and Issues for Congress*, Congressional Research Service, R41424 November, p. 3.

PART II

People of Pakistan

5

THE PEOPLE OF PAKISTAN: RELIGION, ETHNICITY, LANGUAGE AND CULTURE

The people of Pakistan form a segment of the vast population of the Indian subcontinent that stretches from the Hindukush in the northwest to Arakan Mountains in the east. Occupying barely 4 percent of the world area, the subcontinent is home to almost a fourth of humanity. With a population over 1.7 billion, South Asia is one of the great human agglomerations that has witnessed the rise of some very early civilisations, like the Indus valley civilisation that has its cities and monuments spread all across Pakistan.

The Pakistan society and people, numerically a tenth of the population of South Asia, have many common traits, in their social organisation, belief system, language and culture, art and architecture and share common history for long as an integral part of Mughal Hindostan or British India. Yet, undeniably the country and especially its people have many unique features derived from their adaptation to environments marked by a variety of terrain, contrasting climate, history of migrations and, most important, their ethnic-linguistic differences.

The people of Pakistan may have common nationality and common religion (Islam is the faith of 98 percent of Pakistan's population), a constitution and a common official language, but similarity ends there. The different and remote regions of the vast territory of about 800,000 km^2 area of Pakistan, present very contrasting social, cultural, linguistic landscapes. As aptly described by the Pakistan Tourism Development Corporation,

> *Pakistan is a multicultural, multi-ethnic and predominantly Muslim society. Ninety-seven percent of the 140 million (now 200 million) people in Pakistan are Muslims. The population is made up of different ethnic*

DOI: 10.4324/9781003510147-7

groups, such as dominant Punjabis, Saraikis, Sindhis, Pakhtuns, Baloch, Brahuis, Kashmiris, Hazaras, Urdu speaking immigrants from India or Mohajirs, Gojars, Kohistanis, Chitralis and a dozen or so Dardic languages-speaking lingo-ethnic groups.[1]

The number of lingo-ethnic groups mentioned above are not all; their number will swell, if still smaller groups and tribal groups from Balochistan and Khyber Pakhtunkhwa are added. Long settled in a niche or a small isolated region, a specific group has developed a subculture of its own and tends to maintain a separate identity. The Medes of Makran coast or the Kalash tribe of Chitral would, at no cost, like its identity effaced.

5.1 Racial Distinction

There are no apparent racial types, though an anthropologist may find traces of one race or another in a regional group. What is clearly understandable is that much of Pakistan lies in the Indo-Aryan racial domain speaking Indo-Aryan languages. Some of the isolated groups like the Pakhtuns may have retained a greater element of ethnic purity because of their relative isolation, while some other regions like Punjab with layers of humanity superimposition one over the other, have a greater chances of hybridisation. The northern east–west axial region that stretches from Kabul to Lahore carries a far greater impact of successive invasions and migrations. Yet the Indo-Aryan element is far strong in northern Punjab and Khyber Pakhtunkhwa. On the other hand, the Baloch, who arrived in Balochistan quite early claim to be of Arab descent and may have some Semitic (southwest Asian) element in their genetic make-up. The population in the coastal region, subjected to the impact of migration from Africa, along the coast may have some occasional negroid element. In contrast, the people of Sindh, the original Jats and Rajputs of Indo-Aryan descent, have, over the centuries, experienced the impact of conquests from the Arabs, Afghans and others, thus presenting a mixture of various social elements. Being a coastal state, Sindh has traditionally, and even during the last couple of centuries, much greater immigration. The Sindh society is a melange of original native population of Sindh before the Arab conquest in the early 8th century and groups that arrived subsequently, Arabs, Baloch, Afghans, Pashtuns and specifically the Punjabis. The latest group added to the Sindh society is that of the Urdu-speaking refugees from India following the partition of the country in 1947. Called Mohajirs, the Urdu-speaking immigrants have a strong presence in Sindh without having been integrated in the Sindhi society.

Thus, the Pakhtuns, the Punjabis, the Baloch and the Sindhis may differ from each other considerably, but it is difficult to say if there is a racial element in their distinction, or the distinction is more linguistic and a reflection of the historical antecedents of the region.

5.2 Religion, Principally Islam, in Pakistan

Pakistan is an Islamic Republic and a majority of its inhabitants are Muslims. On the basis of religion, population is grouped among the following denominations, as of 1998 and 2017 (Table 5.1).

There has been no substantial change in the religions over the last two decades, except that the population of the Qadians (Ahmadis) has suffered a decline. Their total number in Pakistan is 191,737; at the same time, there is an increase in the relative percentage of scheduled castes.

The dominance of Islam is pervasive and imposes limitations on the rights of minorities. To give an example, as mandated by the Constitution, no non-Muslim can occupy the position of the President or the Prime Minister of Pakistan. Islam has always dominated the thinking of the lawmakers in Pakistan, and the latter's constitution either in 1956 or 1973, accords primacy to Islam, though it has stopped Islamists from taking the country to hard core *Wahhabi* Islam, as advocated by Ahle Hadith Movement. The role of religion in Pakistani society and politics finds its most visible expression in the Islamic Assembly (Jamāat-e-Islami) party. Founded in 1941 by Abu-al-A'la Maududi, the party has always advocated, unsuccessfully though, a refashioning of Pakistan as a chaste Islamic or theocratic state.[4]

A majority of Pakistani Sunnis belong to *Hanafi* School, which is one of the major *madhhabs* (schools) of Islamic jurisprudence. The two reformist schools of India, *Deoband* and *Barelvi* (named after places where these schools are physically located in North India), are widely followed by Sunnis in Pakistan. Another group, known as *Tablighi Jamāat* (founded in 1926) with its headquarters in Lahore, preaches an austere way of life and has a large grassroot following. The Tablighi Jamaat launched one of the greatest

TABLE 5.1 Religious Composition of Pakistan in 1998 and 2017

Religious Groups	*Percentage of Population in 1998*[2]	*Percentage of Population in 2017*[3]
Muslims	96.28	96.27
Christians	1.59	1.27
Hindu (Jati)	1.60	1.73
Qadians (Ahmadi)	0.22	0.09
Scheduled Castes	0.25	0.40
Others	0.07	0.02

protests on a slight modification of the oath-taking phraseology involving reference to Khatme Nabuwwat in the National Assembly of Pakistan.

There are several sects in Islam, but the most prominent, both politically and numerically, are the Sunnis and the Shias. The Wahhabi Movement founded in Saudi Arabia is largely confined to Khyber Pakhtunkhwa. Numerically Sunnis are the largest group accounting for more than 80 percent of the total Muslim population of Pakistan. The other groups like Shias and several smaller sects together make less than 20 percent of Muslim population of Pakistan. Sunnis are the most widespread in the country. A majority of the Sunni Muslims follow the Barelvi School that is moderate in its outlook unlike the stricter Deobandis. It is reported that 78 percent of the Sunnis follow the Barelvi School with only the remaining 22 percent adopting Deobandi practices. Both the groups, Barelvis and Deobandi, have their own institutions and seminaries. It is the Sunnis, who are influenced by the extremist ideology of Salafis and advocate purity of Islam. In Pakistan, Salafi School is represented by Ahle Hadith Movement.

The ***Shia*** sect of Islam has several sub-sects. The principal sub-sect is the Ithna Ashariyyah (or Twelvers) who observe austerity in their life and largely resemble the Shias of Iran. Another sub-sect is that of Ismailis (or Seveners), including the Nizaris (followers of the Aga Khan including Khojas and Bohras).

5.2.1 Other Smaller Sects of Muslims in Pakistan

Besides the two principal sects, there are several minor sects, localised and not universally recognised. Some of these are mentioned below.

Zikri – The Zikri sect of Islam is followed by a small group of Muslims on the Makran coast. This is a heterodox sect that developed as a reaction to the decline of Muslim rule in India, following the arrival of British colonial rule.

Quranists – These are Muslims, small in number, who do not accept blindly the authority of Hadith. They derive their inspiration only from the holy Quran. They are also known as Quraniyoon or Ahle Quran.

Besides these, there is a large percentage (believed to be around 12 percent) of Pakistani Muslims who firmly believe in Allah, but don't affiliate themselves with a specific sect and call themselves only Muslims.

The *Ahmadis* Muslims, also known as *Kadiyani*, after the place of their headquarters at Kadiyan, now in Indian Punjab, were counted as Muslims till 1974, when they were categorised as non-Muslims by a constitutional amendment during the regime of Prime Minister Bhutto. It may be mentioned that some of the very outstanding Pakistanis belonged to this sect. The first among them is the first Nobel laureate of Pakistan Abdus Salam, and there have been earlier personalities like Sir Muhammad Zafarulla

Khan, Pakistan's foreign minister representing Pakistan in the UN General Assembly in the late fifties of the last century.

5.2.2 Islamic Institutions

A universal Islamic institution is the mosque, the shrine where the Muslims collectively offer prayers. This is not typical of Pakistan. The country has an estimated 250,000 mosques. These are the sites of Friday congregation for prayers and the sermons, known as *khutbas*. Besides, there are periodic fairs at the mausoleum of saints or *pirs*. The observance of Islamic rituals and festivals in Pakistan is virtually the same as in any other Muslim society. The common observance of a religious rite is *Ramdan*, a period of month-long fasting, in the holiest month of Islamic calendar, followed by the festival of "*Eid*", the most festive occasion in Islam. Another festival is *Milaad un Nabi*, that celebrates the birthday of Prophet Muhammad. The Shias observe "*Muharram*", a period of ten days, mourning the martyrdom of the first Imam Hussain Ibn Ali, the grandson of the prophet.

5.2.3 Sufism and Sufis

A sect of Islam, originating in Persia, Sufism has many followers in Pakistan. They try to find truth with divine love and knowledge through direct personal experience of God. It has an element of mysticism and the prayers involve singing and dance, unlike the hard core Wahhabi Islam. The sect has spawned the Sufi literature and mystical love poetry.

The most important Sufi orders in Pakistan, known as *silsilas,* are Qaderiyya, Naqshbandiya, Chishtia and Suhrawardya. There are many Sufi shrines in Pakistan. The most well-known are the following.

1. Baha-ud-din Zakaria in Multan – 12th century
2. Shahbaz Qalandar in Sehwan (Sindh) – 12th to 13th centuries
3. Rehman Baba in Khyber Pakhtunkhwa – 17th century
4. Sultan Bahoo (in Sharkot – Jhang – Punjab) – 17th century
5. Shah Abdul Latif Bhitai, at Bhit (Sindh) – 17th to 18th centuries
6. Data Ganj Baksh (Ali Hijweri) in Lahore – 18th century

Mention must be made here of Kutb-ud-din Bakhtiyar, in whose honour the Kutb Minar of Delhi was erected. His disciples Baba Farid-ud-din, Shakar Ganj of Pak Pattan (Punjab) is perhaps the most widely respected saint in Punjab. His disciple, Khwaja Nizamuddin Aulia of Delhi (India), represent a continuation of this tradition. The disciples of these saints founded shrines in Bahawalpur, Taunsa Sharif in Dera Ghazi Khan and elsewhere.[5]

The Sufism and Sufis in Pakistan are threatened by the hard core Sunni Muslims. The latter do not approve of practices that do not conform to their type of Islam. The Sufi shrines and their festival are now being attacked. The latest attack on the Shahbaz Qalandar Shrine at Sehwan in December 2016 by a suicide bomber killed over 130 people and an equal number were injured. Such attacks are a part of a serious conflict that has broken out in Pakistan in the second decade of this century.

Islam in Pakistan, instead of being a unifier of the faithful and producing a harmonious society, appears to be imploding under its own contradictions. The extreme elements that are promoting violence in the society may be in a small number, but large enough to cause disruption and destruction in the society. This is one more evil, besides economic woes, the Pakistani society has to contend with.

5.3 Minorities in Pakistan

The minorities, as generally understood, and in the present context, mean religious minorities. Since, a large majority of population of Pakistan practises Islam, the minorities consist of the non-Muslim population, following religions other than Islam.

While the Muslims with a total population of over two hundred million (200,368,718 in 2017) constitute 96.5 percent of the total population of Pakistan, the minorities with a total population of over 7 million (7,321,998 in 2017) account for 3.5 percent of the country's population. According to Bureau of Statistics, Government of Pakistan and the Census Bureau of Pakistan, the minorities consist of Christians, Hindus, Scheduled castes, Qadiani/Ahmadis and others.[6] Among the minorities, the Hindus are more numerous than the followers of any other religion, and with a population of over 3.5 million, constitute approximately 50 percent of the total population of the minorities. If the Scheduled caste population is added, their number touches 4.5 million, making over 2 percent of the total population of Pakistan (Table 5.2). The other significant minority group mentioned is that of the Christians with a population of over 3.5 million (3,595,256 in 2017). The only other religious group mentioned in the Census is that of Qadiani, with a population of just around two lakhs (191,737 in 2017).

With the exception of Qadiani, the other religious groups have shown an increasing trend in their population. The Hindus and the Scheduled castes, both, have shown an increase in their population between 1998 and 2017, the two census years, much above the national average.

5.3.1 Increase in the Population of the Minorities

The total number of minorities has increased from around 5 million in 1998 to 7.3 million in 2017, showing an increase of 48.7 percent that is

TABLE 5.2 Population of Minority Communities and Their Relative Strength in Percentage

All Communities	*Population 2017*	*Percentage of Total Population of Pakistan*	*Percentage of Total Population of Minorities*
Total population of Pakistan	207,684,626	100.00	
Muslims	200,362,716	96.47	
Hindus	3,595,256	1.73	49.1
Scheduled Castes	849,614	0.41	11.6
Hindus + scheduled castes	4,444,870	2.14	60.7
Christians	2,642,048	2.27	36.08
Qadians/Ahmadis	191,737	0.09	2.62
Others	43,253	0.02	0.6
Total No of minority population	7,321,908	3.53	100.5

below the national growth rate of 57 percent for Pakistan. This is despite the fact that the population of the Hindus has increased by 70 percent and that of the scheduled castes by more than 150 percent. The fact is that there is a drastic decline in the growth rate of Christians. Their number has increased by just 25 percent, pulling down the overall growth rate of minority population. The only community experiencing a drastic decline in number is that of Qadiani. Their number has declined from 291,175 in 1998 to 191,737 in 2017, a negative growth of 34 percent. Similar is the case with other unspecified minorities. Though not very numerous, having a population of 43,253, their number has declined to half of their number in 1998. In terms of growth of different communities the following order emerges (Table 5.3).

In reality, the minorities in Pakistan have a large spectrum of communities. These include besides Hindus, Christians and Ahmadis, mentioned in the Census Report (2017), Bahais, Sikhs and Zorastrians. These latter communities have been lumped into a single category "others".

5.3.2 Regional Distribution of Minorities

Of the total minority population of 7.3 million in Pakistan, more than 60 percent are settled in Sindh. Sindh has 2.5 million or 63 percent of the total minority population (Figure 5.2). Next, in order, is Punjab which accommodates about a third (33.43 percent) of the minority population of Pakistan. The other provinces, Balochistan, Khyber Pakhtunkhwa and Islamabad have each less than 100,000 of the minorities' population, which is just around

TABLE 5.3 Population Growth of Different Religious Communities (1998–2017)

		% of Growth (1998–2017)	*% of Total Population*
All communities		+ 56.91	100.0
Muslims		+ 57.23	96.47
	Scheduled castes	+ 156.8	0.4
Minorities	Hindus	+ 69.7	1.73
	Christians	+ 25.4	1.27
	Qadiani	- 34.15 %	0.09
	Others	- 46.6	0.02

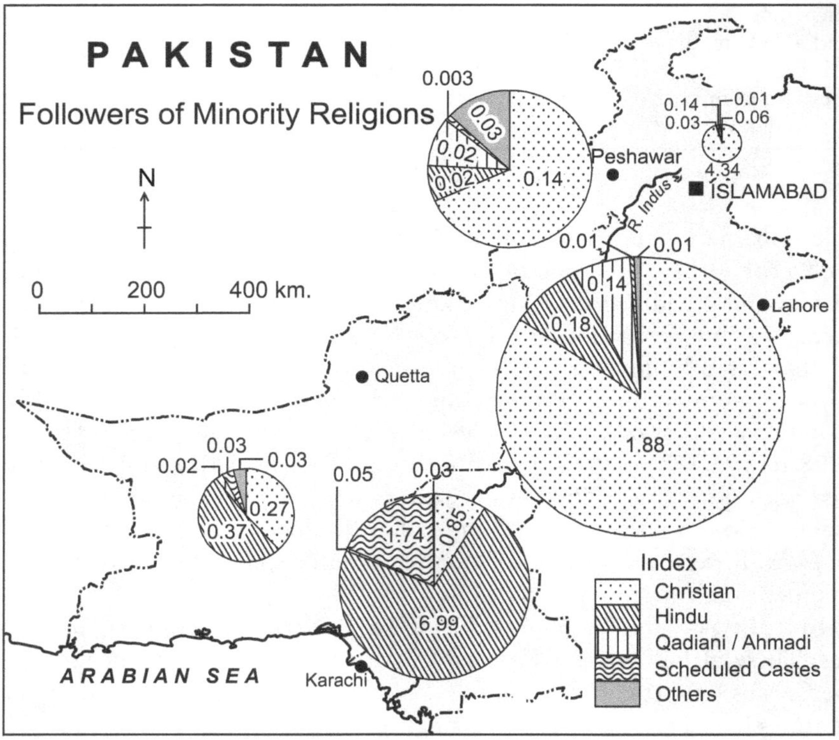

FIGURE 5.1 Followers of Minority Religions in Each Province of Pakistan

1 percent of the total minorities' number in the country (Table 5.4). What is important is the disproportionate concentration of minority communities in Sindh. Almost 10 percent of Sindh's population is composed of one minority community or the other. Punjab has over 2 percent (2.22 percent) of the population composed of minority communities.

TABLE 5.4 Distribution of Minorities in Different Provinces of Pakistan

Pakistan/Provinces	*Population of Minorities*	*Population of Minorities as % of the Total Population of the Province*	*Percentage of the Total Minority Population*
Pakistan	7,321,908	3.52	100.00
Khyber Pakhtunkhwa	73,107	0.20	0.99
Punjab	2,488,053	2.26	33.43
Sindh	4,620,403	9.65	63.12
Balochistan	88,854	0.72	1.21
Federal Capital	91,491	4.56	1.24
Total	7,321,908	3.52	99.99

Source: Final Results of Census 2017, table 9 (The figure of minorities has been obtained by deducing the population of Muslims from the total population of Pakistan and the provinces.)

This disproportionate distribution of the minorities, largely Hindus and Christians, is attributed to the religious composition of each province and their outmigration in the wake, or aftermath, of the partition of British India in 1947.

5.3.3 Distribution of Hindus

The Hindus, as the important minority group numerically, constituting of over 60 percent of the total minority population, are largely concentrated in Sindh. 94 percent of the total population of Hindus is settled in Sindh. Today, the Hindus including the Scheduled Castes, with a population of over four million (4,176,986 in 2017), constitute 8.72 percent of the total population of Sindh. The situation is very different from what was in the pre-partition days when the population of Hindus accounted for 26.1 percent of the total population of Sindh.

5.3.3.1 Impact of 1947 Partition of India – An Exodus of Hindus from Sindh

The following table (Table 5.5) illustrates the numerical status of Hindus in the three census years 1941, representing the pre-partition era, 1951 the post-partition census year and 2017, the latest census year of Pakistan.

An important fact about the population of Hindus is their gradual build-up in Sindh after 1951. Thus, while the population of Sindh during a period of 66 years from 1951 to 2017 increased tenfold and the population of Muslims during the same period increased by a similar multiplier, the population of Hindus multiplied 24 times. The excessive increase in Hindu population cannot be explained any other way except by immigration either from

TABLE 5.5 Population of Principal Religious Communities in Sindh Including Khairpur, in 1941, 1951 and 2017[1]

Year	*Total Population of Sindh*	*Muslims*	*Hindus*	*Scheduled Castes*	*Christians*	*Ahmadis*	*Others*
1941	4,404,908	N. A.			–	–	
	100 %	73.2 %	20.9 %	5.2 %			0.7 %
1951	4,925,342	4,457,951	141,387	323,120	–	–	2,884
	100 %	90.1 %	2.9 %	6.9 %			
2017	47,854,510	43,234,107	3,345,424	831,562	408,562	21,661	13,455
	100 %	90.34 %	6.99 %	1.73 %	0.85 %	0.045 %	0.028 %

Source: 1. Census of Pakistan 1951, vol. 6, Sind and Khairpur, p. 6–2 (Main religious groups)

2. Population of Sindh is inclusive of Karachi

3. Percentage for different communities for 1941 is taken from the Census of Pakistan 1951, table 6, p. 33

the neighbouring state of India or from other provinces of Pakistan, which, anyway, did not have much of Hindu population left after 1947. For example, Punjab including Khairpur had a mere 2,756 Hindus left after 1951, though there were over 30,000 Scheduled caste individuals in the province in the same year, largely concentrated in the dry districts of South Punjab, more specially Bahawalpur and Rahim Yar Khan districts. What is evident is that an excessive increase in the population of Hindus in Sindh after the initial exodus between 1947 and 1951 was caused by a reverse migration of Hindus returning from India. The Hindus of Sindh had migrated to India in the wake of partition, but crossed over, back into Sindh after 1951 after the situation normalised. This fact is further corroborated by a large concentration of Hindu population in the districts of Sindh bordering India.

Distribution of Hindus in Sindh – It may be emphasised that the Hindu community, including Scheduled Castes is largely rural based. Over 80 percent of the total population of Hindus is settled in rural areas. A large part is settled in the two divisions of Hyderabad and Mirpur Khas, which together have 75 percent of the total Hindu population of Sindh. Umarkot, the district bordering India has more than half (52.15 percent) of its population composed of Hindus. Similar is the case of Tharparkar where 44 percent of the district population is constituted by Hindus. There are four districts in Sindh where Hindus constitute more than one-fourth of the population of the districts. These are Umarkot (52.15 percent), Tharparkar (43.4 percent), Mirpur Khas (38.74 percent) and Tando Allahyar (34.17 percent).

The Rural-Urban Divide – The Hindu population is predominantly rural with only 20 percent (21.54 percent) living in urban centres. It is only in Karachi division or in Central Sindh that they are found in urban areas. In underdeveloped south-eastern part of the province, where there is a concentration of Hindus, there is no security risk unlike areas where they are miniscule in number and prefer to settle in urban centres.

5.3.4 The Hindu Society and Their Socio-Cultural Interaction

The Hindus of Sindh are part of larger socio-political organised Pakistan Hindu Council, founded in 2005, which organises the Hindus and aims at social reforms and takes up socio-political issues of Hindus with the provincial governments. It runs schools for Hindus and claims to protect the basic rights of the community. They also organise religious functions on festive occasions. The Hindus of Sindh, as of other provinces, have their own temples. The most famous is the Ramapir temple in Tando Allahyar district, where an annual fair is held. The Shiva temple in Umarkot is another important temple where the Hindus converge in large numbers on the occasion of Shivaratri. Not so important, but quite popular among the Hindus of Sindh is their Panchmukhi Hanuman temple in Karachi.

The most visited and the most famous temple of the Hindus in Pakistan is the Hinglaj Mata Mandir in Hinglaj Town on the Makran coast of Balochistan, in Lasbela district, in the serene surroundings of Hingol National park. The place attracts many devotees from all parts of Pakistan and over 250,000 people visit the temple annually. There are many temples in Pakistan, their number running into hundreds, but they are in a state of disrepair and often deserted in the absence of worshippers.

5.3.5 Christians in Pakistan

There are 2.6 million Christians in Pakistan, constituting 1.27 percent of the population of the country. Their distribution has two important characteristics. First, they are largely city based. In fact 65 percent of the Christian population lives in urban centres. Second, more than three-fourths (78 percent) of their population is concentrated in Punjab province, where Lahore and Gujranwala divisions are the principal areas of their concentration.

The Christians in Punjab, as perhaps anywhere else in the sub-continent, are the converts from the lower castes among the Hindus, and still remain poor and underprivileged, many doing menial jobs in the cities. What is noticeable is that despite the conversion to Christianity they remain confined to the lowest rung in the society in Pakistan. During the colonial period, the British promoted the building of a number of cathedrals and churches, the oldest one being the Roman Catholic St Joseph's Church in Lahore, the city that boasts of the imposing Sacred Heart Cathedral. Cathedrals and Churches were established during the British rule in all provincial capitals and major cities, like Lahore, Peshawar, Karachi, Quetta, Rawalpindi, Multan and many others. These religious institutions still exist. Some of them, like Forman Christian College, established in 1864 and now a university, became a reputed seat of learning. Most other cities of Pakistan have mission schools like St Patrick School in Karachi established by Christian missionaries during the colonial period and managed now by a board with memberships from non-Christian communities as well.

Despite a rich educational and cultural legacy, the Christian community in Pakistan, shorn of the patronage of the colonial masters has not been able to adjust itself to the new administrative, social and economic reality. This also explains why their population has not grown at the same rate as the other religious communities in Pakistan.

State, Society and Minorities in Pakistan – The Constitution of Pakistan, through the article 25 (1), guaranties that "all citizens are equal before law and entitled to equal protection of law". Article (5) provides that "adequate provision shall be made for the minorities to freely profess and practice their religions and develop their cultures", and article 33 declares that "it is the State's responsibility to discourage parochial, racial, tribal, sectarian and provincial prejudices among citizens". Despite these constitutional safeguards, the minorities are reported to have suffered, as much as some other sects of Islam. It is beyond the scope of the present geographic text to enter into another domain and talk of faith-based discrimination against minorities. The readers are referred to the report "Pakistan Minorities under Attack: Faith-based discrimination and violence in Pakistan" produced by the International Federation for Human Rights (FIDH) and the Human Rights Commission of Pakistan (HRCP), 2014.

5.4 Linguistic Landscape of Pakistan

The linguistic landscape of Pakistan is not a monochrome. It varies from the east to west and equally from the north to south, sometimes with a transition, developing in the process a language sharing the features of more than one langue.

Urdu – Urdu is the national language of Pakistan, ordained in the constitution of the country, but not even 5 percent of the people of Pakistan speak it as their mother tongue. Now, well developed as a literary language, it was the mother tongue of the Muslims in Bihar and Uttar Pradesh who were the *"avant garde"* in the movement for Pakistan. In Pakistan, the refugees who came from Bihar and Uttar Pradesh, referred to as *Mohajirs,* speak Urdu as their mother tongue. Apart from them, the educated people in all other provinces speak and write in Urdu as a medium of communication. The language is, however, well rooted in Punjab and Urdu has become the vehicle of expression in art, literature, cinema, media and journalism, besides occupying the position of national language. It is, no doubt, the main language. Punjabi, Pashto and Balochi have a place in folk culture and remained confined as a medium of informal conversation in the family or small groups.

The principal regional languages of Pakistan are the following (Table 5.6).

5.4.1 Other Major Provincial Languages

Every Province of Pakistan has a language of its own, besides Urdu, the national language that is used as a medium of inter-provincial communication. Each of these provincial languages occupies the apex of a regional history, folklore, music and culture. And in some of these regional languages, there exists a rich literature that is the literary history of that province corresponding to the population of the province, Punjabi, Pashto, Sindhi, Saraiki and Balochi, stand in descending numerical order. Saraiki, not a specific mother tongue of any province is a dominant mother tongue among the people of South Punjab, to the extent that they demand a separate Saraiki province. The case of Sindhi appears anomalous as Sindh has a much higher population as compared to Khyber Pakhtunkhwa. This is because Sindh has

TABLE 5.6 Principal Languages Spoken as Mother Tongues in Different Provinces of Pakistan (2017)

Language	*Provinces*						*Pakistan*
	K P	*FATA*	*Punjab*	*Sindh*	*Balochi*	*Islamabad*	
Urdu	0.89	0.48	4.86	18.18	0.81	12.22	7.08
Punjabi	0.53	0.25	69.66	5.31	1.12	52.26	38.70
Sindhi	0.09	0.1	0.154	61.6	4.07	0.77	14.56
Pashto	76.86	98.4	1.98	5.46	35.33	18.5	18.25
Balochi	0.08	0.08	0.83	1.99	35.48	0.15	3.02
Kashmiri	0.14	0.07	0.167	0.145	0.144	2.09	0.17
Saraiki	3.72	0.075	20.68	2.21	2.64	2.12	12.2
Hindko	11.48	0.02	0.58	1.57	0–27	6.4	2.44
Brahui	0.2	0.16	0.03	0.74	17.12	0.17	1.23
Others	5.98	0.33	1.02	2.74	2.47	5.29	2.26
Total	99.97	9.96	99.6	99.4	99.45	99.97	99.91

Source: Bureau of Statistics, Final Results of Census 2017, table 11

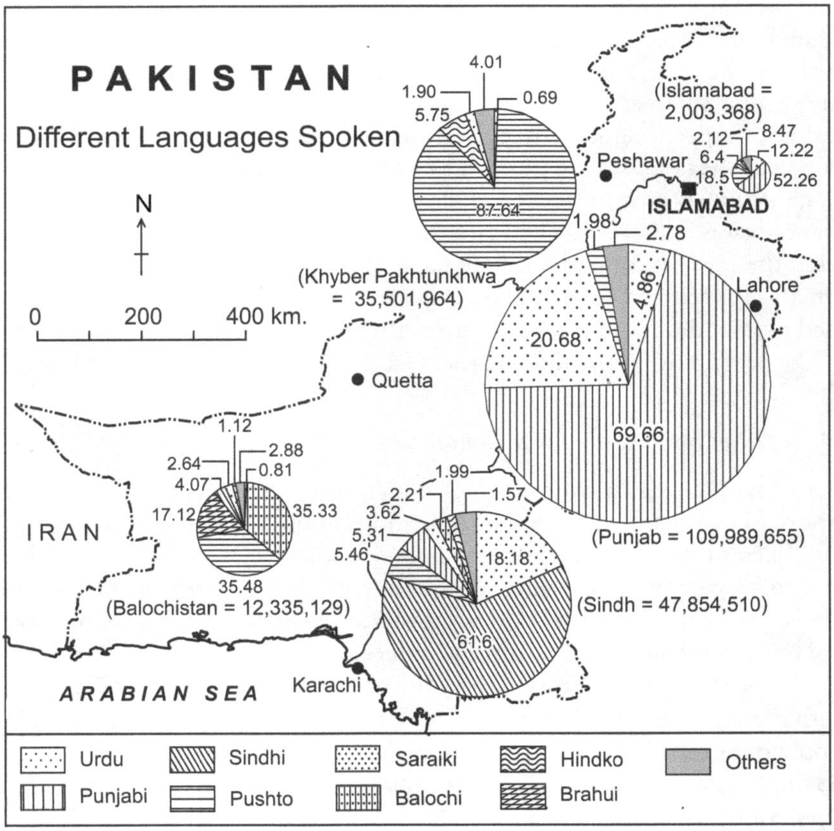

FIGURE 5.2 Percentage of Different Languages Spoken in Each Province

become a multilingual province, thanks largely to the linguistic composition of Karachi. At the other end of the spectrum is FATA, a monolingual area, with more than 99 percent people speaking Pashto. In Khyber Pakhtunkhwa, however, only three-fourths of the population speak Pashto. In the extreme northern parts like Chitral, Kohistan or even Swat, people speak Chitrali, Kohistani and Swati respectively. In fact, Khyber Pakhtunkhwa has the largest number of diversified dialects in the far-flung corners of the territory, which extends from the mountainous Wakhan border in the north to Balochistan border in the south. In contrast, Punjab is an exclusively Punjabi-speaking province; the southern part of the province speaks a variant of Punjabi called Saraiki. Saraiki is also spoken in Sindh, Balochistan and Khyber Pakhtunkhwa, touching the southern border of Punjab.

Sindhi and Sindh – Sindhi an Indo-Iranian language, written in a modified Arabic script, is the mother tongue of 15 percent of the population of Pakistan and the principal language of Sindh. The province of Sindh has a great linguistic variety. It has also the distinction of accommodating a large group of refugees, the *Mohajirs*, whose mother tongue is Urdu. A large

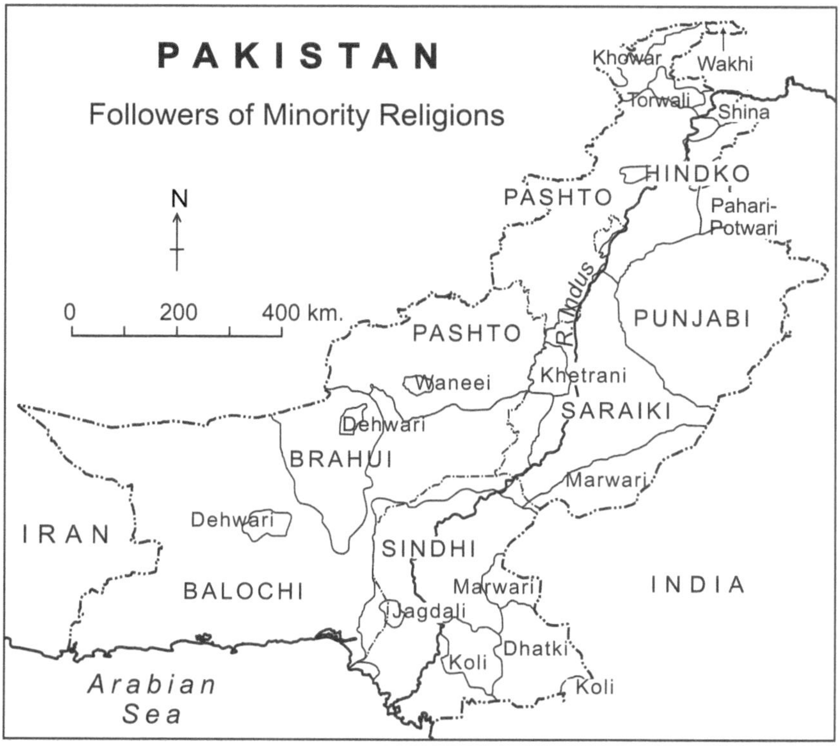

FIGURE 5.3 Regional Distribution of Languages Spoken as Mother Tongue

part of an estimated 15 million Urdu speaking populations – largely consisting of Mohajirs – is settled in Sindh, mostly in urban areas like Karachi and Hyderabad. Sindhi, the regional language of the province, is spoken by around 60 percent of the people. The Mohajirs, constituting roughly 20 percent of the population have retained Urdu as their mother tongue. Yet, Sindhi, the mother tongue of the province, has a third rank in the overall tally of languages in Pakistan after Punjabi and Pashto. But, over 90 percent (92.02) of the Sindhi speaking population is confined to rural areas. Much of the urban spaces of Sindh, and especially Karachi, are occupied by non-Sindhi speaking people. Urdu speaking people form a third of the total urban population of Sindh, while Sindhi speaking people constitute another third, of the remaining one-third Pashto and Punjabi, each claim another 10 percent. While hardly any non-Punjabi language is noticed in Punjab, Saraiki being an offshoot of Punjabi, Sindh has become a multilingual province, with Urdu, Punjabi, Pashto, Balochi and even Saraiki making inroads into the cities of Sindh, pushing Sindhi to the countryside. Sindhi has a number of dialects like Kachhi and Thari (both close to the Indian border), besides Sireli, Vicholi, Lari and Lasi, named after the area where these are spoken.

Balochi – Balochi is the mother tongue of 3.02 percent of the people of Pakistan and 55 percent of the people of Balochistan speak the language as their mother tongue. It may appear paradoxical that only 35 percent people of Pakistan speak Balochi as their mother tongue in view of the fact that the population of Balochistan constitutes 6 percent of the total population of the country. It is clear that the province is inhabited by a large number of non-Balochi speaking people. These are Pashto speaking Pakhtuns, occupying the northern districts of Balochistan, especially Zhob (earlier known as Fort Sandeman), Sherani, Musakhel, Qila Saifullah and Pishin. The Pakhtuns have also extended their dominance in Loralai district. Punjabi and Saraiki-speaking people have spread westward, from the two outposts, Dera Ismail Khan and Dera Ghazi Khan, into Balochistan. A third important language of Balochistan is Brahui spoken by about 18 percent of the population of the state. Once a dominant language is thrown into oblivion; the political landscape is dominated by the Baloch and the Pashtuns.

Balochistan, like Sindh, is a very multilingual province, where Balochi is spoken by only a third of the population as their mother tongue.

5.4.2 Other Minor Languages

According to Raymond, there are more than five dozen minor languages or dialects in Pakistan spoken by people ranging in number from a few hundred to a few thousand. The more commonly recognised minor languages, not enjoying official status but spoken by the people of the region are the following.

1. Brahui – in Balochistan spoken by the Brahui tribe
2. Chitrali – spoken in Chitral district in the northern part of Khyber Pakhtunkhwa
3. Dari – spoken in the border districts touching Afghanistan, north of Peshawar
4. Dogri – spoken in Sialkot and other areas adjacent to Jammu
5. Dhatki (Thari) – spoken in the Thar region of Pakistan
6. Gujarati – spoken by people in Sindh having their origin in Gujarat (India)
7. Hazargi – spoken by the Hazara tribe
8. Swati – spoken in Swat district (K P)
9. Kohistani – spoken in Kohistan province
10. Wakhi – spoken in Wakhan Corridor, northern border of Pakistan

Many of these languages are on the verge of extinction unless genuine efforts are made to save them.

The regional division of the population of Pakistan coincides with an ethno-linguistic division. Each region has a different language and some distinct element of ethnicity. In most cases, these groups coincide with the provinces after which the communities are known. Thus, one may conceive of the population consisting of (1) Punjabis, (2) Sindhis, (3) Balochis, (4) Pakhtuns and (5) some other groups. Most individuals in Pakistan, besides being a Pakistani, have a regional, linguistic and even an ethnic identity. Often regional and linguistic identities overlap or merge into each other. The ethnic identity based on history, group identity and ancestry remains etched in the collective memory of the groups for long. To quote an example, a Baloch or a Pakhtun retains his/her identity regardless of his domicile status.

5.5 The People of Punjab

Punjab has roughly 110 million people, roughly 53 percent of the country's total population, who speak Punjabi or Saraiki as their mother tongue, and can be legitimately called Punjabis. The group includes the Saraiki-speaking people from South Punjab, who form about a fifth of the total Punjabi population. In a population of around 80.5 million Punjabi-speaking people in Pakistan, the Punjabis are (could be) distributed as follows (Table 5.7).

As can be seen, Punjabis, as expected, have not only an overwhelming majority in Punjab, but have also spread to other provinces of the country.

The geographical identity of Punjab was established in the late 16th century during the reign of Akbar, who labelled this land of five rivers (Sutlej, Beas, Ravi, Chenab and Jhelum) lying between Yamuna in the east and the Indus in the west. This was a *Suba* (province) of Punjab and the people spoke a regional variant of an Indo-Iranian language, that later came to be

TABLE 5.7 Distribution of People Speaking Punjabi as Their Mother Tongue in Pakistan (2017)

Pakistan /Provinces	*Total Population in Millions*	*Population of Punjabi-Speaking People in Pakistan /Provinces (in millions)*	*Percentage of Punjabi-Speaking People in the Province*	*Punjabi-Speaking People in the Province as Percentage of the total Punjabi-Speaking Population in Pakistan*
Pakistan	208	80.5	39	100
Punjab	110	77	70	95
Sindh	48	03	06	37
K. P,	35	0.18	0.5	0.2
Balochistan	12	0.14	1.2	0.16
Islamabad	2	1	50.0	1.2

Note: The calculation is based on an estimated population of Pakistan of 200 million. The percentage distribution for each province and each language is based on 1998 Census of Pakistan. a) Decimal points are frequently dropped b) Based on table 11; Census of Pakistan 2017.

formalised as *Punjabi* and the people being in this region, speaking Punjabi came to be identified as *Punjabis.*

Unlike the Pakhtuns or the Baloch, the Punjabis don't have a tribal clan identity. The area was once, no doubt, a heterogeneous collection of tribes and castes, but with a gradual conversion to Islam, and adoption of a common language, a Punjabi identity gradually emerged. Religious differences, the baggage of caste hierarchy, and occupational differentiation between the land-owning class and the landless peasants and the artisans, posed a hindrance in the emergence of a unified distinct Punjabi identity. While a common language brought the religious groups closer, religious identity was a countervailing factor. Even the Punjabi language, a unifying factor suffered from the usage of a range of sub-regional dialects like Hindko, Potohari, Pahari, Chibhali, Dhundhi, Ghebi and Awankari which still hold a niche position in an overall linguistic Punjabi edifice. With the changes of time and a number of competing languages and people in an overall federal setup, Punjabi as a language turned out to be a unifying factor in Punjab.

5.5.1 The Antecedents

Ethnically an Indo-Iranian group, the *Punjabis* often trace their origin to the pre-Islamic caste system and associate themselves with Rajput and Jat castes. A large number still carry the old Hindu titles or honorifics in their names like "*Raja*" and "*Rana*". How important the baggage of the past Hindu social hierarchy has been to the Muslim society is reflected in the 18th century text "*Heer Ranjha* of Waris Shah, one of the important contributors to Punjabi language. The above text, a representative of a genre of Punjabi literature, known as "*Quisse*", tells the story of two lovers – Heer, a Jat girl of Sial (Sayyal) clan and Dheedo Ranjha, a Jat boy of Ranjha tribe. The story has a tragic end when they both commit suicide one after the other. The point to emphasise is that both Heer and Ranjha were born to Muslim parents, yet their caste and clan get prominence in the narrative. In fact, as late as the beginning of the 20th century, the Muslims of Punjab distinguished themselves by appending their caste name, to which they belonged before their conversion to Islam. The 1911 Census of Punjab province gives the numerical strength of Muslims coming from different Hindu backgrounds. They were labelled as Muslim Jats or Muslim Rajputs. In all these expressions, the Hindu ancestry received primacy over their religion. Some of the Muslim Jat clans were Awans, Bains, Bajwa, Khokhar, Kharal, Sial and Tarar. Similarly, some of the Muslim Rajput clans were Khokhars (also a Jat clan), Ghakkars, Khattars, Janjuas, Arains and Gujjars.

The advent of the British rule witnessed a change in the scenario. A revised land settlement, redistribution of land, creation of *zamindars* and *jagirdars* and settlement of irrigation colonies, especially the Chenab colony, changed the social relation from one based on ancestry to one based on land ownership. The array of heterogeneous tribes gradually disappeared and the society was divided into two principal groups, (1) *Zamindars* or *Qoums*, owing large holdings of land and associated with farming and (2) *Moeens*, the traditional artisans. In the category of zamindars were included the Mughals, Rajputs, Jats, Shaikhs, Gujjars, Awans, etc. On the other hand, artisan classes like *lohars* (blacksmith), *khateek* (tending orchards), *chhimba darzi* (tailors), *teli* (oilmen), *qassab* (butchers), *malhah* (boatmen), *dhobi* (washermen), *mirasi* (minstrels), etc. were included in the category of moeens. Even today, *biraderi* is an important influence on individual or group behaviour. Lieven refers to a case of a taxi driver in Rawalpindi, named Mudassar, who discloses his loyalty to Rajput biradari as one of the reasons for him and his entire village to vote for "Alpial Rajput" candidate.[7]

After 1947 and the creation of Pakistan, the Punjabi society is greatly changed. It has accommodated the Mohajirs from Indian Punjab. A redistribution of land following the migration of many landowning Hindus and Sikhs, increasing industrialisation and urbanisation, and above all a participation of people in the democratic process, are the reasons that have given rise to a new class of elites, the politicians. The traditional artisan class has virtually been replaced by city-based professions. Zamindars and big landlords, however, still persist despite land reforms limiting the area of individual holdings during the military regime of General Ayub Khan in late fifties and early sixties and during the regime of Prime Minister Zulfiqar Ali Bhutto during the early seventies.

Though some people from all parts of Pakistan are settled in Punjab, like the Baloch in Dera Ghazi Khan district and like the Kashmiris in Sialkot and neighbouring districts, and a large number of Pakhtuns in the western districts of Punjab, none of these groups has made so much impact on Pakistan as the early immigrants of Kashmir. National poet laureate Mohammad Iqbal, Mian Nawaz Sharif, a former Prime Minister of Pakistan, Saadat Hasan Manto, a distinguished short story writer, all have a Kashmiri origin.

Religion in Punjab has an important place in the life of a Punjabi. In fact, Pakistan being an Islamic Republic, it is important all over Pakistan. The majority of the population of Punjab consists of Sunni Muslims and only a small number practices Shia order of Islam. As for Ahmadiyya, this sect of Islam is delegitimised by a constitutional amendment in 1974, and Ahmadiyyas are no longer considered Muslims. The Sufi order, a mystical Islamic faith has a few shrines in South Punjab and even in Lahore. What is significant is that the hold of religion on the society is greater today than ever before. The impact of ever deepening religious belief and practice in the life of Punjab or for that matter the entire nation will be assessed by the historians in future. What is clear that dogma, ritualism and bigotry have become more important than the essence of Islam that preaches mercy and compassion.

5.5.2 Class Distinctions

A changed economic relation, greater urbanisation, and a decline in the contribution of the primary sector to the national economy, have brought forth a class structure that is different from the traditional society. The importance of the great feudal landowning families (in Punjab) has been greatly reduced in recent decades and its policies are dominated by smaller landowners and new families risen through commercial means or (corrupt) government service.[8] The emergence of a class of "elites" has taken place, a term that "*involved the image of landed feudal group, which was originally part of the ruling elite till 1947. Over the years it has changed with state's civil and military bureaucracy being integrated into the elite*".[9] The elites make a class by themselves, a group of powerful people exercising influence through material resources or political connections. The traditional feudal system based on land is more common in Sindh and not so much in Punjab. The Punjabi society consists of elites, the clerics and the commoners.

5.5.3 Language and Culture of Punjabi People

As observed earlier, a large majority of Punjabi people have Punjabi as their mother tongue. Yet, it is paradoxical that Punjabi as a language has not received the respect and attention it deserves. It has remained the language of communication of ordinary folks in rural society. Though a mother tongue of 60 percent of the Pakistanis, Punjabi (including Saraiki, Hindko and other variants) has no official status in Pakistan Punjab. In a note, Safir Rammah, while evaluating the status of Punjabi in Punjab, makes the following observation.

> *The medium of teaching in government or private schools in west Punjab is Urdu, a lesser extent, English. There is not a single Punjabi medium school in Pakistan. ...Except for a small number of writers and activists, Punjabis are illiterate in their own language.*[10] *The adoption and dominance of Urdu as a medium of instruction in a Punjabi speaking province is paradoxical. It has partly to do with adoption of a national language in preference to a regional language, but equally significant is the adoption of Urdu as a mark of Islamic identity. The spoken language also corresponds to hierarchical level of the society in Punjab, English stays with the elite, Urdu is typically a language of the middle class, and Punjabi is a common spoken language among the uneducated peasantry and labour class.*

In the villages, markets and a majority of the rural and urban homes of West Punjab, Punjabi is the common language of conversation as much as in popular media, but reading and writing is always done in Urdu, the national language.

The Punjabis have a rich literary tradition, starting from Sir Muhammad Iqbal, the poet laureate of Pakistan, and Faiz Ahmad Faiz, the progressive writer and poet, to Saadat Hasan Manto, the unequalled short storywriter and novelist Quratulain Hyder and Nazir Ahmad Dehlvi, there have been a range of writers and literary figures in Punjab. But before that there was a Sufi tradition of writing in Punjabi. Sultan Bahu, Mohammad Baksh, Farid-ud-din Ganjshakar, popularly known as Baba Farid and Shah Hussain, Waris Shah, the author of Heer – Ranjha was a Punjabi, and Bulleh Shah are some of the Sufi poets who are known to have contributed to Punjabi literature.

The elites include the political leaders, rich industrialists and businessmen, army generals and even religious leaders. The traditional land-based feudal class still matters, but not as much as during the colonial period. Most political parties are, however, feudal oriented and according to one estimate, two-thirds of the National Assembly are composed of this class and even key executive posts are held by them. The peasantry in Punjab, except for the big landlords, is relatively poor.

The northern part of Punjab including the Potwar plateau, Islamabad, Rawalpindi and some cantonments in the vicinity, presents a region rich with many institutions of national importance of official Pakistan. Islamabad, the capital, Rawalpindi, the General Head Quarters of Pakistan Army, with Potwar Plateau, the feeding ground of Pakistan army, all these offer great advantage to proud Punjabis who think of themselves to be the core of Pakistani society. Punjabis have a culture of their own, where literature, music, cinema and journalism have deeply struck roots. Known for their hospitality and cuisine, the Punjabi dominates the cultural and political life of Pakistan. Punjab, however, has not recovered fully from gender bias, and Punjab has to tread much ground to create a society where women enjoy equal status as men.

5.6 The Sindhis: A Heterogeneous Society

The people of the province of Sindh are called Sindhis, but the number of actual Sindhis who speak Sindhi as their mother tongue is just over half of the total population.

5.6.1 Evolution of Sindhi Society

In the beginning of the 20th century, over 83 percent of the people of Sindh had Sindhi as their mother tongue. Rajasthani, Balochi and Punjabi were the other languages spoken by immigrants from Rajputana, Balochistan and Sindh–Gujarati was spoken in parts of Tharparkar. The province has been a receptacle of immigrants for centuries. Karachi being for long the principal port of united Northwest India, the capital of Sindh province, and the capital of Pakistan for almost two decades, has attracted immigrants from all parts of Pakistan, similar to Mumbai in India. What has added to the ethnic-linguistic diversity is the post-1947 migration of refugees, the *mohajirs* from India and the arrival of Pakhtuns and Afghans following the war in Afghanistan first against the Soviet Union and lately the invasion by the USA to destroy the Al-Qaida establishment and the Taleban to avenge the bombing of Trade Towers in New York in 2001, by Al-Qaida, a terrorist organisation supported by the Taliban. It must be mentioned that Mohajirs constitute more than a fifth of Sindh population. Sindh has been a home to Baloch immigrants for several centuries. The last ruling dynasty of Sindh, the Talpurs arrived from Balochistan in the late 18th century, conquered it, occupied Sindh, ruled for over a hundred years and distributed a lot of land to their fellow tribesmen. Some of the leading *waderas*, or big landlords, are of Baloch origin. Even Punjabi-speaking people have migrated to Sindh, not all for farming, but to engage themselves in non-agricultural occupation, administration, industry, trade and many other professions. It is significant that while 7 percent of Sindhi population speaks Punjabi as its mother tongue, there is a complete absence of reciprocity, and not even 0.2 percent population of Punjab speaks Sindhi.

The Sindhi-speaking native population of Sindh constitutes roughly 60 percent of the total population of the province. This section of the population traces its ancestry among the Rajputs and Jats of pre-Islamic period. The two main Rajput tribes of Sindh were represented by *Samma* and *Sumra* dynasties. The former occupying the eastern and lower part of Sindh and Bahawalpur, believed to be Yadav Rajputs, and the latter were a branch of Parmar Rajputs. Besides these, the Bhuttos, the Bhattis, the Lakhas, the Sabetas, the Lohanas, the Mohanas and many others also derive their ancestry from Rajput tribes.[11] All these old tribes are together known as "*Sammat*".

The Baloch component of Sindh society has, over the decades and centuries, adopted itself to the Sindhi way of life and appears to have switched over to Sindhi, as, today, only 2.11 percent of the population of Sindh speaks Baloch as its mother tongue. The population of Baloch, classed a tribe in 1941 Census, conducted during colonial British regime, was 17 percent of the population of Sindh and 23 percent of the province's Muslim population. Despite the addition of Urdu-speaking population of Mohajirs (the refugees from India), there cannot be a sudden decline in the population of Baloch group to 2 percent. It appears plausible to believe that a large section of Baloch population has switched over to Sindhi for better integration. The Baloch represent the well-to-do section of the Sindhi society, a larger number of them having large agricultural holdings. They are the dominant group among the *waderas*, the big landlords.

Punjabi- and Sairaiki-speaking People – They are not necessarily immigrants. There is a transitional zone between southern Punjab and northern

Sindh where Saraiki is spoken as a language. But many in this group prefer to declare Punjabi as their mother tongue as a mark of Punjabi pride. There are other Punjabi- and Saraiki-speaking people in major cities in different professions and administration.

Another important linguistic element in Sindh constitutes the Mohajirs, the refugees who migrated from Uttar Pradesh, Bihar and other states of India before and after partition. This group which guards its independent existence constitutes over one-fifth (21.05 percent) of Sindh's population and over 40 percent of the population of Karachi. The Mohajirs with a large segment coming from the Hindi–Urdu speaking areas retained Urdu, the national language of Pakistan, as their mother tongue. Settled largely in large towns like Karachi, Hyderabad and Sukkur, the Mohajirs have benefitted from their educational background and their knowledge of Urdu, and secured jobs in government, public enterprises and local administration at different levels and have turned professionals in large numbers.

The Mohajirs floated a political party, Muttahida Qaumi Movement (MQM), in 1984 under the leadership of Altaf Husain. The party grew phenomenally and was not hesitant to resist any domination by any other ethnic group, like the Pathans or Afghans. The MQM has become a significant factor in the political life of Pakistan as well as Sindh and has a substantial number of seats in the provincial legislature. The group virtually controls the Metropolitan Corporation of Karachi. An unsavoury aspect associated with the group is the confrontational stand on many issues, which leads to violence. Unlike the Punjabi refugees who are almost completely integrated in Punjabi population, the Mohajirs of Sindh maintain a distinct identity in their language, culture and literary achievement. The Mohajirs are not only a social group, but have, over the decades, morphed into a political force. They are, however, confined to only urban areas and their penetration in the countryside is insignificant.

Afghans and Pathans – Afghans and Pathans are relatively recent immigrants in Sindh. They are not mentioned as tribes in Sindh in 1941 Census. The only tribes mentioned in Sindh are Baloch, Brahuis, Bhils, Kolis and Jats. A population of about seven million Pashto-speaking Pathans and Afghans has arrived in Sindh during the last four decades, first following the Soviet invasion of Afghanistan, and subsequently after the invasion of Afghanistan by the USA in the aftermath of 9/11 (the bombing of the Twin Towers in New York by the agents of Al-Qaida). Afghans as well as a much larger number of Pashtuns, following disturbed conditions in Khyber Pakhtunkhwa, descended in Sindh, but especially Karachi, their favourite destination.

The main elements of Sindh's population are the Sindhis, the original inhabitants of the province, the Mohajirs, the Urdu-speaking group of refugees that dominates the affairs of urban Sindh, especially Karachi, Hyderabad and Sukkur and recent arrivals, the Pathans.

5.6.2 Religious Belief of Sindhis

Like other provinces of Pakistan, Sindh is largely inhabited by Muslims, and the Hindus, the largest minority in the province, constitute around 6 percent of the population of Sindh. In fact, much of the Hindu population of Pakistan is concentrated in Sindh, especially in cities like Hyderabad. The Hindus also live in large numbers in the southern district of Tharparkar.

A majority of the Muslims in Sindh follow the *Sunni Hanafi Fiqh* and a substantial number also follow *Shia Ithna Ashariyah* order. Sindh has traditionally been home to the Sufi tradition. The most famous Sufi shrine of *Lal Shahbaz Qalandar* at Sehwan (of 12th century) attracts devotees from all over Pakistan. Besides, there are other Sufi shrines in Sindh like that of Shah Abdul Latif Bhitai in Bhit. The *Urs* (death anniversary) of Sufi saints have the largest gathering at Sufi shrines. Islamic fundamentalists are critical of the Sufi tradition of singing and dancing. The traditional syncretic religious culture of Sindh is under severe strain, first because of sectarian conflict between various sects of Islam, but more because of the infusion of Pashto and Afghan elements who have brought with them a culture of violence and intolerance.

The Sindhis, as a society in Pakistan, had the reputation of being tolerant. There are about three million Hindus in Sindh, and a majority, around 85 percent, are settled in the countryside. Besides, the number of scheduled caste people as defined by 1998 Census of Pakistan is just about 17 percent of the total Hindu population, but regardless of their status, all the Hindus live largely in the rural areas. The depletion of the Hindu population, which constituted 27 percent of the total population of Sindh in 1941, has resulted from a large-scale migration to India following the 1947 partition of India.

5.6.3 The Rural Society of Sindh

The rural society of Sindh is dominated by big landlords, who have retained their large holdings despite the land reforms effected during the military regime of general Ayub Khan (1958–69), and then during the prime ministership of Zulfiqar Ali Bhutto (1973–77). It is believed "*that 44 percent of the farmland in Sindh is controlled by two percent of the rural households*".[12] In Sindh, the big landlords are known as *waderas* who act as a patron to thousands of villages. Lancaster in his write-up in Washington Post cites the case of Ghulam Murtaza Jatoi owning 30,000 acres of prime land. According to him, "*the system Jatoi represents is as exploitative as paternalistic, trapping many sharecroppers who borrow from landlords to pay for seeds and fertilisers, is a form of indentured servitude.*" Another case of a rich wadera is quoted by Aryn Baker (2008) under the title "Landowner Power in Pakistan Election[13] and cites the example of Mumtaz Ali Bhutto, a cousin of assassinated Benazir Bhutto. Mumtaz Ali owns 10,000 acres of land cultivated through thousands of sharecroppers. Such big landlords are far more in number in Sindh and Balochistan than in any other province of Pakistan.

Bhuttos are one of the richest families and are believed to own 40,000 acres of land. The landlords, the waderas of Sindh, have enormous influence on the politics and economics of the province. Writing in *Dawn* on October 10, 2013, Tanveer Arif, the Executive Officer of the Society for Conversion and Protection of Environment, remarked that 5 percent of the landholders possess 66 percent of the total land and 65 percent of the small farmers hold 15 percent of the land and 50 percent of the rural households are landless. The *waderas* of Sindh owning thousands of acres of land, cultivated by deprived class of sharecroppers, illustrate the stark inequality, besides poverty that exists in the countryside in Sindh. The rural society has still not liberated itself from the age-old feudal system, despite a democratic mode of governance of the country and the provinces.

The people of Sindh are resentful of domination of Punjab over the provinces. The shifting of the federal capital from Karachi to Islamabad, a capital territory, yet very much in a Punjabi socio-economic realm, has deprived Sindh of many advantages that the province enjoyed before the capital was shifted from Karachi to Islamabad. With Karachi as an international port and the industrial hub of Pakistan, Sindhis considered themselves as the vanguard of Pakistani society.

5.7 The People of Balochistan

As expected, the most important ethno-linguistic group of people in Balochistan is constituted by Baloch, a tribal group that arrived in Balochistan from Iran over the centuries with an organised administration in the 17th century. The Khanate of Kalat was established in 1666.

Besides the Baloch, the Pashto-speaking Afghans and Pakhtuns, occupying the northern part of Balochistan, form the second important ethnic group. Other minor groups include *Med* along the Makran coast, *Jats* in Las Bela, Sibi and Kachhi areas. One may add *Brahui,* an ethnic group that claims an indigenous status and pre-Baloch existence. Added to these native groups are the immigrants from Sindh and Punjab, especially in the area bordering these provinces.

5.7.1 The Relative Strength of different Ethnic Groups

Language spoken as a mother tongue is a good indicator of the strength of different ethnic groups. The relative strength of different linguistic groups is as follows (Table 5.8).

TABLE 5.8 Languages Spoken as Mother Tongue in Balochistan

Language	*% of People Speaking Different Language as Their Mother Tongue in Balochistan*
Balochi	35.58
Pashto	35.33
Brahui	17.12
Sindhi	4.07
Punjabi	1.12
Saraiki	2.21
Others	8.64

Source: Pakistan Bureau of Statistics, Final Result of Census 2017, table 11

The Baloch and the Pashto-speaking Afghans and Pathans make almost 90 percent of the total population while Sindhi, Punjabi and Saraiki speaking immigrants, settled in the border region of the province constitute around 10 percent of the population of the province. Balochistan is the least populated province of Pakistan, and its population is only 6 percent of the total population of Pakistan. At 60 percent Baloch component of the province's population would mean that Baloch as an ethno-linguistic group have a population of roughly seven million, including many tribes and sub-tribes. The Baloch, as an ethnic group, are spread all over the province, except the northern part of the province, which is occupied by Pashto-speaking Afghans and Pathans.

There are multiple theories about the origin of Baloch. There is a tradition among Baloch that they came from Aleppo (now in Syria). Their immediate past habitat, before they arrived in Balochistan, appears to be Iran. In fact, Dames (1904) suggested that Baloch are Iranians.[14] But whatever their original habitat, they had taken up a position in close proximity to Makran, early in the 7th century.[15] The Baloch had pastoral habits and moved from one place to other in search of fresh pastures. With the discovery of minerals like coal and natural gas the occupational structure of the family is getting diversified, though pastoralism still persists. Many of the tribal names bear the names of localities they once occupied. They were more firmly organised in a state like structure only in the 17th century, yet there were many tribal chiefs, to which the tribes owed their loyalty.

5.7.2 The Tribes of Balochistan

The *Baloch* are divided into a large number of tribes and sub-tribes, and they number several hundred. In fact, over the centuries, the tribes have gone on multiplying in smaller groups. The principal tribes of Balochistan are named after the children of Jalal Khan, the legendary chief of the Baloch chiefs. The four sons and a daughter of the Khan were named Rind, Hot, Lashari and Korai, and the daughter Jatoi. Thus the principal tribes of Balochistan are:

1. Rind	4. Korai
2. Hoth	5. Jatoi
3. Lashari	

Besides these, some other important tribes are Mirwani and Rakshani. Of all the tribes, the Rind and the Lashari are numerically the most important. It may be noticed that *Bugti* and *Marri*, the most militant tribes of Balochistan fall under *Rind*, one of the major tribes. The Khan of Kalat come from Mirwani tribe, named after Mir Mirwani Baloch. Many of the Baloch tribes are a sub-tribe of *Rind* or *Lashari*. To illustrate, one may cite Jardaris and Legharis, the two well-known names, representing two subgroups in the larger group *Rind*.

The geographical distribution of the tribes is very impermanent and often a number of tribes are found in close proximity. Perhaps, the only exclusive domain is that of the Afghans and Pathans settled in the northern part (north of the Dera Ghazi Khan–Quetta-Pishin line). The Baloch, as a tribal group are distributed in two disparate areas, eastern and western. The eastern Baloch area consists of Bugti and Marri hills, the former around Dera Bugti and the latter around Kohlu, and Kachhi plain. The western Baloch group extends westward beyond 66° E. Between the two areas of Baloch concentration, there is a wedge of Brahui dominated land that includes the Kharan and Jhalawan country. The Sarawan country is the tract between Kalat and Quetta and the Jhalawan country is the area south of Khuzdar. Brahui, though a different tribe, are Baloch to the core, unlike the Afghans and Pathans of northern Balochistan.

Baloch, by tradition, have been a nomadic tribe, living in small hamlets of a dozen households owning cattle, sheep, goats and camel and moving from one area to another looking for better pastures. Those engaged in agriculture are relatively better settled in Kachhi plain or in the Las Bela region in south, the two moderate agriculture regions. One may even say that the Baloch are an ever migrating tribe. In the past, they have migrated to Sindh, of which roughly 40 percent population claims Baloch ancestry. They have migrated and settled in the border region of Punjab in Dera Ghazi Khan district and Derajat area, known for immigrant Jat population from Balochistan. Baloch, like many other sub-Himalayan communities, practice a modified form of transhumance, moving between the mountain territory and the agricultural areas. The Brahuis, in particular, have a regime of movement that runs as follows.

Months	*Routes*	*Area of Migration and Type of Work*
September	Downward	Movement from the hills
October – November		Harvesting rice in Sindh
December – January		Harvesting jowar
March – April		Spring harvest of wheat
June	Returning to the hills	
July and August		The rest period for the Brahuis

The Baloch as an ethnic group have an above average height, often very courageous (an important virtue of the Balochs) and relatively truthful, and genuinely hospitable. Being good riders, Baloch spend much time outdoors.

Balochi as a language has many variants with multiple dialects of which *Rakshani* is a major dialect. Yet, eastern and western Balochi are the two principal categories, being spoken in eastern and western Balochistan. Originally an unwritten language, Balochi has adopted Urdu–Persian script and a modern literature is developing in Pashto. It must be added that most Baloch are bilingual. Besides speaking Balochi and some other dialects, they speak Urdu and some are even acquainted with Sindhi.

The Baloch as a community and as a provincial group are a bit disenchanted with the Government of Pakistan, after the forced annexation of the state of the Khan of Kalat in 1948, despite an assurance to the contrary by Quaid–i-Azam Mohd. Ali Jinnah, the first Governor General of Pakistan. In protest, there have been a series of uprising and outright revolts led by the successors of the last Khan of Kalat and even other tribal chiefs like those of Bugti and Marri tribes. Their demand is to have a greater share in the resources of the province and withdrawal of the militarisation of Balochistan. The last serious and bloody conflict ended in the killing of Nawab Akbar Khan Bugti, the Oxford educated tribal chief of the Bugtis. A number of militant groups and parties like the Balochistan Liberation Army, United Baloch Army or Baloch Republican Party are in the field, demanding greater share in the resources of the province.

The Afghan/Pathans The Pashto-speaking Afghans and Pathans occupy a quarter of Balochistan territory. They are not a recent immigrant group but are believed to have settled even before the Baloch arrived in Balochistan. Their number is not small; they constitute over 30 percent of the population of Balochistan. They are settled west of Takht-i-Sulaiman and northeast of the Quetta-Pishin line. The districts of Sherani, Zhob, Qila Saifullah, Pishin, Musakhel and part of Loralai are the strongholds of Pashto speaking Pathans. They speak a softer version of Pashto, unlike the Pathans of Peshawar region.

The Pathans settled in Balochistan are divided into a number of tribes. The chief among them are the four tribes, viz., Kakar, Tarin, Pani and Sherani. Each tribe is concentrated in a specific area.

Tribe	*Area of Concentration in Balochistan*
Kakar	Zhob, Quetta, Pishin and Loralai
Tarins	Loralai, Sibi and Quetta
Panis	Musakhel and Sibi
Sherani	Close to Takht-i-Sulaiman (Sheran district)

The Pathans of Balochistan are known for managing the road transport of the region. They have a virtual monopoly of goods transport by roads. They are also graziers and a small number have specialised in *karez* building for irrigation.

Jats One of the earliest community, they are confined to only agricultural regions like Kachhi area, the triangular area with its apex at Sibi and the base formed by the interprovincial boundary between Balochistan and Sindh. The other area where Jats predominate is Las Bela, another plain, between Bela and the Arabian Sea coast, drained by the river Porali. The Jats have another name called *Lasis*. The *Medes*, one of the oldest human groups, possibly of African origin, largely engaged in fishing, are found on the Makran coast and speak Makrani as a language of communication.

Brahuis The Brahuis, a large group, speaking Brahui as a language, believed to be of a Dravidian family, are an independent ethnic group. Consisting of 29 tribes, some of which are on the verge of extinction the Brahuis number over two million in Balochistan. Historically, they owed allegiance to Brahui-speaking Khan of Kalat. Since the end of *Khanate*, the small subgroups of the tribes have their own chief. They occupy the Kharan and Jhalawan counties, the former being the hilly region lying between Quetta and Kalat, on an stretch of roughly 150 km^2, and the latter being the area of Khuzdar. In fact, Brahuis occupy the hilly eastern part of Balochistan, south of Quetta and west of Kachhi plain and the Brahui Hills. Brahuis are not a keen agriculturist community and often engage themselves in non-agricultural occupations. Most of them are pastoralists and goat breeders while others are wage earners, migrating long distances specifically in Sindh in search of jobs. Most Brahuis are bilingual.

Balochistan has several tribal communities and many languages and dialects; it is the Balochis who are politically more enlightened and active. They have not yet shed the memory of their recent history, especially the forced annexation of the Khanate of the Khan of Kalat or even Makran and the forced eviction of the two principal tribal chiefs, the Bugtis and the Marris. The Baloch have a perception that the mineral-rich Balochistan is exploited without giving them their due share. Second, they resist the persecution of such Balochi people as dare to criticise the federal government or organise movements putting forth their demand and resisting persecution by federal armed forces. Though writing a political commentary is beyond the scope of this text, we feel tempted to comment on the situation. The strength of Baloch resistance is the geographical location and remoteness of Balochistan from the federal capital. Its case is fundamentally different from that of East Pakistan that seceded and became an independent state. Balochistan is a vast desolate country with small population, though rich in minerals. It still practices subsistence farming and can hardly support its own population. Its association with mainstream Pakistan society and politics, based on an understanding and reconciliation is a better option, than localised eruption of violence that harms individuals, creates sensational news, but hardly does harm to federal or provincial administration. One has to look at the problem from a more practical perspective. With a population of just over 20 million (12,246,275 in 2017) and an area of 345,190 km^2, that gives a density of 35 persons per km^2, any resistance, based on force, is an unequal fight. Reconciliation, on best terms, is a better option for Balochis and Balochistan. After all, they are a now part of Pakistan.

5.8 The People of Khyber Pakhtunkhwa

As the name suggests, Khyber Pakhtunkhwa (KP) is predominantly inhabited by Pakhtuns/Pashtuns, commonly known as Pathans and some nomadic tribes like in the north-eastern part of the province. Caroe (1958) observes that

> *the Afghans of Afghanistan and the Pakhtuns of Khyber Pakhtunkhwa are part of the same stock. The Afghan of the plains and low altitude open plateau, representing a senior branch are entitled to Afghan name, while the Afghans occupying the highlands are the "Hill Pathans', a deformation of Pukhtanah, a plural of Pakhtun.*[16]

The Afghans have two branches: (1) Western Afghans and (2) Eastern Afghans, with several clans in each category.

The *western Afghans* include: (1) Abdalis (also known as Durranis) and (2) Ghilzai. The western Afghans have been subjected to Persian influence, like the Durranis who picked up Persian language, as they came in contact with Safawid Empire of Persia. Herat and Kandahar are their cultural capitals. The *eastern Afghans*, on the other hand, were more influenced by the Mughals of Hindostan who ruled even over Peshawar and Kabul. The eastern Afghans are known as *Yuzufzai* and also include a hundred clans of Peshawar plain and the valleys in the north. The Pakhtuns or Pashtuns, the Highlanders, occupied the land between the east and west Afghan territories.

Though Khyber Pakhtunkhwa as much as the western extension into erstwhile Federally Administered Tribal Areas (FATA), now merged in Khyber Pakhtunkwa, are predominantly inhabited by Pakhtuns, the sub-mountainous northern areas have some of the non-Pakhtun tribes. The erstwhile Hazara division, incorporating Abbottabad, Haripur, extending further into Swat have a substantial population of Gujjars, a pastoral tribe, which still practices transhumance, moving to higher regions in summer in search of pastures, and descending down to lower levels during winters. Dera Ismail Khan, another district of Khyber Pakhtunkhwa is largely inhabited by Jat Muslims.[17]

5.8.1 Principal Pashtun Tribes and their Geographical Distribution

Of the total population of Khyber Paktunkhwa, about 75 percent speak Pashtu as their mother Tongue. The remaining 25 percent is constituted of non-Pashtu-speaking people. These are the people away from the core areas, in Hazara division, Chitral and even in Swat Valley.

There are roughly 60 Pashtun tribes, but the number multiplies to hundreds if sub-tribes are taken into account. The Pakhtuns are divided among several tribes. Numerically the most notable are the Yusufzai, Afridi, Orakzai, Bangash, Khattaks, Wazirs, Marwats, Mehsud and Turi.[18] The following is the locational distribution of these tribes as suggested by Caroe (Table 5.9).[19]

TABLE 5.9 The Pakhtun Tribes of Khyber Pakhtunkhwa

Tribes	*Location (districts of Khyber Paktunkhwa)*
Yusufzai	Kohistan, Swat and Dir districts of K P
Tarklanri	Malakhand and Lower Dir
Mohmand	Charsadda, north of Kabul River
Shinwari	North of Safed Koh Mountain
Afridi	South of Safed Koh, extending to the north of Kohat
Orakzai	East–west band south of Afridi territory
Bangash	South of Orakzai (Hangu and Kohat area)
Khattak	W of Indus and along it; Nowshera in the north to Bannu in the south
Wazir	Bannu, North Waziristan
Darwesh Khel	Along Afghanistan border in South Waziristan
Mahsud	A square patch northwest of Wana (Umar pocket)
Marwat	West of Indus along the river, south of Khattak territory
Turi	Kurram district of Khyber Pakhtunkhwa

Source: Map in ch. XXIV of Olaf Caroe (1958)

The Pathans have some unique elements in their conduct, collectively known as Pakhtunwali, which has three important components.[20]

Each of the tribes is divided into sub-tribes, locally called "*Khel* or "*Zai*. A *khel* often includes an entire village and may sometimes consist of several villages, all coming under a specific *khel*. A *khel* is usually named after the patriarch, the earliest migrant of the area. Division of land has led to the multiplication of clans. The social affairs of a tribe are managed by "*jirga*", the assembly of the elders, and their decision is binding on the community and individuals. The Federally Administered Tribal Area (FATA) and PATA (Provincially Administered Tribal Area) have since been abolished through a constitutional amendment, these have been reorganised into districts. The *Jirga* still exists, but its importance has been diminished and regular district administration, as in other parts of Pakistan, has taken roots.

Most of the Pakhtuns are Sunni Muslims, but there are a few Shias whose number is less than 1 percent of the total population. Turis and some of the Bangash tribes in Kurram Valley are avowed Shias as are many of the Orakzai Pathans in Tirah.[21] Shias also occur in Kohat and further south in Shirani country (in Balochistan). Before 1947, there were some Hindu traders and Sikhs, but they have since migrated to India. There is a semblance of some non-Muslim community in Peshawar where an insignificant number of Sikhs and Christians are still living.

The Pakhtuns speak Pashtu, a language with many variants, of which two are well-defined. These are usually labelled as soft and hard Pashto. These two variants are associated with certain tribal groups and are spoken in specific regions. The softer variant, *Pashtu* is spoken in the southern part of Khyber Pakhtunkhwa, in north and south Waziristan, Bannu, Lakki, Marwat, Hangu and Kohat and Orakzai agencies by tribal groups that include Durranis, Ghalzis, Wazirs. The hard variant, written as *Pakhtu* is spoken in the northern and north-western part of Khyber Pakhtunkhwa, coinciding with the region incorporating Peshawar, Swat, Buner and Bajaur. These districts are inhabited by Afridis, Orakzais, Shinwaris, Bangash and Turish tribes. There is, in fact, a language divide. If on draws a SE–NW line from Attock on Indus passing through Thal (Tall) up to Kurram River, the area north of this speaks Pakhtu while the tribes, south of this line, in Bannu, Tank, Lakki Marwat and North and South Waziristan speak a softer version of the language. The softer version extends further south in North Balochistan, which has a large population of Pathans.

1. Honour, called *nang*
2. Revenge, known as *badla*
3. Hospitality, known as *melmastia*
4. Giving refuge to one who asks for protection.

While the honour relates to the chastity and reputation of one's female relatives, revenge means avenging the damage of one's honour. Hospitality to guests has certain well-established norms and etiquettes. Accommodation of a huge influx of refugees from Afghanistan in 2001 and thereafter in Khyber Pakhtunkhwa, an unprecedented situation, is ascribed partly to the extension of hospitality as a deep-rooted element in the character of Pakhtuns.

Besides these aspects of personal conduct typical of Pathans, the latter have a tendency to pick up arms. The Pathans are generally recognised as a martial tribe. With hardly a fifth of the area, confined to the valleys, under cultivation, Pathans have been traditionally a migrating community. A large number of Pathans migrate from their home region seeking employment opportunities elsewhere.

A culture of violence has pulled the region back from modern development. But recent efforts to mass-spread education among the tribes may change the socio-cultural landscape of the region. Khyber Pakhtunkhwa has over two dozen universities, half of them centred at Peshawar and many schools and colleges. These may bring about a change in the social outlook of the people. These educational and cultural institutions are located in the central axial region of the province. The problem lies in the remote hilly areas, isolated for centuries.

Other Tribes of Pakistan – Besides the generally accepted ethno-linguistic groups, there are smaller groups, especially the tribal groups in the far corners of the country.[22] The fishing communities like *Kihals* and *Mors*, living along the bank of the Indus in Dera Ismail Khan and Dera Ghazi Khan are the examples of obscure neglected tribes. Then, the pastoral tribes like the *Ode* or *Ode Rajbuts* living in the Thar Desert. Other tribes mentioned by Wagha include *Ihabels, Kehals, Mores* and *Kutanas*. Mention must be made of *Kalash* tribe of Chitral, which speaks *Kho* or *Chitrali*, different from Pashtu.

To conclude, despite an apparent homogeneity, under the Islamic umbrella, there is a striking diversity in the population of Pakistan. These diversities are based on a primordial loyalty of individual groups to their tribal past, pre-Islamic heritage or even regional distinctions and pride. These subtle differences get accentuated when there is some crisis of confidence and a specific ethno-linguistic or regional group perceives discriminated against by federal government or federal institutions, or dominated by a numerically strong and materially prosperous group. This is, however, not only the case of Pakistan and is common to most democracies across the world.

How to retain and respect regional and cultural diversity, without letting it be suppressed in the name of development, has to be one of the objectives of a functioning democracy, including Pakistan.

Notes

1 Pakistan Tourism Development Corporation, http.//www. tourism.gov.pk. peoples-of-Pakistan (accessed on 29–12–2016).
2 Population Census Organization, Government of Pakistan, http//www.census.gov.pk/religion.htm (accessed on 24–6–2013).
3 PBS, Final Census results 2017, Table 9
4 Encyclopaedia Britannica (2016) The Section on Islam or Islamic Population is Largely Based on Encyclopaedia Britannica, *Encyclopaedia Britannica 1985*, vol. 11, p. 353.
5 *Imperial Gazetteer of India*, Punjab (1908) vol. I. p, 50, Government of Printing Press, Calcutta.
6 In Census of Pakistan, the Scheduled Castes are categorised separately from Hindus and the Hindus till 1998 census were categorised as Hindu (Jati).
7 Lieven, Anatol (2011) *Pakistan: A Hard Country*, Allen Lane, London, p. 123.
8 Lieven, Anatol (2011) *Pakistan: A Hard Country*, Allen Lane, London, p. 278.
9 Siddiqa, Ayesha (2016) Pakistan's Elite, *The Express Tribune.*
10 Safir Rammah (2017) Punjabi in Pakistan, http://www.Punjabics.com/exclusive article (accessed on 2–5–2017).
11 Abdullah Hamid (1973) The Historical Background of Pakistan and its people, Tanzeem Publications, Karachi, reproduced in parts by Sani H. Panhwan (2009) under the title "The People and Land of Sindh".
12 Lancaster, John (2003) Pakistan's Modern Feudal Lords, *The Washington Post*, 8 April 2003.
13 Baker, Aryn (2008) Landowner Power in Pakistan Election, *Time Magazine,* 18 February.
14 Dames, M. L. (1904) An Historical and Ethnological Sketch of the Baloch Race, quoted from Imperial Gazetteer of India, 1908, vol. VI, p. 289.
15 Imperial Gazetteer of India (1908) *Provincial Series- North-West Frontier Province*, Imperial Gazetteer of India, Calcutta, vol. VI, p. 285.
16 Olaf Caroe (1958) *The Pathans, 550 BC – AD 1956*, Macmillan, London.
17 Imperial Gazetteer of India (1908) *Provincial Series- North-West Frontier Province*, Imperial Gazetteer of India, Calcutta, p. 23.
18 Olaf Caroe (1958) *The Pathans, 550 BC – AD 1956*, Macmillan, London, p. XIV.
19 Olaf Caroe (1958) *The Pathans, 550 BC – AD 1956*, Macmillan, London, map entitled "Tribal Locations of the Pathans".
20 Amin, Mohamed, Duncan Welletts and Graham Hancok (1982) *Journey through Pakistan*, Vikas Publishing House, New Delhi.
21 Imperial Gazetteer of India (1908) *Provincial Series- North-West Frontier Province*, Imperial Gazetteer of India, Calcutta, p. 34.
22 These tribes are mentioned by Wasim Wagha (2012) in a note "*Country Technical Note on Indigenous People's Issues - Islamic Republic of Pakistan*" published by International Trend for Agricultural Development.

6

POPULATION OF PAKISTAN

Pakistan is the sixth most populous country of the world, ranking after China, India, the USA, Indonesia, and Brazil. The population of Pakistan according to the latest Census of 2017 is 207,684,626 though the latest estimate (July, 2022) by the United Nations Organisation raises this figure to over 229 million, with a sex ratio of 1,050 males per 1,000 females. Distributed over an area of 756,056 km^2, the population density of the country works out to 261 persons per km^2. While this low density of population, thanks largely to the empty spaces of Balochistan and Khyber Pakhtunkhwa, compares favourably with its South Indian neighbours (India 382 persons/km^2 and Bangladesh 1,265 persons/km^2), it is the growth rate of population of Pakistan that is demographically a worrying phenomenon. Pakistan has an average annual population growth rate of 2.1 percent, against the world average of 1.03 percent and 1.1 percent for both India and Bangladesh. Some of the vital statistics of populations, like fertility, infant mortality and life expectancy and the median age of the country`s population further clarify the state of the demographic health of the country. Pakistan has a fertility rate, defined as "the average number of children, born to a woman during her reproductive years" of 3.3 that compares adversely with the world fertility rate of 2.4. It has an infant mortality (defined as the number of infant deaths per 1,000 births during a year) of 54, one of the highest in the world; and it has a life expectancy at birth of 68 years, not far behind the world-population life expectancy of 72 years. The median age of the population of Pakistan is 21 years suggesting that it is a very young population, with 43% of the population below the age of 16 years.

The population of Pakistan has grown not only steadily, but even fast, during the last 75 years. From the first Census in 1951 to the last one in 2017, the population has grown from 32 million to 208 million, a six and a

DOI: 10.4324/9781003510147-8

half fold increase. The level of urbanisation during the period (1947–2017) is doubled from a mere 18 percent in 1951 to over 36 percent in 2017.

6.1 Primary Information about the Population of Pakistan

The population of Pakistan as recorded in 2017 Census was 207,684,626, spread over the four provinces and the Federal territory, Islamabad, with 36.4 percent of the population living in urban areas and the remaining 63.6 percent living in rural areas. The country has a sex ratio of 104.91 males per 100 females. The overall density of population for the entire country is 261 persons per km^2. The average household for the entire country is 6.39 persons. And, the country has experienced an average annual growth rate of 2.4 percent during the census period 1998–2017. The estimated population of Pakistan for the year 2022 is 220 million.

Sources of Information: The main source of information for the population of Pakistan and all related aspects like detailed district wise distribution, sex ratio, age-structure, density of population, level of urbanisation and many other qualitative aspects like literacy, religious and linguistic composition, household details and many other aspects, is the Census of Pakistan, Pakistan Bureau of Statistics. Starting from 1951, Pakistan has conducted population censuses, though at irregular intervals. The population censuses in Pakistan were conducted in 1951, 1961, 1971, 1981, 1998 and 2017. The last of the series, the 2017 Census, was conducted after a gap of 17 years. Thus, the population figures relating to the post 2017 years are, at best, an approximation, unless backed by some government agencies. Such estimates have been made by several international agencies like the United Nations Population Fund, the US Bureau of the Census International Data Base, the World Bank, the Population Division of the Department of Economic and Social Affairs of the United Nations and several others. These have been appropriately quoted as sources wherever necessary (Table 6.1).

6.2 Distribution of Population in Pakistan

The spatial distribution of population in Pakistan is very unequal with great diversity, presenting a density pattern ranging in density from the very sparsely populated districts of Balochistan like Awaran, Jhal Magsi and Chagai district, each having a population density of less than 10 persons per km^2, to some of the most densely peopled districts of Punjab, like Faisalabad and Gujranwala, each having a population density of over 1,300 persons per km^2. Such contrasts are not only inter-provincial but even in the same province such contrasts in density are observed. The contrasts are not only limited to population density but are also reflected in the level of urbanisation, linguistic composition and several other qualitative aspects of their life and culture.

TABLE 6.1 The Area, Population and Population Density of Pakistan, and Its Constituent Provinces, with the Share of Area and Population in percentage terms, of each Province and Islamabad, the Federal Capital

Administrative Units Pakistan / Provinces	*Area in km²*	*Area in % of the Total Area of Pakistan*	*Population as per 2017 Census*	*Population as % of the Total Population of Pakistan*	*Density of Population, Persons per km²*
Pakistan	7,960,961	100	207,684,626	100	261
K. P.	74,521	9.36	30,508,920	14.7	409
FATA	27,220	3.42	4,993,044	2.4	183
Punjab	205,345	25.8	109,989,656	52.95	536
Sindh	140,914	17.7	47,854,510	23.04	340
Balochistan	347,190	43.61	12,335,129	5.93	36
Islamabad	906	0.11	2,003,368	0.96	2,211

Source: Population Census Organisation, Government of Pakistan, 2017; Final Results of Census 2017, T-1; Area, Population by Sex, Sex Ratio, Population Density, Urban Population, Household Size and Annual Growth Rate.

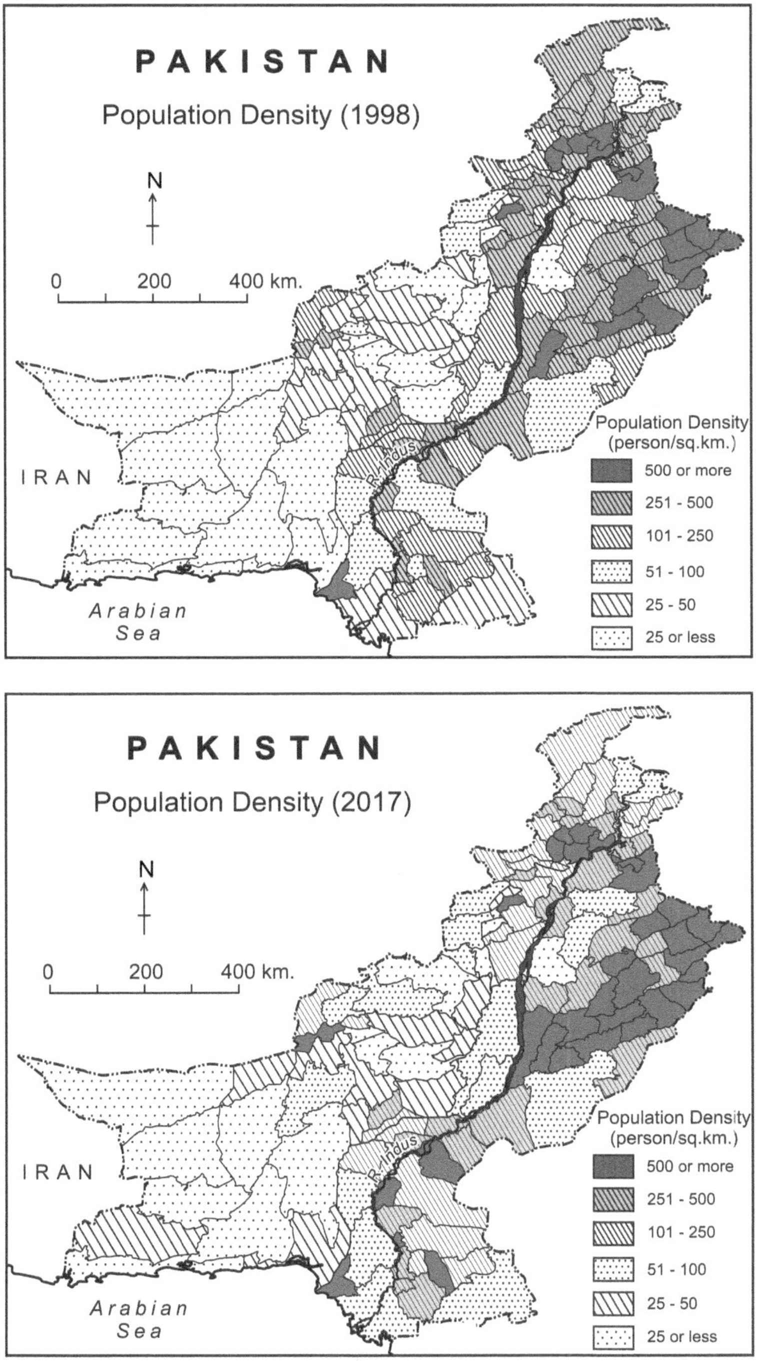

FIGURE 6.1 District-wise Population Density of Pakistan (1998 and 2017)

Press release of the Pakistan Bureau of Statistics on August 5, 2023 mentions that the population of Pakistan reaches 241.49 million as the Digital Census.

In terms of inter-provincial distribution, the two most striking facts are a very low density of population of Balochistan (5.93 person/km^2), lying on one end of the density spectrum and Punjab with a population density (536 persons/km^2), more than double the national density, occupying the other end. The two provinces viz., Sindh and Khyber Pakhunkhwa with a population density of 340 and 409 persons per km^2 have a population density above the national average of 261 persons per km^2. More striking is the fact that the largest of the provinces, Balochistan, occupying over 43 percent of the area of Pakistan has less than 6 percent (5.91 percent) of the latter`s population. In contrast, Punjab occupying a quarter (25.8 percent) of the area of Pakistan has more than half (52.95 percent) of its population. Sindh has an above average density of population (340 persons/km^2), covering 18 percent of the area of Pakistan but accounting for over 23 percent of its population. Punjab is the most densely populated and highly urbanised province of Pakistan, followed by Khyber Pakhtunkhwa (409 persons km^2), Sindh (340 persons/km^2), Federally Administered Tribal Area (FATA) (183 persons/km^2) and Balochistan (36 person/km^2) in that order. It may be mentioned that after the Census of 2017, the Government of Pakistan merged the two provincial units of Khyber Pakhtunkhwa and the FATA into one single administrative province of Khyber Pakhtunkhwa. Thus, the combined province of Khyber Pakhtunkhwa has an area of 101,741 km^2, i.e., 12.78 percent of the area of Pakistan having a total population of 35.5 million, giving a population density of 349 persons per km^2.

Distribution Pattern in Each Province Another important aspect of population distribution is not just inter-provincial, but even intra-provincial contrasts and large spatial variations in each of the provinces. This is attributed to terrain conditions, harsh climate, especially desert or desert-like conditions and low or almost insignificant agricultural productivity and overall economic poverty.

6.2.1 Population Distribution Pattern in Balochistan

With a population of 12 million spread over an area of 3,47,190 km^2, Balochistan has an overall population density of 36 persons per km^2, the lowest density among the provinces of Pakistan. Even, this low density is highly variable, ranging from 4.13 persons/km^2 for Awaran district, in the arid southwest region, to 658 persons/km^2 for Quetta district, in the high altitude axial zone of the province. Most of the districts with the exception of Quetta, Jaffrabad and Nasirabad on the border of Sindh, and Qilla Abdullah, have a population density of less than 100 persons per km^2, a large number have a density of less than 50 and a few have a single digit density of less than 10 persons per km^2. The south-western part has desert conditions, a sandy rocky wasteland with very low rainfall without any perennial river, resulting in the absence of any agricultural land, and very small population. The northern part of Balochistan is a mountainous plateau with heights lingering around 2,000 m ASL, bordered on the northwest by the Toba Kakkar Range. Despite the winter rainfall that ensures soil moisture, the relief conditions permit only horticulture. The Zhob River that traverses the entire northern part of Balochistan makes only a narrow valley, devoid of a wide expanse of alluvial plain. Only small settlements are strung along the river and the main transport routes.

It is to be noted that only 1.05 million hectares (10,500 km^2) of land is classed as the Net Sown Area (NSA) which is just around 3 percent of the geographical area of the province, limiting agricultural prospects and productivity. The province is, no doubt, rich in mineral resources but these occur in patches and employ a few hundred or at best a few thousand people, promoting a mining or industrial settlement, as in case of Sui gas area. The mining of chromite, copper and gold has not yet fully realised. Quetta–Pishin is the area of high density, largely as result of the impact of Quetta, the capital city of the province. The Zhob region, in the extreme north, has a density of around 50 persons/km^2. The central linear zone along the railroad connecting Jacobabad with Quetta and Chaman is another area of moderate population density of around 50 persons/km^2.

Despite a low density of population, Balochistan has a relatively high level of urbanisation, with 28 percent urban population. This is attributed to a few cities like Quetta, Gwadar, Turbat, Lasbela and some others which house a substantial part of the district in the district headquarters.

6.2.2 Population Distribution in Khyber Pakhtunkhwa

Despite being a hilly frontier region, Khyber Pakhtunkhwa has a relatively high population density of 410 persons per km^2. This is higher than the national average of 261 persons/km^2, and second only to Punjab which has a density of 536 persons/km^2. Even if, one combines the two administrative units of Khyber Pakhtunkhwa and the Federally Administered Area (FATA), as was done in 2018 by a constitutional amendment, making a single province of Khyber Pakhtunkhwa, the density comes down to 349 persons/km^2, which is still higher than the population density of Sindh (349 persons/km^2) and Balochistan (36 persons/km^2). It may be noted that no district in Khyber Pakhtunkhwa has a population density of less than 100 person/km^2, with the sole exception of partly glaciated mountainous and the northernmost district of Chitral having a population density of 39 persons/km^2.

The population distribution and density pattern in Khyber Pakhtunkhwa follow a pattern that corresponds with the suitability of land for agriculture and agricultural intensity. Thus the valley regions of Kabul, Swat and their tributaries have better prospects of agriculture and have attracted large population. The trans-continental route, the Grand Trunk Road that cuts across the province from south–east to north–west, coincides in its alignment with the famous Kabul Valley, and the low level rich alluvial plain has attracted population over the centuries.

One can identify three broad population density zones:

1a) A very low density mountainous region of northern Khyber Pakhtunkhwa. This zone includes the mountainous districts of Chitral and Kohistan, having a population density of 39 and 104 persons per km² respectively. Lying north of 35 degrees N Latitude, these districts with their high mountainous relief and partly glaciated terrain with heights ranging from 2000 to 7000 m ASL, cover around 30 percent of the area of the province and are responsible to bring down an overall density level of the province.

1b) A second area of low density is the zone of frontier districts contained in what was earlier known as the Federally Administered Tribal Area (FATA) before 2018. Being mountainous with a dissected terrain, though not as cold as northern Khyber Pakhtunkhwa, home to a number of tribal agencies like Kurram, Mohmand, North and South Waziristan, Orakzai and a number of tribal frontier districts, have a low density of population, ranging from 102 persons per km² to 209 persons per km² for Mohmand district. Small tribal areas between the provincially administered district and the frontier tribal agencies, earlier known as frontier districts, have a very low density of population. Some of these districts like frontier Tank district, frontier Dera Ismail Khan and frontier Bannu district have a population density of 30, 34 and 57 persons per km², respectively. It has to be understood that these districts stood as a first line of defence between the British administered districts and the tribal agencies occupying the frontier region and were not very popular with the people as a safe area for settlement.

2a) An Intermediate Density Zone of river valleys – South of the mountainous region, the north–south river valleys like those of Swat and Panjkora not only provide some narrow alluvial plain but also some sources of irrigation. Districts like Swat (Mingora), Buner and Sangla have a medium density of 400 to 800 persons per km², though some of the districts in this zone, like Lower Dir, benefitting from irrigation and consequently more productive agriculture, have a little higher density of 900 to 1,000 persons per km².

2b) The mid-eastern part of the province, the Hazara division – The trans-Indus eastern part of the province, consisting principally of the districts of Abbottabad, Mansehra, Haripur and Battagram, has a medium level of density, ranging from 400 to 600 persons to a km^2. The northern part of the region, consisting of Battagram and Mansehra districts, because of its sub-mountainous nature, has a lower density than the southern part covered by Abbottabad and Haripur. Abbottabad and Haripur have a population density of 678 and 581 persons per km^2, while the northern parts have a density lower than 400 persons per km^2.

While the northern part of the region is occupied by Indus Kohistan and the mountainous Kaghan Valley, the southern part occupies a plateau like relief and has a salubrious climate for which Abbottabad is known. The southern part, with better climatic conditions, has the advantage of being traversed by the Karakoram Highway that has not only improved connectivity but opened the prospects of industrialisation of the region. The region, especially, Abbottabad, is home to some important public institutions and also serves as a hill station. It is still attracting ever more population, for tourism, industry and even for reasons of its liveability.

3) The region of highest population density The area of highest density in Khyber Pakhtunkhwa is the linear zone extending between Peshawar in the west and Swabi in the east, spread over Peshawar, Charsadda, Mardan, Nowshera and Swabi districts. Four of these five districts have a population density of over 1,000 persons per km^2. The population density of these five districts is as follows (Table 6.2):

As can be observed, this high density zone occupying only 9.6 percent of the area of the province is home to 44 percent of the population of the province. Each of the districts, mentioned in the table, has a population density that ranges from over two times in case of Nowshera, to more than eight times the population density of Khyber Pakhtunkhwa (409 persons/km^2) in the case of Peshawar. The highly urbanised district of Peshawar

TABLE 6.2 Districts of Khyber Pakhtunkhwa with High Population Density

Districts	*Area (km^2)*	*Population*	*Density of Population*
Peshawar	1,257	4,267,196	3,395
Charsadda	996	1,610,960	1,617
Mardan	1,632	2,373,399	1,454
Nowshera	1,748	1,521.00	870
Swabi	1,543	1,625,477	1,063
Total	7,176	11,398,027	1,588

accounts for roughly 14 percent of the population of the province. Being the capital of the province, and having a population of over 2,000,000, (Population 1,970,042 in 2017), Peshawar is the nodal point of the province attracting ever more population.

This high density, linear zone, often referred to as Peshawar valley or Kabul valley, occupies a mid-location, between North and South Khyber Pakhtunkhwa and is the economic and cultural core of the province. Aligned to the historic Grand Trunk Road and the main railway line of Pakistan, running from Lahore to Peshawar, the region is agriculturally most productive and industrially better developed. Enjoying the benefit of irrigation, from the river Kabul and its tributaries, the region is also the main agricultural belt of the province. Thus, the agricultural productivity of the region has traditionally attracted population to this region.

6.2.3 Population Distribution in Sindh

With a population of over 47 million (47,854,510 in 2017), Sindh is the second most populous province of Pakistan, next only to Punjab. It has a population density of 340 persons per km^2. In terms of density, the province is placed lower than Punjab as well as Khyber Pakhtunkhwa. This appears paradoxical as the province is economically and industrially better developed than other provinces of Pakistan. It is also true that some of the most densely populated districts of Pakistan like Karachi and Hyderabad are in Sindh, yet the province has a population density much lower than expected. The explanation lies in a very sparsely populated large area, consisting of arid districts of eastern Sindh, which more than counterbalance the high density of Central Sindh, pulling down the overall density of the province.

Unequal Distribution of Population – Sindh has a very unequal distribution of population. On the one hand, it has a compact Metropolitan Karachi showing a population density of over 43,000 per km^2, the highest in Pakistan, while on the other hand, there are large areas like the districts of Tharparkar bordering India, Sujawal, the deltaic district and Jamshoro, a mountainous district, each having a population of less than 100 persons per km^2. The high density central corridor of Sindh is bordered on the two sides with areas of very low population density. The eastern part of Sindh is partly occupied by the Thar Desert with very sparse population, reflected in the population density of Khairpur (151 persons/km^2), Sanghar (191 persons/km^2), Umarkot (191 persons/km^2) and Tharparkar (84 persons/km^2). With less than 2 percent of the area of the province, Karachi and Hyderabad together account for roughly 40 percent of the population of Sindh. Could there be a greater contrast in population distribution? It may be mentioned that, despite a low population density and its unequal distribution, more than half the population of Sindh is urban.

6.2.4 Population Distribution in Punjab

With a population of about 110 million (109, 989, 655 in 2017), Punjab accounts for over half the population of Pakistan. Having just over a quarter (25.8 percent) of the area of Pakistan, the province is home to more than half the country`s population. This explains a population density (536 persons/km^2) that is more than twice the density of Pakistan, at 261 persons per km^2.

Like other provinces, Punjab, too, has an unequal distribution of population, but the inequality is not so great. There is no district of Punjab with a population density of less than 100 persons per km^2, and barring a few districts like Bahawalpur, occupying a large part of the Cholistan Desert, and Ranjanpur and Dera Ghazi Khan, in the trans-Indus zone, no district in Punjab has a population density of less than 200 persons /km^2. In fact, over 60 percent of the districts have a population density of over 500 persons per km^2.

The most densely populated and the most urbanised district of Punjab is Lahore with a population density of 6,275 persons per km^2. At the other end of the spectrum, are the districts in the drier southern part of the province, such as Bahawalpur (148 persons/km^2) and Bahawalnagar (335 persons/km^2), and those in the trans-Indus region in the south, like Dera Ghazi Khan (241 persons/km^2) and Ranjanpur (162 persons/km^2).

Punjab has a relatively high population density, supported by rich agriculture and an equally high level of industrialisation. The entire province, except a small sub-mountainous northern part, is a sedimentary basin covered with the alluvium of the flood plains of the tributaries. The river basins of these tributaries, in Punjab, are imperceptibly flat, proving to be a very suitable agricultural land. Supported by canal irrigation from these rivers, Punjab has 11 million hectares of Net Sown Area which is 55 percent of the geographical area of the province. Having the benefit of irrigation, more than 60 percent of the area enjoys multiple cropping, and the total cropped area amounts to over 17 million hectares. The above details define the agricultural productivity of Punjab. Producing over 20 million tons of wheat which is 80 percent of the total wheat production of Pakistan, Punjab is the bread basket of Pakistan. The province has also developed agro-based industries like textiles and sugar, giving it an additional advantage.

Being part of a larger British province of Punjab during the colonial period, the province received all the attention of the British for exploitation of the resources; and in the process the province enjoyed an early momentum of development. The natural advantage of a vast agricultural plain, with the benefit of irrigation, attracted large population from other areas of what is now Indian Punjab, and the population of Punjab, and continued growing with increasing economic activities. It must be emphasised that industrialisation has led to large concentration in certain areas, increasing the population of industrial centres or even creating large metropolitan centres, but it has not seriously influenced the general population density pattern.

6.2.4.1 Population Density Pattern in Punjab

Determined primarily by agricultural productivity, the density of population, there is a general decline in the density of population from the north to south, primarily because of aridity, declining quality of soil and less effective utilisation of irrigation. A more effective way of looking at the density pattern is to evaluate doab-wise variation density of population. The following distinct population density zones can be recognised.

1. A moderate density Bari Doab Region – Though the river Beas is nowhere in the picture, the Sutlej–Ravi Doab is often referred to as Bari Doab. The upper part of this doab incorporating in its fold the districts of Kasur, Okara, Sahiwal, Pakpattan and Vehari districts, has a moderately high density of 600 to 800 persons per km^2. This is above the average population density of Punjab. The region lying in the upper south-east corner of the province has the benefit of irrigation and high agricultural productivity. As one moves south-westward, one encounters the arid and semi-arid Bahawalpur region having large tracts of uncultivated areas and sparsely settled population.
2. Rachena Doab: the Zone of highest density in Punjab – This is the most densely people part of the province, taking in its fold Lahore, Sialkot, Gujranwala, Sheikhpura, Faisalabad, extending further south into Khanewal and Jhang. This doab is home to the famous Chenab Colony with its centre at Faisalabad, the erstwhile Llyalpur district of British Punjab. The region enjoys a little higher rainfall than other parts of the province and has enjoyed an early benefit of irrigation from the Lower Chenab Canal. The area also carries the impact of metropolitan Lahore and another million-city Gujranwala. The north-eastern part of the area has the privilege of being connected by the Grant Trunk Road, between the two million-cities, viz., Lahore, in the south-east, and Rawalpindi, in the northwest corner of the province. Most of the districts in the upper Rachena Doab have a population density exceeding, or close to, 1,000 persons per km^2. To quote a few examples, Lahore, Gujranwala, Faisalabad, Sialkot and Sheikhupura have a population density of 6,275, 1,384, 1,346, 1,291 and 924 persons per km^2, respectively.
3. The Chej Doab: An area of moderately high density – With Sargodha as the focal point, the Chej Doab is much smaller in its areal extent as compared to Rachena Doab. It has certain constraints of relief in its upper part and not a very fertile land in the central and lower part. Most of

the districts in this zone have a density close to, or above the average for the province. The three main districts in this zone viz., Gujrat, Mandi Bahauddin and Sargodha have a population density of 865, 598 and 631 persons per km^2, respectively.

4. The low density North-western and Western Punjab The north-western part of Punjab, unlike the rest of the province, lies in the piedmont zone of the Himalayas, consisting of the Potwar Plateau and the Murree Hills in the north and the Salt Range in the south. The Potwar Plateau, badly dissected, with no benefit of canal irrigation, and practising only rain fed agriculture is not densely peopled. Equally, the Salt Range girdling the Potwar Plateau, from the south, hardly provides any scope for agriculture, except in some depressions. The trans-Jhelum area covered by the arid Sind–Sagar Doab with desert-like conditions, and very little, indeed very limited, rain-fed cultivation, has not attracted much settled population. The result is a low to very low density in the Potwar Plateau, Salt Range and the Sind–Sagar Doab.

   The exception is the Rawalpindi district. The city of Rawalpindi, home to General Headquarters of the Pakistan Army, has boosted the density of the district of the same name despite agricultural poverty in the region. The fourth largest city of Pakistan, with a population of over two million (2,098,231 in 2017), with a slew of activities associated with the army establishment, Rawalpindi has attracted huge population, pushing up the overall population density of the district.

   All other districts, coinciding either with Salt Range like Jhelum, Chakwal, or Khushal, or in the dry Sind Sagar Doab region, like Mianwali, Bhakkar and Layyah, have a low density of population, often less than half the national average. While Jhelum, Chakwal and Khushal have a population density of 341, 229 and 197 respectively, the districts of Sind Sagar Doab are equally depleted of population. Attock, Mianwali, Bhakkar and Layyah have a population density of 275, 264, 202 and 290 persons per km^2, respectively, much below 535 persons per km^2, the average population density for Punjab.

5. The Trans-Indus Area of very low density – The Trans-Indus area of Punjab, sandwiched between the Indus and the outliers of the Sulaiman Range, is formed by an infertile piedmont zone and a part of the terraces of the Indus and its flood plain. The two districts in this zone, Dera Ghazi Khan and Rajanpur have a population density of 241 and 162 persons per km^2, respectively.

The Impact of Large Cities: The large cities have, invariably, their impact on the population density of the districts in which they stand. Punjab has five million-cities. These are Lahore (P-11,126,285), Faisalabad (P-3,203,846) Rawalpindi (P-2,098,231), Gujranwala (P-2,027,001) and Multan

(P-1,871,843). Besides, there are cities with a population close to a million, like Bahawalpur, Sialkot and Sargodha. These cities have contributed to raising the density of the district. In the population spread, these cities stand like poles, around which many economic activities gravitate, raising the population of the district.

According to the 2017 Census, as given in more than three-fourth of the population is concentrated in the two provinces of Punjab and Sindh, the former having more than half and the latter nearly a quarter of the population of Pakistan. Khyber Pakhtunkhwa, having as much density of population as Sindh, has a share of only 13 to 14 percent in the population of the country, because of its smaller area. As a matter of generalisation, one may identify three major zones of population concentration in Pakistan.

1. A southeast-northwest belt running from Lahore to Peshawar, north of Salt Range and Potwar Plateau, following the Grand Trunk Road
2. Central Punjab, aligned along Faisalabad–Multan Corridor
3. Longitudinal belt along the Indus in Sindh, characterised by irrigated land, good agricultural productivity and a string of towns

6.3 Population Distribution between Rural and Urban Areas

In its geographical distribution, much of the population of Pakistan – attached to land and practising agriculture – lives in 47,290 rural localities. A rural locality, a term used by the Pakistan Census in 2017, is more or less equivalent to a *mouza*, the lowest revenue unit of administration. A *mouza* includes several rural settlements. The number of rural settlements is quite large and numbered 170,246 in 1998 (Table 6.3).

TABLE 6.3 Rural Population Living in Mouzas and Total Number of Settlements of Rural Population

Pakistan/Provinces	*Rural Populated Mouzas*	*Total Number of Settlements*
Pakistan	47,482	170,246
Punjab	23,941	78,218
Sindh	5,573	52,058
Khyber Paktunkhwa	11,540	23,359
Balochistan	6,428	16,611

Source: For the number of Mouzas and rural settlements - Census of Pakistan 1998. The mouzas of FATA are included here in Khyber Pakhtunkhwa (Rural Population- Pakistan: 132,013,769; Khyber Pakhtunkhwa – 29,626,670, Punjab - 69,442, 450: Sindh- 23,021, 876; Balochistan - 8,928,426 and Islamabad -994,365)

6.4 Urban Localities and Urban Population in Pakistan

There are over 600 urban localities in Pakistan. The actual count is 626 distributed in different population size groups. According to their population, the urban localities are grouped in the following categories (Table 6.4).

Rural and Urban Population – The urban population is distributed in these 626 urban centres. It must be noted that majority of the towns lie in the 10,000 to 50,000 group, though they don't have so much of population. The total population of Pakistan is divided into rural and urban, where there is a dominance of rural population. According 2017 Census, the rural urban population in Pakistan is split into 63.56 percent rural and 36.44 percent urban population. The 1998 Census of Pakistan gives the following ratio of rural urban population.

Today, 75.67 million that is more than a third (36.44 percent) of Pakistan's population lives in urban localities, more commonly known as urban centres, i.e., cities and towns. There is a significant increase in terms of the total number of urban population which increased from 43 million in 1998 to over 75 million in 2017, an increase of 74 percent. In terms of the level of urbanisation, it has increased from 32.5 percent in 1998 to 36.44 percent in 2017 (Table 6.5).

Like Pakistan, the provinces have also shown a higher level of urbanisation as seen in the Table 6.22. Punjab has added more than 17 million people to its urban population, rising from 23 million in 1998 to 40 million in 2017, raising the level of urbanisation from 31.3 percent in 1998 to 36.8 percent in 2017, an overall increase of 5 percent. A similar increase is observed in other provinces, with the sole exception of the Federal Territory, Islamabad. Though in absolute terms, the urban population of Islamabad increased from a little over 500,000 in 1998 to over a million in 2017, the rural population more than tripled with a large mass of people settling in the rural part of the Federal Territory, causing a decline in the percentage of urban population.

TABLE 6.4 Total Number of Urban Localities and Their Population Size Groups (2017)

Population Size Groups	*Number of Urban Localities*
1. Population over a million	13
2. 500,000 to 1,000,000	08
3. 200,000 to 500,000	27
4. 100,000 to 200,000	59
5. 50,000 to 100,000	113
6. 25,000 to 50,000	178
7. 10,000 to 25,000	192
8. 5,000 to 10,000	28
9. Under 5,000	8
Total No. of Urban Localities	626

TABLE 6.5 Rural/Urban Population Ratio of Pakistan (1998 and 2017)

Pakistan / Provinces	*Rural*		*Urban*	
	1998	*2017*	*1998*	*2017*
Pakistan	67.49	63.56	32.56	36.44
Punjab	68.70	63.14	31.3	36.86
Sindh	51.25	48.11	48.75	51.89
Khyber Pakhtunkhwa	83.12	81.20	16.88	18.80
FATA	97.31	97.20	2.69	2.80
Balochistan	76.11	72.38	23.89	27.62
Islamabad	34.29	49.63	65.71	50.37

Source: Census of Pakistan *1998, Demographic indicators, and Census of Pakistan 2017, table-1 of Final Results of Census 2017

The increase in urban population will continue with increasing industrialisation and greater economic growth, as elsewhere in the South Asian countries.

6.5 Growth of Population: A Historical Perspective

Though Pakistan came into existence only in 1947, its Census Organisation has worked out the population of the territory corresponding to present day Pakistan since 1901. Thus, the figures given in the following table present the entire sequence of decadal population and its growth from 1901 to 2017.

It is observed that in a span over hundred years, the population of the area, now known as Pakistan, multiplied over 12 times, from 16.5 million in 1901 to 207 million in 2017. In the first half of the 20th century, the growth was rather slow. It took 50 years, from 1901 to 1951, to double the population of the area, coinciding with present day Pakistan. From 16.5 million in 1901 the population grew to 33.7 million in 1951. After 1951, Pakistan experienced an accelerated growth. While in the beginning of the 20th century, doubling of the population took 50 years (1901–51), it took just two decades for another doubling when the population of Pakistan grew from 33.7 million in 1951 to 65.3 million in 1972 (Table 6.6).

Until 1951, the population of Pakistan, as much as the entire South Asia, grew at a rate ranging between 1 and 2 percent annually. The year 1951, usually considered a watershed in the growth history of population of South Asia, witnessed an upward trajectory of population growth. After the mid-twentieth century, the population of Pakistan began experiencing a much higher growth rate reaching beyond 3 percent per annum. The most spectacular feature of post 1951 growth pattern is the exceptional growth of population seen in 1972, when in a period of 11 years, the population

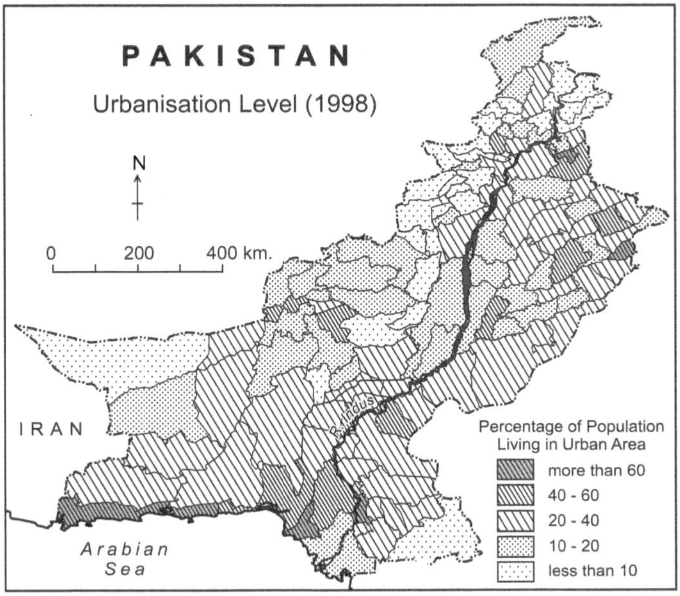

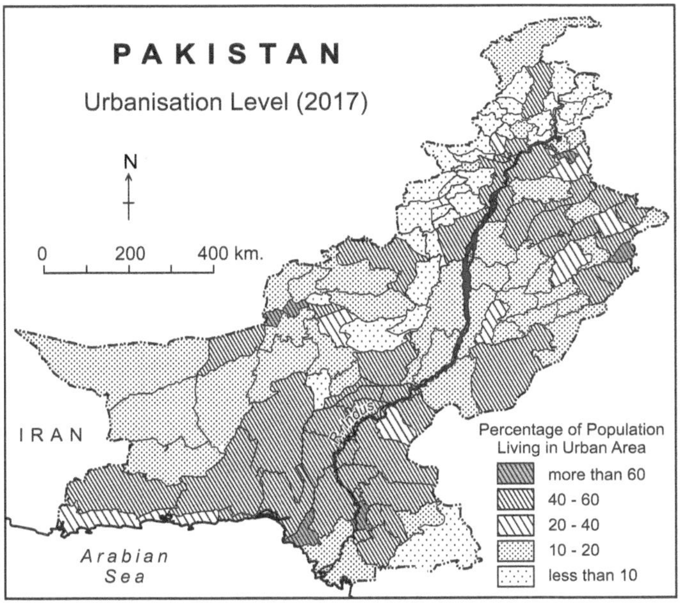

FIGURE 6.2 District-wise Percentage of Urban Population in Pakistan (1998 and 2017)

TABLE 6.6 Decadal Variation of Population in Pakistan Since 1901

Year 1	*Population 2*	*Change 3*	*Growth in Inter-census Period in Percentage 4*	*Crude Annual Average Growth in Percentage 5*
1901	16,576,593			
1911	19,381,532	+2,804,939	+16.92	1.69
1921	21,108,779	+1,727,247	+8.91	0.89
1931	23,542,071	+2,433,292	+11.52	1.15
1941	28,282,064	+4,739,993	+20.13	2.01
1951	33,779,555	+5,497,491	+19.43	1.94
1961	42,880,378	+9,100,823	+26.94	2.69
1972	65,309,340	+22,428,962	+52.30	4.74
1981	84,253,644	+18,944,304	+29.01	3.22
1998	132,352,279	+48,098,635	+57.08	3.36
2017	207,684,626	/75,332,347	+56.91	2.4

Source: 1. Population figures for the years 1901-51 are from the Census Organisation 1951, by E.H. Slade, Vol. I, Report and tables – Table 3, Decadal Changes in Pakistan, since 1901: Adjusted to apply to present boundary wherever necessary. 2. Population figures for 1961 to 1998 are from Population Census Organization, Government of Pakistan 1998, Summary table 2.1, Population of Pakistan 3. Population figures for 2017 are from, Census of Pakistan, Final Results of 2017, Table 1, Area, Population, by Sex, Sex Ratio, Population Density, Urban Proportion, Household Size and Annual Average Growth Rate. 4. The crude Growth Rate is the total percentage growth during the inter Census period divided by the number of years in the inter Census period, and not the compounded average Annual Growth Rate.

of Pakistan increased by over 50 percent (52.3 percent). What contributed to this abnormal growth was the large scale immigration after the secession of Bangladesh from Pakistan in 1971–72. Thousands of Biharis, the Urdu-speaking population of East Bengal, arrived from Bangladesh producing an exceptional rise in the population of Pakistan. Almost all the provinces showed a population increase of over 45 percent; with an exceptional growth in Sindh and Punjab. The population of Sindh rose by 80 percent and that of Punjab by about 70 percent, between 1961 and 1972. The immigrant population especially swelled the population of cities such as Karachi, Lahore, Faisalabad and Rawalpindi, which experienced a growth of 84, 67, 93 and 80 percent, respectively in their population.

In the demographic history of Pakistan, there has never been a decline. There has always been an increase during all the inter-census periods. The only exception is the Federally Administered Tribal Area (FATA), which showed a marginal decline between 1972 and 1981, attributed to the disturbed conditions in neighbouring Afghanistan, following the USSR invasion of Afghanistan in 1979. All other provinces have shown consistent increase. Between 1951 and 2017, the population of Pakistan has increased 6.15 fold. The provinces also follow the same pattern though on a variable scale. Among

TABLE 6.7 Growth of Population of the Provinces of Pakistan between 1951 and 2017, in Thousands

Provinces/ Regions	*Census Years*					
	1951	*1961*	*1972*	*1981*	*1998*	*2017*
Pakistan	33,740	42,880 (+27.08 %)	65,309 (+52.3 %)	84,254 (+29.0 %)	132,352 (57.08 %)	207,684 (+56.9 %)
Punjab	20,641	25,484 (+23.9 %)	37.607 (+47.6 %)	47,292 (+25.5 %)	73,621 (+55.9 %)	109,989 (+49.3 %)
Sindh	6,04	8.367 (+38.3 %)	14,156 (+69.2 %)	19,029 (+34.4 %)	30,440 (+59.9 %)	47,854 (+57.2 %)
NWFP	4,557	5,731 (+25.7 %)	8.389 (+46.3 %)	11,061 (+31.8 %)	17,44 (+60.9 %)	30,508 (+72.3 %)
Balochistan	1,167	1,353 (+15.9 %)	2,429 (+79.5 %)	4,332 (+78.3 %)	6,566 (+51.5 %)	12,335 (+87.8 %)
FATA	1,332	1,847 (+38.6 %)	2,491 (+34.8 %)	2,199 (-11.7 %)	3,176 (+44.4 %)	4,993 (+57.2 %)
Islamabad	96	118	238	340	805	2,003

Source: For 1951 to 1998: Census of Pakistan, Summary tables, Table 2.1- Population of Pakistan and Provinces: For 2017 Census of Pakistan, P. B. S., Final Results of Census Table-1, The figures in the bracket represent the percentage growth of population during the inter-Census period.

the provinces, Balochistan shows the highest relative population growth, multiplying 10.5 times between 1951 and 2017. During the same period, the population of Punjab grew 5.35 times, that of Sindh 7.91 times and the population of North West Frontier province (now Khyber Pakhtunkhwa) multiplied 6.7 fold. Among the more populous provinces, Sindh has grown fastest (7.9 times) and Punjab the slowest (5.3 times) (Table 6.7).

6.5.1 Change in Density of Population and Strain on the Environmental and Economic Resources

Given the fact that the area of Pakistan (796,096 km^2) remained constant; any increase in population would mean increase in population density and strain on the economic and environmental resources. A comparison of the density of population of Pakistan and the provinces in 1951 and 2017, the earliest and the latest Census years, is quite informative (Table 6.8).

The impact of increase in population between 1951 and 2017, a period of 56 years, can be seen in the increase of population density. The area of the country as well as the provinces remaining unchanged, it means increasing load on the land and the land-based resources, specifically agriculture. An increase in the density of population from 43 persons/km^2 to 261 persons/km^2, a six-fold increase, entails one-sixth availability of agricultural land and water per person in 2017 as compared to 1951. This would also multiply the need of essential commodities six times. The immediate impact of this increase in population is reflected in the gap between the domestic production consumption of food grains, especially wheat, turning Pakistan from a food-surplus to a food-importing country. Many other adverse impacts of uninterrupted growth can be recounted, like making ever increasing provision for public services like health, education and transport. But the most pressing problem that faces a developing country, – and in no less a measure Pakistan – is to generate employment for a growing population. The country is quite alive to this problem but this is a sensitive issue which involves the masses. One hopes that the country is able to bring down its fertility rate to below two, as Bangladesh has done.

The population of Pakistan continued to rise at a relatively high rate between 3 and 3.5 percent till the end of the 20th century. The country witnessed an unprecedented fall in its population growth, especially in its birth rate in the 21st century. The birth and death rate and the natural increase as given by the Bureau of Statistics, Government of Pakistan, show unfailingly a steep decline in the growth rate.

Before 2000, the population growth rate of Pakistan lingered around three percent; after 2000, the birth rate began to decline drastically, from over 35 per 1000 during the period 1995–97 to 26 per 1000 in 2005 and further to

TABLE 6.8 Population Growth and Increase in Population Density (1951 to 2017)

Territory	*Area in km²*	*Population (1951)*	*Density (1951) Persons/km²*	*Population (2017)*	*Density(2017) Persons/km²*
Pakistan	796,096	33,740,167	42.83	207,684,626	261
Khyber PK	74,521	4,556,545	61.14	30,508,920	409
Punjab	205,345	20,540,782	100.0	109,989,656	536
Sindh	140,914	6,047,748	42.9	47,854,510	340
Balochistan	347,190	1,167,167	3.36	12,335,129	36
FATA	27,220	1,332,005	48.9	4,903,044	183
Federal Capital	906	95,940	105.9	2,003,368	2,211

TABLE 6.9 Average Birth Rates, Death Rates, Infant Mortality Rates and Natural Growth Rates of Pakistan, According to Pakistan Demographic Survey

	Years							
	1989–94 Average	*1995–97 Average*	*1998–2001 Average*	*2003*	*2005*	*2006*	*2007*	*2020*
1. Crude birth rate per 1000 persons	39.5	35.2	29.0	26.5	26.1	25.9	25.6	27
2. Crude death rate per 1000 persons	10.1	9.0	7.8	7.0	7.1	7.0	6.8	7
3. Natural growth rate per cent per annum	2.94	2.62	2.12	1.95	1.9	1.89	1.88	2.2
4. Infant mortality per 1000 live birth	102.8	88.2	79.6	76.2	76.7	76.2	75.2	54

Source: 1. Pakistan Bureau of Statistics Govt. of Pakistan, 2008 Pakistan Demographic Survey, p. 328
2. Figures for 2020 are from United Nations Population Division, World Population Prospects 2020

24.3 per 1000 in 2012. During the same period, death rate declined from almost 40 per 1000 during the five-year period of 1989–94 to 24 or 25 percent around 2010 but not in the same proportion as the birth rate, bringing down the overall growth rate. The growth rate of less than 2 percent is an achievement and a further decline is optimistically anticipated.

6.6 Phases of Population Growth in Pakistan

One should recognise several phases of retarded and accelerated population growth in the demographic history of Pakistan. These may be elaborated as

1. A transitional phase following the partition of India in 1947 and large-scale migration on both sides, Pakistan and India, as refugees
2. A phase of stabilisation (1951–65) when natural growth was the only component of growth
3. A phase marked by political instability (1965–1972); and the creation of Bangladesh
4. A phase of unchecked natural growth, developing into a crisis situation
5. The last phase of national awareness, family planning and a drastic decline in the population growth rate of Pakistan

6.6.1 The Period 1941–51

This was the most critical period for both countries, India and Pakistan. Besides the transitional instability associated with the transfer of power from the British Crown to the two independent countries, there was an added problem of refugees (Mohajirs) of which 699,000 Mohajirs had migrated to East Bengal, leaving on 6,527,000 refugees to be settled in what is the present Pakistan. But, this is not the entire story; an almost equal number of Indian refugees, largely Hindus and Sikhs, migrated from Pakistan to India, thus neutralising the sudden burden of increased population due to immigration. A simplified numerical picture of migration from both sides, India and Pakistan, in each other's territory is as follows:

Total No of Mohajirs (refugees) in Pakistan (1951)	7,226,000[1]
No of Refugees in East Bengal (1951)	699,000
No of Mohajirs in what is present-day Pakistan	6,526,900

Slade, E.H. Census of Pakistan, 1951, Vol. 1, Reports and Tables, Statement 2-E, Manager of Publication, Government of Pakistan, Karachi, p. 31.

As recorded by the Census of Pakistan (1951), more than seven million refugees arrived from India between 1947 and 1951, of which 5,785 million were from Punjab and Delhi region.[2] This arrival of over six and a half million refugees was compensated by over 4.5 million refugees who left Pakistan for India. In fact, the total number of refugees who arrived in India was 8,229,699 including those that came from what is now Bangladesh. This figure is arrived on the basis of "Place of Birth" criterion employed by the Indian Census.

Total number of immigrants in India born in Pakistan (1951)	8,229,699
Total number of immigrants born in the present-day Bangladesh	3,693,785[3]
Total number of immigrants in India born in present-day Pakistan	4,535,914

Thus, again a strength of over six and a half million refugees are four and a half million left Pakistan, leaving still a burden of about two million Mohajirs which added to Pakistan population, and the country recorded an annual growth of almost two percent.

It may be mentioned in passing that the population of the Mohajirs, in the initial stage in 1951, was largely concentrated in Punjab and Bahawalpur states, which together accommodated over 80 percent of the refugees. Punjab had in 1951 over five million refugees (5.28 million), Karachi over six lakh (<9 percent) and Sindh and Khairpur state over half a million (>8 percent) refugees. Balochistan and NWFP had 28,000 and 51,000 refugees respectively.

6.6.2 The Growth Trend between 1951–61

This was the phase of demographic stabilisation and an apparent trend of rising population growth rate. During 1951–61, the second decade of Pakistan's existence, Pakistan experienced an annual growth of 2.1 percent, a growth rate much higher than the one between 1931 and 1941 or during the 1941–51 periods. The jump from 1.7 (1931–41) to 2.7 percent between 1951 and 1961 is reflective of a universal rise in population growth in South Asian countries, following the end of the Second World War, and the discovery and spread of antibiotics and advances in medical facilities. In fact, 1951–61 was a relatively stable period in the history of Pakistan when the Nehru–Liaquat Pact, between the prime ministers of India and Pakistan, for the protection of the refugees and the Indus Water Treaty (1960) for sharing the water of the Indus and its tributaries, were signed. These agreements, and especially the latter, have stood the test of time, forming the basis for the settlements of water related disputes. This period was one of socio-economic stabilisation, recovery from the shock of 1947 partition. Within a decade, the people felt more or less settled.

The political instability, abrogation of the Pakistani constitution and the imposition of martial law by general Ayub Khan, the then president of Pakistan, and the Indo-Pakistan war of 1965 don't appear to have influenced the population growth rate. The revolt in East Bengal and the secession of the territory as an independent country, Bangladesh, following the 1971 war also do not appear to have affected the growth trend of population. What, on the contrary, is intriguing is an abnormal annual population growth rate of 4.75 percent between 1961 and 1972. One of the reasons advanced is the large-scale migration of non-Bengali Muslims from Bangladesh to Pakistan for fear of reprisal at the hands of Bangladeshis and biased treatment of local administration. Indeed, the 1961–72 census period has been the decade of the highest growth rate in the demographic history of Pakistan. And, though all the provinces of Pakistan experienced a high growth rate and bore the brunt of this unexpected exodus from Bangladesh, the maximum impact was felt by Sindh and Balochistan. Sindh added nearly 70 percent to its population, as its population jumped from 8.3 million in 1961 to 14.2 million in 1972. Balochistan also showed an exceptional growth around 80 percent during this period. The population of Karachi increased from less than 2 million in 1961 to 3.5 million in 1972. One may justifiably call 1972 as a landmark in the population of Pakistan characterised by a Second Refugee Wave, the first being from India in 1947 and the second from Bangladesh in 1972.

After 1972, the population of Pakistan showed an unchecked annual growth rate of 3.3 percent during the last quarter of the 20th century (1972–1998). The population almost doubled in a space of 26 years, rising from 65.3 million in 1972 to 132 million in 1998. Though, this was a period of military rule followed by a succession of transient governments, nevertheless the population of the country grew. It appears political instability has not much to do with population growth. The burden of this high growth of population began to be experienced by the close of the century, leading to a greater demand on the national resources and causing strain to the economy. The need to moderate the population growth was realised at the beginning of the 21st century, the growth rate of population began to slide down. A high fertility rate and a high growth rate was a subject of alarm among the political leaders and political planners. This led to a high-pitch campaign of awareness of the adverse consequences of a high growth rate of population. The results have been quick and spectacular, bringing down the growth rate substantially. Though the average growth rate between 2002 and 2012 is estimated to be 2.6 percent per annum, the figures for the last few years show a further decline, coming close to, or lower than, 2 percent, as suggested by the estimates of population growth presented in the "Pakistan Economic Survey[4], which broadly concur with the estimates of the US Bureau of Census International Data Base.

TABLE 6.10 The Latest Estimate of Population Growth, during 2009-10, 2010-11 and 2011–12

Year	*Birth Rate Per 1000 Population*	*Death Rate Per Thousand Population*	*Natural Growth Per Thousand*	*Total Fertility Rate*
2009-10	28.4	7.6	20.8	3.9
2010-11	27.5	7.3	20.2	3.9
2011-12	27	7	20	3.3
2016-17	28.7	7.1	21.6	3.6
2022-23	26	6.7	19.3	3.2

Source: 1. Pakistan Economic Survey 2012, Govt. of Pakistan, p. 156 2. US Bureau of Census International Data Base, quoted from Encyclopaedia Britannica 2013–14, p. 691.

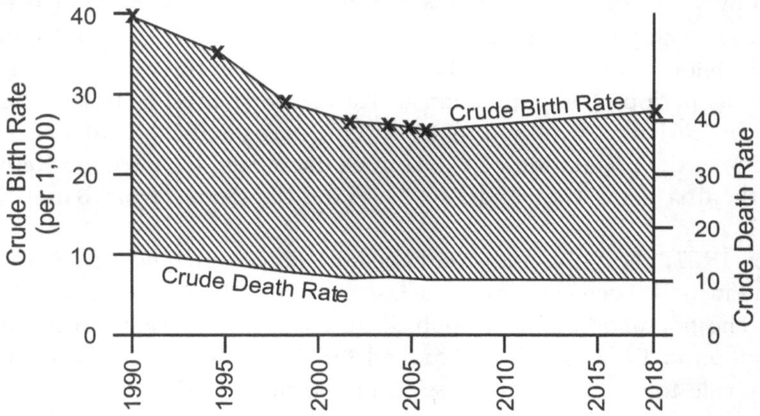

Source: Census of Pakistan 2017 and Pakistan Economic Survey

FIGURE 6.3 Average Birth Rates, Death Rates, and Natural Growth Rates of Pakistan

Besides the decline in growth rate, there is a significant decline in total fertility on the one hand and infant mortality on the other. On the latest count, the total fertility as in 2012 was 3.07, and the infant mortality was 61.3, though lower than earlier years, yet far above the expected levels.

6.6.3 Provincial Variation in the Growth of Pakistan's Population

As it is clear from the above table, Balochistan has experienced the highest growth of population in percentage terms, as its population multiplied ten times between 1951 and 2017. The growth was especially sharp immediately after the emergence of Bangladesh that continued till the end of the 1971–81 decade. On the contrary, FATA showed the lowest growth. A very high

TABLE 6.11 Population Growth in Provinces of Pakistan between 1951 and 2017

Pakistan / Provinces	*Year*						*Ratio Population 2017/1951*
	1951	*1961*	*1972*	*1981*	*1998*	*2017*	
Pakistan	33,740,167	42,880,378	65,309,340	84,253,644	132,352,379	207,684,626	6.16
a. Population		+2.7	+4.75	+3.2	+3.4	+2.4	
b. Annual growth rate in %							
Punjab	20,540,262	25,463,974	37,607,423	47,292,641	73,621,290	109,989,655	_ 5,35
a. Population		+2.39	+4.32	+2.85	+3.27	+2.13	
b. growth rate %							
Sindh	6,047,748	8,367,065	14,155,909	19,028,666	30,439,893	47,854,510	7.9
a. Population		+3.8	+6.28	+3.82	+3.52	+2.41	
b. growth rate %							
KP	4,556,545	5,730,991	8,388,551	11.061,321	17,743,645	30,508,920	6.7
a. Population		+2.58	+4.21	+3.54	+3.55	+2.89	
b. growth rate %							
Balochistan	1.167,167	1,353,384	2,428,678	4,332,376	6,565,885	12,335,129	10,6
a. Population		+1.6	+7.2	+8.7	+3.02	+3.37	
b. growth rate %							
FATA	1,332,005	1,847,195	2.491,230	2,198,547	3,176,331	4,993,044	3.8
a. Population		+3.86	+3.16	-1.3	+2.6	+2.4	
b. growth rate %							
Islamabad					805,236	2,003,368	4,9

Source: 1. Census of 1998, for the period 1951 to 1998, Final Results of Census of Pakistan 2017, table 1

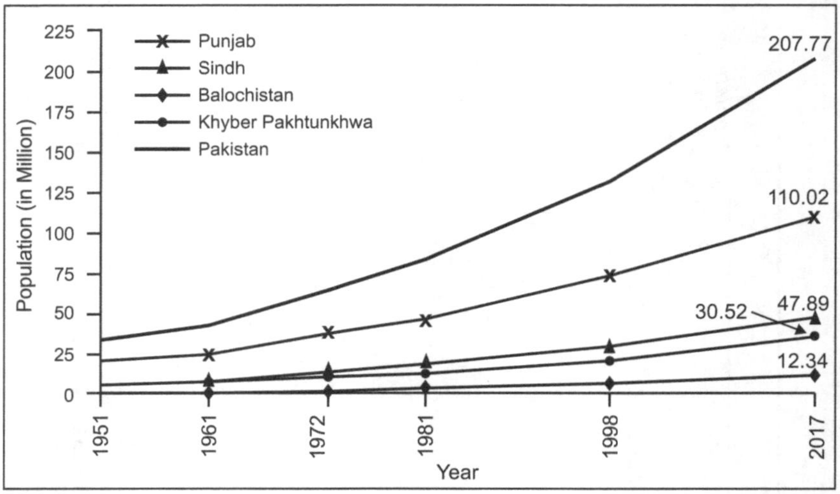

FIGURE 6.4 Census-wise Population of Pakistan (1951 – 2017)

growth rate in Balochistan is firstly accounted for its low base population that was just over a million in 1951 and any addition of refugee population, besides its own natural growth, resulted in a spectacular high growth rate, so in terms of absolute number the increase is not so significant.

Both, Punjab and Sindh, had a high influx of refugees during 1971–72, adding an unusually high number to their population. It must be emphasised that all the provinces with the exception of FATA, showed a high growth during 1961–72; the growth in Sindh was exceptionally high. Karachi, itself, added 1.6 million to its population, between 1961 and 1972, as its population jumped from 1.9 million in 1961 to 3.5 million in 1971, a growth of 84 percent, unimaginable by natural growth, and speaks of massive influx of refugee population. In all the decades since 1951, Sindh experienced a higher growth rate of population than Punjab and Khyber Pakhtunkhwa, the two provinces that matter demographically. The impact of this enormous growth in Sindh had been more adverse than favourable. With the limited agricultural resources and malfunctioning of important industrial units like Pakistan Steel, the state had also to contend with the problems of a larger metropolis like Karachi.

The annual average growth of population for Pakistan for the period 1998–2017 is 2.4 percent, with Punjab recording the lowest annual growth rate of 2.1 percent and Balochistan the highest at 3.34 percent. The growth rate in Sindh has kept level to the national average of 2.4 percent. Khyber Pakhtunkhwa has a shown a growth rate of 2.9 percent. The growth rate is believed to be declining, and according to some estimates, it is veering around an annual growth rate of 2 percent (Figure 6.5).

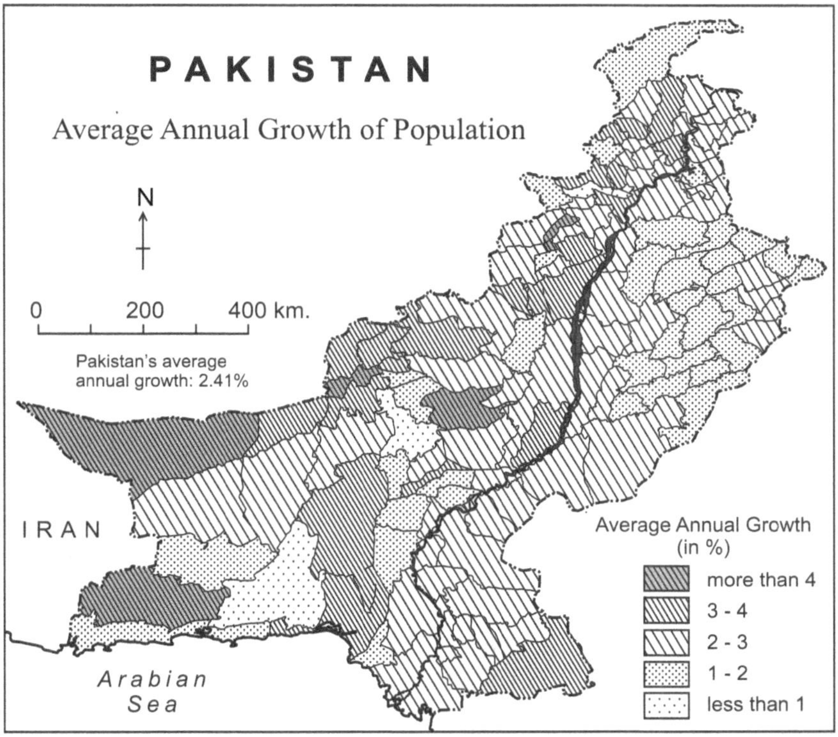

FIGURE 6.5 District-wise Average Annual Growth of Population (1998 to 2017)

6.7 Some Qualitative Attributes of Population

6.7.1 Age Structure of Population

Like many other developing countries, Pakistan with its higher fertility and birth rate has a relatively young population. The age structure is very significant in assessing the strength of working population. In developing countries of South Asia, the longevity of the people being not as high as in the developed countries, the age pyramid takes a different shape. While in developed countries the top, representing the old age is quite large, more so because a corresponding increase in the natural growth of the population has not occurred. As can be seen from the following table, there are clear indications that the population of Pakistan is relatively young (Table 6.12).

The detailed age structure of the population of Pakistan according to the latest Census of Pakistan is as follows.

The age structure given in the Pakistan Census 2017 shows a clear bulge between 15 and 50 years of age (Fig. 6.6). Little over a quarter (28.5 percent) of the total population of Pakistan is under 10 years. About a fifth of the population is covered by the age group 15 to 24 years. And, the population

TABLE 6.12 Population of Pakistan and its Provinces in Selective Age Groups, in Millions (2017)

Administrative Unit	*Age-groups (in years)*					
	< 5 years	*< 10 years*	*< 16 years*	*15-64 years*	*65 and above*	
Pakistan	29.16	59.18	88.15	116.25	7.71	78.44
K. P,	4.63	9.34	13.96	16.19	1.18	93.5
Punjab	14.38	29.22	44.08	63.58	4.61	87.07
Sindh	6,82	13.97	20.52	26.83	1.51	83.58
Balochistan	2.19	4.36	6.27	6.01	0.32	96.79
Islamabad	0.23	0.47	0.71	1.25	0.073	68.42

Source: Population Census Organization, Government of Pakistan 2017, Final Results table-5

TABLE 6.13 The Detailed Age Structure of the Population of Pakistan

Number	*Age Interval in Years*	*Population*	*Percentage of Population*
1.	All Ages	207,684,626	100.00 %
2.	Under 1	5,074,735	2.44
3.	Under 5	29,162,990	14.04
4.	under 10	59,189,308	28.49
5.	under 16	88,150,064	42.44
6.	05 to 24	94,416,019	45.46
7.	15 to 24	39,862,513	19.19
8.	15 to 29	50,263,645	24.20
9.	15 to 49	100,022,887	48.16
10.	15 to 64	116,255,314	55.97
11.	18 to 60	99,803,261	48.39
12.	18 and above	110,896,945	53.39
13.	60 and above	12,024,677	5.78
14.	65 and above	7.712,815	3.71

Source: Pakistan Census 2017, Final Results, Table 5. Population by selected Age Group, Sex and Rural/Urban

below 24 years of age makes 60 percent of the total population of Pakistan. The median age of Pakistan's population is 20.5 years, i.e., half the population of Pakistan is below the age of 21 years. Only 9.5 percent of the population of Pakistan is above the age of 60 years, of which those above 65 years of age make 3.7 percent of the total population of Pakistan.

6.7.2 Dependency Ratio

The census considers population in 15–64 year age group as working population and the population under 15 and over 64 years of age as the dependent

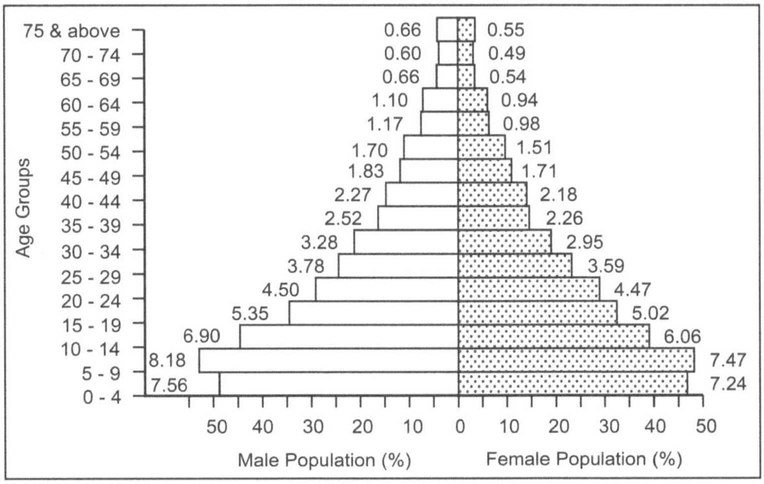

FIGURE 6.6 Age–Sex Structure of the Population of Pakistan (Census 2017)

population. Pakistan has over 116 million (116,255,314) of its population in the working age (15–64 years) group. This amounts to 56 percent of the total population. The remaining non-working population representing a sum of the population below the age of 15 and above the age of 65 years of age, numbering over 91 million (91,429,312) or 44 percent of the total population is the dependent population. This means 44 percent of the country`s population, in the non-working group depends on 56 percent of the working population in the 15–64 year age group.

$$\frac{(\text{population below } 15 + \text{population above } 64) \times 100}{\text{population in age group } 15 \text{ to } 64}$$

$$\frac{(83{,}716{,}490 + 7{,}712{,}816\ ) \times 100}{116{,}255{,}320}$$

$$= 78.5\%$$

Thus counted, the dependency ratio in Pakistan is 78.5 percent. The most significant demographic fact of Pakistan is that 42 percent of the country's population is below the age of 16 years, and more than 80 percent below 45 years of age. The median age of Pakistan's population is 20.5 years. This shows how young the population of Pakistan is, a phenomenon often believed to be holding bright prospect, usually described as "*demographic dividend*. But whether this existing age group distribution offer a "demographic dividend" is a fact that needs evaluation.

6.7.3 Comparison of Age Structure of Pakistan with Some Other Countries

For comparison, the age structure of some developing and some developed countries are taken. The age-group categories and the details are sourced from a reliable reference book.

This elaborate age group structure, each group at 15 years interval, presents a comparable picture of some of the developing as well as developed countries of the world and reveals some significant features. While the developing countries like Pakistan, India and Bangladesh show a heavy concentration of population in the younger age group, the developed countries have a relatively balanced and more uniform distribution, with lower percentage in the younger age group and a larger percentage in the older age group. The expected decline in percentage (the tapering of the top of the population pyramid) of population in the older age group, in the developed countries, is not as much as in the developing countries. Pakistan for instance, has 65 percent of its population concentrated in the age group below 30 years of age. The situation in India and Bangladesh is not very dissimilar, the former having 55 percent and the latter 61 percent of its population in the same age group, i.e., under 30 years. In contrast, countries like the UK, Germany and the USA have only 35 to 40 percent of their population under 30 years of age. In contrast, the developed countries have a higher proportion in the old age group. For instance, in the age group above 75 years of age, Pakistan, India and Bangladesh, each, has less than 1.6 percent of their total population, whereas the developed countries like the the UK, Germany and the USA have 6 to 8 percent of their population in this old age group.

This suggests a lower infant mortality, higher survival rate and a higher life expectancy, resulting in a much higher percentage of older people in the

TABLE 6.14 Age-group Break-up for Pakistan and some Developing and Developed Countries

Age groups	*Countries with Relevant Reference Years – Percentage of Population in Each Group*						
	Pakistan	*India*	*Bangladesh*	*U.K.*	*Germany*	*USA*	*China*
<15 years	34.7	30.9	34.8	17.4	13.6	19.8	19.5
15-29	30.4	26.9	26.6	20.1	17.5	21.0	24.3
30-44	17.9	21.2	19.5	20.2	21.2	19.8	24.9
45-59	10.7	13.1	12.0	19.6	22.2	20.9	24.9
60-74	4.9	6.4	5.5	14.8	16.9	12.5	9.3
75-84	1.2	1.3	1.2	5.6	6.4	4.2	2.5
85 and above	0.2	0.2	0.4	2.3	2.2	1.8	0.5

Source: Encyclopaedia Britannica, Book of the Year 2014, Age break-up for Pakistan p. 691 and for other countries from the relevant pages.

population of developed countries. The developing countries of South Asia have to plan and put in efforts to reduce infant mortality, attain a higher survival rate and a higher life expectance. Education, hygiene and healthcare provide the answer. The causes and consequences of an age structure, with a heavy concentration below 30 years of age have to be interpreted and evaluated in the light of total national growth rate, total fertility, infant mortality and the life expectancy.

The most striking dissimilarity in some of the vital statistics given above is seen in the rates of infant mortality and natural fertility. The wide, almost gaping, contrast in the value of these vital statistics, between the developed and the developing countries, is a measure of societal health as it exists between the developed and under developed countries. Infant mortality, not only as a demographic variable, but even a humane aspect of social existence, is of great significance and reflective of the care a society takes of the newborns. This adverse aspect of a country`s population is much higher in South Asian countries like Pakistan and India than in the developing countries. The infant mortality in Pakistan (54 per 1,000 live births) is double the world average of 27 per 1,000 live births and ten times the average for the UK and Germany. Like the UK and Germany. Infant mortality needs to be avoided by augmenting the measures for pre- and post- natal care.

No doubt, some of the developed countries like Germany, France and the UK face the prospects of a declining population as is evident from their low total fertility rate. A fertility rate of 2.1 is needed to maintain a stable population: a fertility rate lower than 2.0 is the indication of a declining trend of population. This is the case of the UK, the USA, France and Germany. All these developed countries face the prospect of a decline in their population, if not compensated by immigration. Yet one finds in these developing countries a higher percentage of population in the lower age group, especially the group under 15 years of age. This, obviously, reflects a high birth rate and even a higher natural growth rate. Two things which influence the age structure of population more than any other factor are the birth rate and infant mortality rate. While affecting a decline in the birth rate requires sagacious family planning, the elimination or maximum deduction in infant mortality is as much a moral issue as it is a demographic one. Both, these aspects, curtailment of birth rate and the reduction in death rate need organised and motivated effort on the part of developing nations and Pakistan is no exception (Table 6.15).

6.7.4 Sex Ratio

It is a widely shared view among the biologists that every 100 female child born in the world, there are 107 births of male child. This ratio, however, changes with advancing age group. The present sex ratio of world population

TABLE 6.15 Vital Population Statistics for Pakistan and Selected Countries

Countries	*Birth Rate Per 1000*	*Death Rate Per 1000*	*Natural growth Rate in % per Annum*	*Total Fertility*	*Infant Mortality per 1000 Live Birth*	*Life Expectancy at Birth*
Pakistan (2020)	27	7	+2.0	3.46	54	73
Bangladesh(2020)	18	6	+1.2	2.0	24	80
India (2020)	17	7	+1.0	2.2	28	75
Indonesia	17	7	+1.0	2.3	20	80
Malaysia(2020)	16	5	+1.1	1.98	7	87
U.S.A. (2020)	11	10	+0.1	1.6	5.0	88
UK	10	10	+0.00	1.6	4.0	92
Germany	09	12	-3.0	1.5	3	93
France	11	10	0.01	1.8	3	93
World	18	8	+1.0	2.4	27	73

Source: World Bank Data 2020.

TABLE 6.16 Sex Ratio of Pakistan in Different Age Groups

Age Groups in Years	*Sex Ratio (No. of males per 100 females)*
All Ages	104.9
0-01	106.3
0-05	105.1
0-10	105.9
0-18	108.4
18-60	101.08
60 and above	109.2

Source: Calculated from table 5; Population by selected Age group, Sex and Rural Urban: Final Results of Census of Pakistan, 2017

is 101 male for every 100 females, i.e., the male–female ratio of the world population is 50.25 to 49.75.

Sex Ratio in Pakistan – The sex ratio of the total population of Pakistan, according to 2017 Census, is 104.9 males per 100 females. There has been a substantial change in the sex ratio during the last two decades. The sex ratio in 1998 was 108.5 males per 100 females. The present sex ratio of 104.9 males per 100 females is a marked improvement in the sex ratio of 1998, suggesting an increase in the number of females in relation to males. The sex ratio in different age groups of sex ratio is available, as given in table 6.16 Table 6.22

The overall sex ratio of Pakistan is 105 (104.9), i.e., 105 males per 100 females. As can be seen the number of males per 100 females progressively increases till the age of 18. In the higher age group, especially 18 to 60 years, the number of females appears to be catching up with that of the males, after which their number declines and in the old age group, the group above 60 years, there is an obvious dominance of males, showing 109.2 males per 100 females.

This kind of sex ratio, as seen above in the case of Pakistan, where there is a dominance of males over females is typical of developing or underdeveloped countries. In the developed countries, it is just the reverse. The pattern of sex ratio in different countries is highly influenced by their socio-economic development. As a rule, the number of females is higher than the number of males in the developed countries, and the situation is just the reverse in developing countries where the males dominate as a whole and in all the age groups except the old age.

The contrast in sex ratio between the two sets of countries is very apparent. Situation in Pakistan, India and Bangladesh represents the case of traditional societies where the status of women in society is yet to attain a full measure of equality with men. The male dominated sex ratio is also explained by the neglect of female children and women. Things are gradually

TABLE 6.17 Sex Ratio in Pakistan and in some Developing and Developed Countries

Countries	*Males (%)*	*Females (%)*
Pakistan	51.5	48.5
Bangladesh	50.07	49.93
India	52.0	48.00
China	51.25	48.75
France	48.4	51.6
U.S.A.	49.5	50.5
U.K.	49.23	50.6
Germany	49.03	50.5
Netherlands	49.51	50.2
Japan	48.64	51.2

Source: World Bank data based on the figures of the United Nations, 2021

TABLE 6.18 Sex Ratio in Pakistan and the Provinces of the State

Province / Pakistan	*Sex Ratio Males (%)*	*Sex Ratio Females (%)*	*No. of Males for 100 Females*	*No. of Females per 1000 Males*
K. P.	50.63	49.37	102.54	975.26
FATA	51.12	48.88	104.54	956.87
Punjab	50.85	49.15	103.41	967.05
Sindh	52.00	48.00	108.29	923.46
Balochistan	53.40	47.43	110.82	902.36
Islamabad	53.55	46.45	110.68	903.48
Pakistan	51.2	48.8	104.91	953.2

Source: Census of Pakistan 2017, Final results, Table.1.Area, Population by sex, Sex Ratio and Density

improving in all the South Asian countries, where in not too distant a future, the sex ratio showing parity between males and females may be observed. Inside Pakistan also, the sex ratio differs from one province to another.

Among the provinces, a more prosperous Punjab has a less distorted sex ratio than the other provinces. It is difficult to point out which area is socially or culturally better evolved, but a better balanced sex ratio could be an appropriate measure for the socio-cultural evolution of a society.

6.7.5 Educational Status of Population

The age structure of the population marked by a high degree of concentration in the young age group following a high birth rate, a shocking high rate of infant mortality and a sex ratio dominated by male population may be attributed to the socio-cultural evolution of the society and the level

TABLE 6.19 Percentage of Literate Population above the Age of Ten Years

Territory	*Gender*	*Rural*	*Urban*	*Total*
Pakistan	Males	61.31	77.98	67.79
	Females	38.57	68.14	49.69
	Total	50.05	73.23	58.91

Source: Census of Pakistan 2017, Final results.

of literacy. Only 55 percent of the people of Punjab are literate. Further resolved into literacy in male and female population, it emerges that only 67 percent males in Pakistan are literate, while the percentage of literates among female is only 43 percent. By any standard, this is a low level of educational attainment.

The level of literacy, a measure of educational attainment, appears to have a direct bearing on several parameters of population like sex ratio, birth rate and infant mortality. The overall literacy rate in Pakistan, measured and evaluated in the population above 10 years of age, is 58.91 percent. It is not a bright picture. It is much higher (67.79 percent) among males than among females having a literacy rate of 49.7 percent. There is also a variation between the rural areas. T

6.7.6 Literacy Rate in Percentage Terms in Pakistan

Among the provinces, Punjab has the highest literacy rate of 64.01 percent, Sindh has an overall literacy rate of 54.57, Balochistan has 43.56 percent, Khyber Pakhtunkhwa has 54.02 percent, and Islamabad has 81.5 percent of its population evaluated as literate. In all provinces, the literacy among women is lower than among men. In Balochistan and Khyber Pakhtunkhwa, the percentage of literacy among women is roughly half of what is found among men. Also, in the urban areas, the gap between the male and female literacy area is much less than in rural area. To take the instance of Punjab, the most prosperous and urbanised province of Pakistan, while the male and female literacy percentage in urban areas is 80 and 73 percent respectively, in rural Punjab, this is 65 and 46 percent respectively. In rural areas, not only the literacy percentage is lower but it shows a wide gap between the literacy ratios of the two sexes.

There is a universal perception that the education of women as mothers is a positive factor in regulating birth rate, upkeep and care of children and keeping infant mortality at a minimum. Perhaps, more than anything else, the South Asian countries have to pay great attention to educating the women folk as an important part of their education policy and programmes.

TABLE 6.20 Number of Speakers of different Languages in Pakistan and the Provinces Where These Are Spoken

Languages Spoken as Mother Tongue	*Number of Speakers*	*Percentage of Speakers*	*Areas of their Dominance*
1. Punjabi	80,538,390	38.78	Punjab
2. Pashto	37.885.441	18.24	Khyber Pakhtunkhwa, Balochistan
3. Sindhi	30,256,372	14.56	Sindh
4. Saraiki	25,324,637	12.19	South Punjab
5. Urdu	14,706,159	7.08	Sindh, Punjab
6. Balochi	6,281,087	3.02	Balochistan, Sindh and Punjab
7. Hindko	5,065,879	2.43	Khyber Pakhtunkhwa
8. Brahui	2,587,677	1.24	Balochistan, Sindh
9. Kashmiri	359,669	0.17	Punjab. Islamabad
10. Others	4,701,315	2.26	Khyber Pakhtunkhwa. Sindh. Punjab

Source: Census of Pakistan, Final Results of Census 2017, Table-11, Population by Mother Tongue, Sex and Rural/Urban

6.7.7 Ethno-linguistic Composition of the Population of Pakistan

An important element in the diversity of Pakistan's population is its ethnolinguistic composition, attributed largely to its diverse relief, mountainous barriers and the impact of centuries-old successive waves of immigration. The four regional identities – the Punjabis, the Sindhis, the Baloch and the Pashtuns – subsumed in the larger national identity of Pakistan, are so strong and deeply entrenched that they persist despite many political changes and the emergence of a democratic setup in the country. It is beyond the scope of this chapter to discuss the ethnicity of the individual group, as a separate chapter is dedicated to this theme, but one can have some idea by looking as the linguistic makeup, especially the mother tongue of the people. The languages, spoken as mother tongue, most often though not invariably, speak of the ethnic affiliation of a group.

Broadly speaking, there are four major languages that are spoken as mother tongues in the four provinces. Though in some areas of each province, the people speak some kind of patois derived from the principal language of the region or a combination of a number of regional languages in the province. The languages, spoken as mother tongues in Pakistan, are given below in the order of the numerical strength of the speakers (Table 6.20).

6.7.8 Spatial Spread of Different Languages in Pakistan

Punjabi, as the mother tongue, is spoken by the largest group of people, constituting 39 percent of the population of Pakistan. More than 90 percent of the people speaking Punjabi are from Punjab, the province after which the people and Punjabi, as a language, are named. A spill over effect of the language is seen in northern Sindh, contiguous to Punjab, where more than two and a half million people speak the language as their mother tongue. Numerically, the second important language, spoken as mother tongue, is Pashto, spoken predominantly, almost exclusively, in Khyber Pakhtunkhwa. Besides, Khyber Pakhtunkhwa, the core area of Pashto, more than a third of the population of Balochistan, composed of Pashtuns settled in northern Balochistan, speaks the language as its mother tongue. Sindhi, as the name suggests, the principal language of Sindh, is spoken by about 60 (61.6 percent) of its inhabitants.

The linguistic composition of the population of Sindh is a very complex phenomenon which has appeared after 1947, following a huge influx of Urdu speaking population refugees, the *Mohajirs* from India. The latter group constitutes about 18 percent of the total population of Sindh. More puzzling is the Pashto speaking group of Pashtuns who now constitute more than 5 percent of the population of Sindh. It may be noted that, unlike Punjab, there is no common border between Sindh and Khyber Pakhtunkhwa, yet there is a huge influx of Pashtuns, largely in Karachi and large urban area. Good connectivity, promoted by Peshawar–Karachi Railroad, and the attraction of Karachi, as an economic and industrial magnet, have facilitated the inter-state movement of over 2.5 million Pashtuns to Karachi, reflected in the linguistic composition of Sindh.

Saraiki, numerically the fourth important mother tongue is spoken by over 26 million people. Known earlier as Multani, the language covers a large swath of land in south-west Punjab. Believed to be an old language sharing partly the properties of Sindhi and Punjabi, it has acquired eminence lately. It has generated a distinct identity of its speakers in South Punjab, a phenomenon that has acquired a political hue in the demand for a separate province of South Punjab.

Urdu, the national language of Pakistan, is spoken by only seven percent of the population of Pakistan, as its mother tongue. The language is largely confined to the two provinces of Sindh and Punjab. Around 60 percent of Urdu-speaking people is settled in Sindh. These are *Mohajirs*, the Urdu-speaking refugees from India, who have proudly retained Urdu as their mother tongue. Being a large group, about 18 percent of the population of Sindh, they have very proudly preserved Urdu as their mother tongue, besides maintaining a separate social and political identity. Over five million people of Punjab, speaking Urdu as their mother tongue, are the elites who promoted Urdu not only as a national language but also its growth in a Punjabi-speaking society, as a medium of education and culture.

Balochi and Brahvi (also written as Brahui), the two languages spoken largely in Balochistan, are spoken roughly by four percent of the population of Pakistan. Balochi has over six million speakers, of which over 4 million are in Balochistan and is the principal language of Central and South Balochistan. Besides, Balochistan, a million people, each in Punjab and Sindh speak Balochi as their mother tongue. This has resulted from a continued migration, over the centuries, of Baloch population in Sindh and trans-Indus region of Punjab. Brahui is a very old language, probably the language of the first settlers of the area, and its aerial space has slowly narrowed to a very small area. Spoken by 2.5 million people, it is largely confined to Balochistan, with some Brahui speaking people settled in Sindh.

Another significant linguistic group having an independent identity is the group of Hindko-speaking people. Representing a group of Indo-Aryan dialects, the language combines several dialects, spoken largely in the south-western region of Khyber Pakhtunkhwa, especially the Hazara division, around Abbottabad.

6.8 Perspective on the Population of Pakistan

What is the optimal population for a country? It is a difficult question, as the optimality will vary with the quality of manpower, availability of resources, technology and the overall socio-economic development. The capacity to support a large mass of people will depend on how well a country has developed its economic and human resources.

The first question that is always asked, to evaluate *a priori* the pressure of population on land, is the density of population, especially so in the developing and underdeveloped countries. Taken all the countries together, population density of the world is 60 persons per km^2. Yet there are extremes of high and low densities. The developed countries of the world fall in different categories of population density. The USA with a vast expanse of land has a population density of only 36persons /km^2, but Japan, once the Asian economic giant, has a relatively high density of 328 persons/km^2. The developed countries of Europe like Germany and the UK have a population density of 233 and 287 persons/km^2 respectively. This shows that the

TABLE 6.21 Population Density of Some South Asian Countries (population/km^2) – 2021

Country	*Density of Population per km^2*	*Growth Rate between 2012 and 2021*
Pakistan	292	2.08
India	469	1.09
Bangladesh	1278	1.09
Sri Lanka	358	0.83
South Asia	394	

Source: World Bank 2021.

population density has little relation with economic prosperity in the developed countries with a large industrial base. The countries can be rich or poor regardless of their density. But, in case of underdeveloped or developing countries, density of population takes precedence, as primary sector of the economy, primarily agriculture, is the main support base, and population density reflects the pressure on land, and in turn, the per capita share in the economic pie of a nation. In such situations, a lot of national effort goes in looking after the huge mass of population, in providing education, health care and employment. It takes decades, if not centuries, to transform the unskilled population into required and useful manpower to add to the productivity of a nation. In such situations, as in the countries of South Asia, the pressure of population becomes a problem.

Among the major South Asian countries, Pakistan is the least densely peopled country, and one may argue that there is no reason for alarm. But the growth rate of population has been rather high. Pakistan has a comparatively lower population density as compared to some of its South Asian neighbours, what is alarming is a high population growth rate, a high total fertility rate, and several adverse vital statistics in case of Pakistan. The pressure of population in Pakistan is not as much as some of its neighbours. It is even lower than the combined population density of South Asian countries. In terms of per capita available arable land as well, Pakistanis is favourably placed as compared to the main countries of South Asia and even China.

While Pakistan is not unfavourably placed in terms of per capita Net Sown Area, the agricultural productivity of the country is not the highest. The crop yields in Pakistan are still low and need improvement. The mean yield of wheat (kg/hectare) for the world is 4,071 kg per hectare. Pakistan's main cereal, wheat, has an yield of 2,844 kg/ha. This is rather low. The European countries like the UK, Germany and France produce more than 7,000 kg of wheat per hectare. The wheat yield in China is 5,409 Kg/ha, and even in Bangladesh it is 3,029 kg/ha. This comparison shows, Pakistan's agriculture needs added attention, more so, because of its underdeveloped

TABLE 6.22 Percentage of Arable Land and Per Capita NSA in ha for Different Countries of South Asia

Countries	*Percentage of Arable Land*	*Per Capita NSA in Hectares*
Pakistan	26.9	0.12
India	45.0	0.13
Bangladesh	50.6	0.059
China	10.3	0.07
Sri Lanka	19.1	0.06

industrial base. With an annual population growth above 2.0 percent, the country may face food shortage, unless the situation is rapidly remedied.

Water Problem: As compared to most of the countries of South Asia, Pakistan receives relatively less rainfall, which is variously estimated at an average of 350 to 450 mm for Pakistan as a whole. But, no country in the world receives so much of glacial melt water as Pakistan does. It is estimated that 50 per cent of the total water in Pakistan is derived from the glaciers. Since the lower part of Indus basin – Punjab and Sindh – has a mantle of deep alluvium, there is a considerable storage of ground water. However, with global warming and the melting of the glaciers, the country may face a shortage of water. One may theorise that the rainfall amount and intensity of rainfall may increase and redress the water problem. In any case, like other monsoon countries, Pakistan has undertaken comprehensive water conservation measures.

Besides the strain on the natural resources, like land and water that a growing population may bring, it neutralises the modest growth that the country achieves. A growing population, with a corresponding increase in increased productivity, perpetually brings down the per capita Gross Economic Product (GDP). If one considers the basic resources, Pakistan is not overpopulated, yet, with a crude density of around 261 persons/km^2, and limited arable land, it has to regulate the population growth rate. One my say, Pakistan has reached an optimal level of population density; a continued high growth rate of population may prove detrimental to its prosperity. What the country has to emphasise is the improvement in the quality of population by education and making it better skilled.

Ballooning Population and Sub-optimum Economic Growth For a variety of reasons Pakistan's economy has not grown adequately to keep pace with its population. While the population has ballooned from a mere 34 million in 1951 to 208 million in 2017, there is no corresponding growth of the economy. In fact, the economy in a developing country like Pakistan has to grow much faster, not only to take care of the growing population but also enhance the per capita income of the populace that could usher in a better standard of living.

Pakistan, as a country, is mindful of its population problem, but like many other countries has achieved only a moderate success. It has certainly brought down its population growth rate, total fertility, infant mortality and has an increased life expectancy at birth. With a total fertility rate of 3.46, an average annual growth rate of 2.4 percent (2017 Census), estimated to be around two percent in 2021, an unbearably high infant mortality rate of 54 per 1,000 live births (World average is 54 and developed countries like the USA have around five, and the European countries have brought down to around three), the country has to struggle to bring these vital parameters of population, to a tolerably acceptable level. No doubt, its young population would prove a productive fore and would produce what is termed as demographic dividend. The most important vital statistics need attention are lowering of birth rate, infant mortality and total fertility (2.1 total fertility rate ensures population stability). The final objective is to create a demographically stable population.

The Government and the people of Pakistan are alive to these problems and would be able to influence the numerical growth of the population towards stability, with healthy attributes like a balanced sex ratio and age structure and a highly literate, healthy and forward-looking population.

Notes

1 Census of Pakistan (1951) vol. I, Report and tables. Statement 2.E, p. 31.
2 Main stream of Mohajirs, Map 2.4, p. 29.
3 The total number of immigrants from East Bengal is approximate and is based on the immigration in the neighbouring Indian States of West Bengal, Assam, Tripura, Manipur and Orissa. Thus the number of refugees leaving present day Pakistan is a close approximation.
4 Pakistan Economic Survey (2019–20) Finance Division, Government of Pakistan.

7

RURAL SETTLEMENTS IN PAKISTAN

Like most other South Asian countries, a large part of the population in Pakistan lives in villages, large and small, home to 63 percent of the country's population. The remaining 37 percent of the population is classed as urban. The distinction between a rural and an urban centre in most South Asian countries, as inherited from the colonial past, is essentially the size of the population of the settlements. While the settlements depending largely on agriculture for their livelihood were classed rural, others, with a population of over 5,000, with 70 percent of their population engaged in non-agricultural occupation, and having a local self body, like a corporation or a municipality, were classed as urban. The terminology adopted by Pakistan Census is a little different. According to 2017 Census, rural and urban centres are called rural and urban localities.

7.1 Definition of Rural and Urban in Pakistan

The Government of Pakistan has compiled "*Mouza Statistics 2008*"[1] for planning of the settlements, through its Agricultural Census Organisation. A *mouza*/deh

> is a traditional unit with a separate name, definite boundaries and areas, precisely measured and divided into plots/*khasras*/survey numbers. Each *mouza* is a revenue estate, and has a cadastral map maintained in the land revenue record with a *hadbast* number except in Sindh province. *Mouza*, *deh*, village, *killi* and *chak and* goth are the names commonly used for rural settlements. There may be one or more settlements in a *mouza*, like *abadis*, *bastis*, *dhokas*, *goths* etc. in the territory of a *mouza*/*deh*. The

DOI: 10.4324/9781003510147-9

mouza/deh may also have scattered inhabitation while there may be some *mouzas* without population as well.

What is clear is that the word *mouza* in Pakistan is generally a revenue village, which may have one or several settlements or, as in a few cases, no settlement at all.

The total number of *mouzas*/revenue village, spread over 132 districts of Pakistan is as follows (Table 7.1):

The urban *mouzas* appear to be virtual urban centres without the official tag of an urban locality. Similarly, rural-urban *mouzas* are the ones combining the agricultural villages with many urban functions. *Rural Mouzas are the actual villages, classified as rural localities, with one or several settlements.* The unpopulated *mouzas* are the ones with no settlements. Such *mouzas* are either deserted villages or have agricultural land cultivated by the farmers of adjacent villages.

The number of settlements is, however, much larger, more than three times the number of *mouzas*. The total number of rural *mouzas* including 1,576 partly urban *mouzas*, but excluding urban, forest and unpopulated *mouzas,* is 47,482. These, along with the number of settlements as distributed in different provinces are as follows (Table 7.2).

A *mouza* in Pakistan, as defined earlier, is a revenue entity. It has a fixed boundary, demarcated on a cadastral map. The rural settlements, on the other hand, are the habitat, a collection of houses, cattle sheds, streets, a mosque, wells and a common ground, in the midst, or on the margin of agricultural fields. Thus, a *mouza* has cultivated fields, one or several settlements, besides roads, canals and pump houses and water channels.

In short, a Mouza is a revenue entity, with or without settlements, maintained in the land records of the Government, while a rural locality, as defined in 2017 Census, is also a revenue entity but always with a human habitat, universally known as a village, with or without one or several satellite settlements.

TABLE 7.1 Number of *Mouzas* in Pakistan 2008

Type of Mouzas	*Number of Mouzas*	*Percentage of the Total No of Mouzas*
1. Urban Mouzas	1,099	2.1
2. Urban/Rural Mouzas	1,576	3.00
3. Rural Mouzas	45,906	87.64
4. Forest Mouzas	439	0.83
5. Unpopulated Mouzas	3,356	6.4
Total number of Mouzas	52,376	99.97

Source: "Pakistan 2008, Mouza Statistics", Govt of Pakistan, Statistics Division, Agricultural Census Organization.

TABLE 7.2 Distribution of Mouzas and Rural Settlements in the Provinces of Pakistan

Provinces	*Area* in km²*	*Rural** Population (1,000) 1998*	*No of Rural Mouzas*	*No of Rural Settlements*	*No of Rural Settlements (1998) per Mouza*	*Rural Population per Rural Mouza*	*Rural Population per Rural Settlement*	*No of Rural/ Mouzas per 1,000 km²*	*No of Rural Settlements per 1,000 km²*
Punjab	205,345	50,602	23,941	78,218	3.27	2,114	646	117	381
Sindh	140,914	15,600	5,573	52,058	9.34	2,800	300	40	370
KP	74,521	14,749	11,540	23,359	2.03	1,278	631	155	313
Balochistan	347,190	4,997	6,426	16,611	2.58	778	301	19	49
Total of the four Provinces	767,970	85,948	47,480	170,246	3,58	1,810	505	61.82	222

Source: Pakistan (2008) Mouza Statistics, Govt of Pakistan, Statistics Division, Agricultural Census Organization.

*Area refers to the entire area of the province and not only the rural area.

**The figures as given in the table omit the rural population of Bajaur, Orakzai, Kurram, N Waziristan, S Waziristan agencies and frontier region Kurram. The figures for the latest available number of rural localities as per 2017 Census is given in the sequel.

7.2 The Rural *Mouzas* and the Rural Settlements

In Pakistan, there are around 47,480 rural mouzas having 170,246 rural settlements. Roughly, half the total number of rural *mouzas* have only one settlement each, about 20 percent have two to three settlements, and the remaining have more than three settlements. Around eight to nine percent of the total *mouzas* have ten or more settlements. The numerical relationship between a *mouza* and the number of rural settlements is not uniform. This can be seen from Table 7.3 that specifies the number of *mouzas* and the varying number of rural settlements in each province.

There is a wide variation in the number of rural settlements owned by each *mouza* in each province. This can be seen from Table 7.2 above that specifies the number of *mouzas* and the varying number of rural settlements in each province.

While Khyber Paktunkhwa has, on an average, two rural settlements per *mouza*, Sindh has as many as nine settlements per rural *mouza*. Punjab presents a moderate case where each rural *mouza* has, on an average, 3.2 rural settlements, which means roughly 300 to 400 rural settlements per 1,000 km^2 of the area of a province, and each settlement occupying 2 to 3 km^2 (200 to 300 hectares) of land. The case of Balochistan is very different; there are only 50 rural settlements in an area of 1,000 km^2, each having a population of around 300 people.

The mean population of a rural settlement in Punjab and Khyber Pakhtunkhwa is roughly 600. In Sindh, because of there being too many rural settlements in a *mouza*, the mean population of a rural settlement is around 300. About 40 percent of the *mouzas* in Sindh have 10 or more rural settlements each, a case rare in Punjab or even Khyber Pakhtunkhwa.

An Exceptional Case of Sindh with fewer mouzas and a large number of rural settlements – The number of *mouzas* in Sindh is relatively small, a fact attributed to a higher level of urbanisation (around 50 percent) compared to other provinces. The province has only 5,573 rural mouzas, which works out 40 mouzas for every 1,000 km^2 area of the province. Thus, for each 25 km^2 of Sindh territory, there is only one *mouza*.

What is intriguing in the case of Sindh is that unlike other provinces, the number of rural settlements per *mouza* is higher, thus bringing the average area of a settlement comparable to that of Punjab and Khyber Pakhtunkhwa. With only 5,573 rural *mouzas*, the province has an aggregate of 52,058 rural settlements, commonly known as the *goths* in Sindhi, giving an average of over nine rural settlements per *mouza*. In other words, in a revenue village, on an average, there are nine rural settlements.

TABLE 7.3 Rural *Mouzas* with Varying Number of Settlements

	Rural Mouzas *with Only One Settlement*	*With 2 to 3 Settlements*	*With 4 to 5 Settlements*	*With 6 to 9 Settlements*	*With 10 Settlements and Above*	*Total*
No of mouzas	24,494	9,354	6,215	3,178	4,241	47,482
Percentage of *mouzas* having variable number of settlements	51.5	19.7	13.08	6.69	8.93	99.90

7.2.1 The 2017 Census and the Introduction of the Terms "Rural Localities" and "Urban Localities"

The Census of Pakistan and the Bureau of Statistics classified all the settlements of Pakistan, into two categories, as 1. Rural localities and 2. Urban localities.

According to the Census of Pakistan (2017), the total number of rural localities (Villages) in Pakistan is 47,290. This figure closely corresponds to 47,482, the number of rural *mouzas* as per the Pakistan Mouza Statistics (2008). It is likely that some of the rural *mouzas* qualified as urban mouzas, during the decade 2008–2017 and are not included in the list of rural localities in 2017.

For understanding the number, size and distribution of villages, the Census Statistics of 2017 is adopted in this text. For all purposes, a "rural locality" is considered as a village. The total number of rural localities (villages) in Pakistan as per 2017 Census is as follows (Table 7.4)

It may be mentioned that in the present text, the three terms, viz., rural mouza, rural locality and a village, are used interchangeably.

As can be seen from the above table, Pakistan has 47,290 rural localities (villages), with a mean population of 2,791 per village. With 25,321 villages, out of a total of 47,290 for Pakistan, Punjab has more than half the total number of villages in the country. It is quite in consonance with the population of the province which is also 52 percent of the population of Pakistan. There are 117 villages per 1,000 km^2 of territory in Punjab. The mean size of a village in Punjab being 2,742, it lingers around the national average of 2,791 persons per village. Each village in Punjab, on an average, has three to four satellite settlements in its orbit.

The villages in Khyber Pakhtunkhwa are relatively large, though not as large as in Sindh. Khyber Pakhtunkhwa has 7,696 villages, though with a larger average population size of 3,219 persons per village. Surprisingly, the density of rural localities in Khyber Pakhtunkhwa is higher than any other province in Pakistan. The number of secondary settlements linked to a village is just around two. For reasons of security, the villages in Khyber Pakhtunkhwa, it appears, have not developed isolated secondary settlements.

Balochistan is an extreme case of a small population scattered in a large number of hamlets or small cluster of houses. The average size of a village in Balochistan is 1,390, and even this is scattered in a number of secondary settlements. A settlement is usually a cluster of five to ten houses attached to a specific village. Each village, on an average, has 30 to 35 small clusters of houses.

Sindh is a case of a fewer large villages, each with multiple secondary satellite settlements. The province has only 5,726 villages (rural localities), the

TABLE 7.4 Number of Rural Localities (villages) in Pakistan and Their Provincial Distribution

	Number of Rural Localities	*Percentage of Total Rural Localities*	*Total Rural Population*	*Percentage of Total of Rural Population*	*Average Population of a Village*
Pakistan	47,290	100	132,012,864	100.00	2,791
Khyber P	7,696	16.27	24,773,492	18.76	3,219
Punjab	25,321	53.54	69,441,410	52.60	2,742
Sindh	5,726	12.10	23,021,876	17.44	4,020
Balochistan	6,419	13.57	8,926,426	6.76	1,390
FATA	1,999	4.22	4,853,178	3.67	2,427
Federal Capital	129	0.27	994,365	0.75	7,708

Source: Census of Pakistan 2017, Table-3. Number of Rural localities By Population Size and their Population by Sex.

lowest among the provinces. It has, however, a large number of settlements, large and small, attached to a specific village. Sindh has over nine settlements in each rural *mouza*, The average size of a village is roughly 4,000 persons, the highest in Pakistan.

7.2.2 Population Size Groups of the Villages in Pakistan and the Provinces

Most of the villages in Pakistan have a population of less than 5,000. The number of villages with a population of over 5,000 is not more than 15 percent. The province of Sindh is an exception where about one-fourth of the villages have a population of over 5,000. In Sindh, a village has a large cluster of satellite settlements which add up to a higher population of the villages (Table 7.5).

In most of the provinces, the largest number of villages falls in the population group of 2,000 to 5,000. In Pakistan, more than 25 percent of the villages lie in this group. So is the case in other provinces. In Punjab, 28 percent of the villages lie in this group. In the case of Khyber Pakhtunkhwa, the villages appear to be distributed more evenly over a wide range of sizes. Even the number of small villages with a population of 200 to 500 is fairly large. In Sindh, 26 percent of the villages have a population of over 5,000.

The median size of the population, indicating half the number of villages having a population below, and the other half above this number, is 1,600 for Pakistan as a whole. It is highest in Sindh. The median village size in Sindh is 3,300; in Punjab, it is 1,660; in Khyber Pakhtunkhwa, it is 1,760; and in Balochistan, it is less than 500.

7.2.3 Density and Spacing of Rural Settlements

It must be emphasised that, while discussing the density and spacing, all the settlements, in the sense of habitat, commonly known as *basti* or *abadi* in Pakistan, are taken into account. The number of settlements is many fold the number of villages, or the rural localities. As seen in Table 7.2, the density of settlements varies between 300 and 400 per 1,000 km^2 of territory. It is 381 in Punjab, 370 in Sindh and 313 settlements in Khyber Pakhtunkhwa per 1,000 km^2 of area. Balochistan stands at a different footing: it has only 49 settlements per 1,000 km^2 of territory. Besides the general density, the distribution and spacing of rural settlements are not uniform. In Punjab, except the Thal Desert, it is relatively uniform, and so is the case with Sindh. In the mountainous Khyber Pakhtunkhwa, it is irregular largely, confined to valleys, leaving large parts of high ranges, absolutely rocky or snow covered areas, uninhabited.

TABLE 7.5 Distribution of Rural Localities (Villages) in Different Size Groups in the Provinces of Pakistan and Their Relative Number in Percentage

Population size	*Pakistan*	*Khyber Pakhtunkhwa*	*Punjab*	*Sindh*	*Balochistan*	*FATA*	*Islamabad*
Over 5,000	7,103	1,356	3,670	1,504	300	235	38
Percentage	(15.02)	(20.25)	(14.49)	(26.26)	(4.6)	(11.75)	(29.45)
2000–4,999	12,333	1,420	7,386	2,417	6.53	463	21
Percentage	(26.07)	(18.45)	(29.16)	(45.21)	(10.17)	(21.81)	(16.27)
1,000–1999	9,468	1,417	5,619	988	963	461	20
Percentage	(20.020)	(18.41)	(22.19)	(17.25)	(15.00)	(23.06)	(15.5)
500–999	7,028	1,353	3,679	385	1,264	327	15
Percentage	(14.86)	(17.58)	(14.52)	(6.72)	(16.69)	(16.35)	(11.62)
200–499	5,270	1,196	2,200	207	1,362	292	13
Percentage	(11.14)	(15.54)	(8.68)	(3.61)	(21.21)	(14.6)	(10.07)
< 200	4,590	858	1,728	153	1,594	157	19
Percentage	(9.7)	(11.14)	(6.82)	(2.67)	(24.83)	(7.75)	(14.72)
Uninhabited	1,579	91	1,039	72	283	91	3
Percentage	(3.33)	(1.18)	(4.10)	(1.25)	(4.40)	(4.55)	(2.32)
Total	47,290	6,696	25,321	5,726	6,419	1,999	129
Percentage	(100)	(100)	(100)	(100)	(100)	(100)	(100)

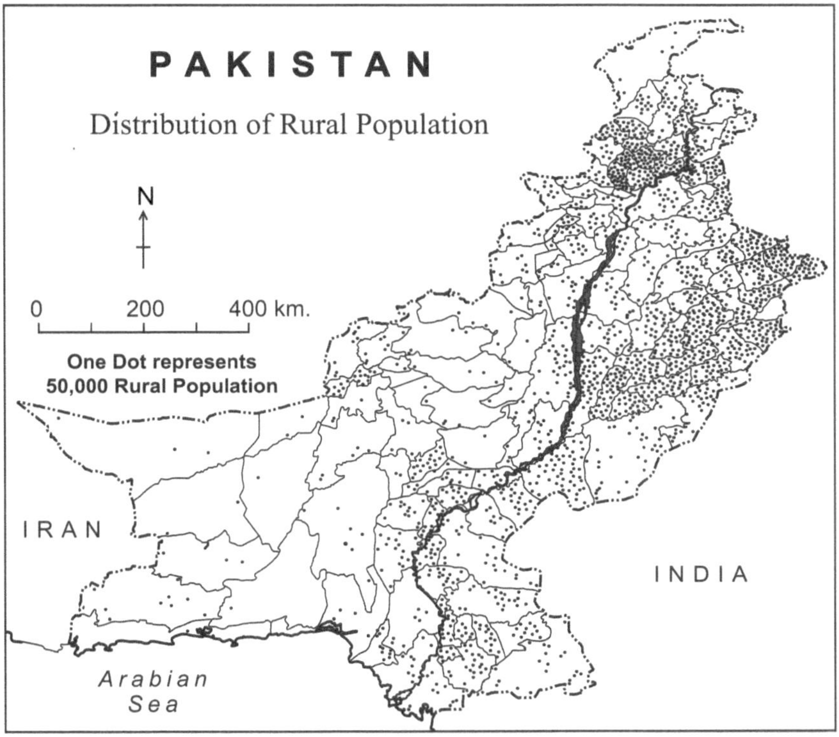

FIGURE 7.1 Distribution of Rural Population in Pakistan (2017)

7.2.4 Accessibility to Village Settlements in Pakistan

About 60 percent of the 47,482 rural Mouzas (villages), described by Pakistan Census as rural localities, fall within one km from the nearest metalled road. The accessibility to road transport is, however, not uniform. Of the four provinces, Punjab is the most privileged to have roughly 80 percent of the villages within a distance of one kilometre from the nearest metalled road. In the case of Balochistan, it is the lowest, and only 20 percent of the villages are located within a kilometre from the nearest metalled road. Sindh and Khyber Pakhtunkhwa have 67 percent and 38 percent of the villages respectively, located within one kilometre of a metalled road.

Though the accessibility to road transport has rapidly increased in Pakistan within the last couple of decades, in the case of Balochistan, which forms 43 percent of the area of Pakistan, the situation is still far from what is needed. The *mouzas* are far apart, sometimes as much as 20 km or more from each other. Only 20 percent of the rural *mouzas* have been provided metalled road in their vicinity.

TABLE 7.6 Number and Percentage of Rural Mouzas Lying within a km from the Nearest Metalled Road

Provinces	*Total Number of Rural Mouzas*	*Rural Mouzas Lying within a km from the Nearest Metalled Road (2008)*	
		No of Mouzas	*% of Mouzas*
Khyber Pakhtunkhwa	11, 540	374	33
Punjab	23,941	19,099	80
Sindh	5,573	3,711	67
Balochistan	6,428	1,276	20
Pakistan	47,482	28,460	60

Source: Pakistan (2008) Mouza Statistics, Op. cit.

7.2.5 Drinking Water Sources and Facilities in Rural Pakistan

The country relies on hand pumps, tube wells, wells, canals and rivers for drinking water. In most cases, the villages utilise more than one source. A village may use both, wells as well as a tube well, and some residents of the village may have even piped water. The use of hand pump is the most common practice in most *mouzas*. It is economical and site is no constraint for it. Some of the villages have even assured piped water supply, but their number is relatively small, Surprisingly, Khyber Pakhtunkhwa has the largest percentage villages with piped supply of water.

Like other South Asian countries, Pakistan has, yet, not been able to provide filtered and piped water to its villages. What is significant to note is that 18 percent of the villages still use well water for drinking, though the situation is fast changing. In 2008, more than half the number of villages (56 percent) were using hand pump to tap ground water as a source of drinking water. Hand pump is the most common source of drinking water in Punjab and Sindh. These two provinces, Punjab and Sindh, have been using hand pump since long. And, even in 2008, 80 percent of the villages in Punjab and 70 percent in Sindh were using hand pump as a source of drinking water. The situation must have improved further during the last one and a half decade since 2008.

In Balochistan and Khyber Pakhtunkhwa, the main sources of drinking water are springs and wells. Thirty-five to forty percent of the villages in both the provinces still use well water for drinking. Hillside springs have a special place in the life of the people of Khyber Pakhtunkhwa. These springs are a perpetual source of water for all purposes. In Khyber Pakhtunkhwa,

TABLE 7.7 Percentage of Villages with Different Modes of Drinking Water Supply

Province/State	*Percentage of Mouzas with different Modes of Drinking Water Supply*					
	Piped Water	*Hand Pump*	*Tube Well*	*Well*	*Canal / River*	*Spring/Stream / Karez*
Punjab	9	82	15	6	3	1
Sindh	7	69	6	9	38	1
K P	20	18	20	38	13	39
Balochistan	9	7	29	39	10	20
Pakistan	12	56	17	18	10	13

Source: Pakistan (2008) Mouza Statistics, Op. Cit (A Mouza may have more than one source of water supply. That explains the addition of different sources to be more than 100 percent).

40 percent of the population uses spring water for drinking. Hand pumps, so common in Punjab and Sindh, are not so common in Khyber Pakhtunkhwa, because of the absence of widespread alluvial terrain and an easy access to ground water.

Topography, lithology and climate are the important determinants in deciding the source of water. The alluvial plains of Punjab and Sindh hold enough water in their underground aquifers which can be easily tapped by sinking a bore hole and installing a simple hand pump that provides secure water. The dry and rocky areas of Balochistan depend largely on deep wells, or some nearby *karez* (hill streams) and some occasional tube wells for their drinking water supply. The ease with which the ground water is tapped in the Indus Plain is not encountered in Balochistan, where a proper site has to be found to dig a well, deep enough to reach a permanent aquifer. Not very dissimilar is the case of Khyber Pakhtunkhwa, though parts of the province, like Peshawar and Mardan Basins, easily tap the ground water through a hand pump. But, spring water is still an important source of drinking water in Khyber Palhtunkhwa.

While hand pumps and tube wells are important for drinking water, canals and tube wells are crucial for irrigation. More than half the villages of Pakistan irrigate their fields using tube wells. Tube wells are especially important in Punjab where 75 percent of the village use tube wells for irrigation, often in conjunction with canals. In Sindh, the use of tube wells, though quite frequent, is not as common as in Punjab. In Sindh, only 30 percent of the villages supplement their irrigation with tube wells. Canals remain the main stay of irrigation in Sindh, where about 90 percent of the villages depend on canal for irrigation and more than one-third of the villages use canal water for drinking.

7.3 Village Nomenclature in Pakistan

A village settlement in Pakistan, also called an *ābadi* or basti has numerous endings of which some are more common than others. Usually, a village name is linked with an important personage, a local landmark, or as in the case of a canal colony, is marked by the number of a chak, which simply means the number of that canal *chak* or village. It is difficult to decipher the chronology of the settlements, nor is it the intention of this chapter to do so. Yet, certain broad hints may be offered, which may be evaluated. Then there are different sizes of villages, large, medium and small, yet in the fully irrigated area of central Punjab, the size and spacing of the villages remain clustered to a mean.

Some of the most common endings are

Suffix:	1. pur	Prefix:	1. Dera
	2. *abad*		2. *Qila*
	3. *basti*	3. *Basti*	
	4. *goth*		4. *Abadi*
	5. *mahal*		
	6. *wal / wala*		

Each has its own significance. Not a small number of settlements carry the ending "*pur*'. The most common example is "Rajanpur" or "Ahmadpur" in Jhang district. Carrying *pur*-ending is associated with some important personage. But why add *pur*, a Sanskrit word, and not *bad* (a Farsi word) as is commonly the case? It is likely that the settlements carrying *pur* as an ending have a great antiquity and have undergone a change of names several times, retaining only *pur* as a suffix. The name Ahmadpur, in Jhang district, appears incongruous. Why Ahmadpur instead of Ahmadabad? It appears "*pur* is inherited from antiquity of another settlement, the root word which may have been replaced several times. It is surmised that the settlement carrying the suffix "*pur*" are very old and have assumed different names at different times. Ahmadpur in Jhang district, Alipur in Muzaffargarh, Deenyapur in Multan district and Nurpur, Madnipur and Khalakpur in Gujranwala district are examples of settlements ending in "*pur*".

Basti and *abadi* are used interchangeably for a habitat or a settlement, more often used for sub-standard housing colonies. A *goth* is a cattle ranch which develops into a regular settlement with age, as more and more people assemble around a "*ranch*" where initially a few farmers or graziers started developing a pen for the night. In due course, the cattle ranch grows into a large settlement and is transformed unrecognisably into a regular village, but the original title *goth* continues. Goths, as villages, are the most common features of Sindh. "Basti Mardanpur" and "Abadi Adda Bosan" in Multan district are examples of "*basti*" and "*abadi*", prefixed before the proper name of a village. It is likely that a "*basti* is a settlement, adjunct to the main settlement, developed much later.

Many settlements have a prefix "*dera*, which means a camp, a temporary resting site, which, in due course, turns permanent. The most prominent examples, though of districts and cities, are Dera Ismail Khan on the right bank of the Indus in Khyber Pakhtunkhwa and Dera Gazi Khan, a district of Punjab, lying west of the Indus. Other prefixes, which occur, though not so frequently, are *Qila* or *Kot*, both implying a fort. These were the defensive positions or garrison sites or some kind of defensive establishments, named after a "chief". During the Sikh rule, a number of sites in Punjab carried defence establishments around a fortress. Qila Man Singh, Qila Didor Singh or Qila Diwan Singh – names of actual settlements – are examples of a defence site where a "*qileddar*" was a chief, and the settlement assumed a name after the custodian of the *qila*.

There are single-syllable names of villages that are difficult to explain like "Duni or "Jalan". In most cases, a village carries in its name the characteristics of the site, the name of some important personage of the settlement, a historical event, a religious event or the name of a "Pir", a saintly person, or in some cases a name difficult to explain.

The *Chak* Villages: The Chak villages, developed after the laying of canals and the distribution of land among the cultivators in the canal irrigated areas, were simply numbered with a prefix "*chak*" like Chak 6/AH. The larger irrigation colonies were well planned and systematically organised and so were the newly established villages, which had no name but only the number. This was a convenient mode of collecting revenue from these villages. Often these *chaks* carried with their number, as suffix, the name of a canal or a branch of it, numbered. In Sindh, unlike Punjab, most villages and settlements carry only the names of the local chieftain, a zamindar. Occasionally, one notices "*goth*" or "*basti*" as a prefix, but there is a preponderance of proper names that villages carry.

The most common ending in Sindh is "*Goth*" and has assumed a new meaning in Sindhi signifying a village, but is always conjoined with some individual. Such individuals may have been pioneers in the area or important leaders or landlords. Settlements often carry the names of individuals and are known after them. Some examples from Larkana district in Sindh are given below:

1 Samsher Goth	5 Gul Mohd Chacha
2 Basti Sugrana	6 Jamal Khan Laghari
3 Meehar Goth	7 Sardar Rais Bhanbho Khan Chamdo
4 Sahib Kalhoro	8 Sir Shahnawaz Bhutto

Names like Jatoi and Laghari often appear as settlements. This shows the dominant influence of local landholders, some of them having large holdings of thousands of acres of agricultural land. To have an idea of the measure of importance of individual names; one has only to look at the alphabetical listing of places, and starting with "Abdul" one may find the number of villages running into hundreds. Beginning with Abdul, one may recount of Abdar Rahim, Abdu Karim, Abdul Aziz, Abdul Bakshi, Abdul Ghafur, Abdul Ghani, Abdul Hakim, Abdul Haq, Abdul Karim ... It goes on, showing the influence of proper names of persons in the naming and identification of settlements.

As a simplistic generalisation, while in South Sindh "goth" is a common suffix like Ismali Khan Goth, Yusuf Khan Goth, Abdul Rana Goth, Haji Saawan Goth, the northern part of Sindh frequently carries "abad" as a suffix like Sultanabad, Umarabad and Nawazabad.

7.4 The Structure and Layout of Village Settlements in Pakistan

The structure and layout of the villages, as much as their distribution in different size groups, vary widely from one province to the other. These depend on the topography and availability of water having their bearing on the availability of land for cultivation and the intensity of agriculture. It has to be kept in mind that the rural economy is largely based on agriculture, and spacing and layout of the villages depend on the type and scale of agriculture in a region.

7.4.1 The Punjab Villages

It may be emphasised that about 60 percent of the total cultivated land of Pakistan lies in Punjab, and a large part of the cultivated land is utilised for multiple cropping. In consequence, a high agricultural productivity of the land supports larger population distributed over a large number of villages. That explains why more than 25,000 villages, out of a total of 47,000 in Pakistan, exist in Punjab.

Secondly, unlike other provinces, the village structure in Punjab has undergone complete transformation following the initiative of the colonial Government that included 1. Survey and settlement of land and 2. Introduction of canal colonies in the canal irrigated areas and the settlement of the Chak Villages. These Chak villages were created to settle the immigrant cultivators in the canal irrigated land. Thus emerged a new class of rural settlements, different from the traditional villages.

It may be mentioned that the traditional Punjabi villages before independence, or even before the colonial British rule, were either clan villages, or landlord controlled villages or the *ryotwari* villages. The clan villages were jointly managed by members of a clan or the community, like the Jat villages of Central and South Punjab. The landlord villages were the ones owned by landlords as in the Salt Range region. Such villages were owned by the village landlords who assembled around them dependent communities of cultivators. The third type was the *ryotwari* villages, found usually in South-western part of the Punjab Plain. In such cases, the village used to be a revenue unit and the actual settlement consisted of a group of isolated home-steads built around a well or some other source of water in the otherwise dry region. After the settlement survey and the opening of canals and the establishment of canal colonies, the villages as institutions and their internal structure are greatly transformed.

7.4.2 Planned and Unplanned Villages

In contemporary Punjab, villages can be grouped in two categories, 1. Planned villages associated with the irrigation system and the establishment of irrigation colonies and 2. The old unplanned villages with irregular layout.

The planned villages – One may call them Chak villages, originally inhabited largely by immigrant farmers from other districts including some like Amritsar, Jullundar and Hoshiarpur now in India. Most of these villages are numbered by Chak numbers followed by a village name. In cases where there were pre-existing core villages, before the arrival of canal irrigation, the village names were also added. To illustrate, take the case of Chak-168-GB-Siraj, a Chak village in Faisalabad district. Here, 168 is the number of the block of irrigated land, given by the irrigation department; GB indicates the source of irrigation, that is Gujera branch of Chenab canal, and Siraj was the original village. These villages are, by and large, compact, surrounded by agricultural fields. The field pattern in such villages is extraordinarily rectangular, parallel to each other. The general dimension of a field is 25 m x 50 m. One seldom observes a field with a width of over 30 m.

The planned villages in Punjab are more often rectangular. An average village settlement may have a dimension of 300 m x 400 m of a rectangular shape. Houses in Punjab are usually rectangular in their outline, made of mud or bricks with a flat roof. Almost all houses are single-storied. The normal size of a house is 10–15 m in width and 20 to 40 m in length, usually with a courtyard, which in most cases is open on two sides. The streets are wholly or partially straight, though they don't form a pattern, yet accessibility to each house is assured. Often, there are a few trees inside the settlements, sometimes even in a courtyard.

In some villages, apparently the older ones, the settlements have fewer central streets and present a jumbled arrangement. The village named *Qatarpur* (30°39'N and 71°55'E) has a very irregular outline. Such villages appear to have grown gradually by accretion on their peripheries. The village appears to have had a water pond in the centre, which was filled up after the arrival of canal irrigation, and is now a green patch, plump in the centre of the village. Another village *Duni* (32°05'N and 73°43'E approximately) which extends 500 m north–south and 600 m east–west, is irregular in shape. The village has good houses with a mosque to the west of the village.

The featureless Punjab Plain is criss-crossed by canals, their distributaries, sub-distributaries and service roads, and canals sometimes make convenient boundaries between the villages.

In Sindh, the situation is different with large areas of *mouzas*, each *mouza* carrying as its annexe several settlements of a relatively smaller size. The village settlements, unlike those in Punjab, have more irregular shapes. Almost every compound of a house has a tree. Often the settlements are linear, following a road, a canal or a distributary canal. The field patterns are not well-defined and are often quite irregular. The houses, with usually a north–south orientation, are thinly spread and not huddled together. There are compact villages tending to be circular, as is the case with *Bhaan* village, 10 km west of Naushahro Firoze (26°45' N and 68°30'E). The village is surrounded by greenery and a large mango garden to the southwest of the village.

The villages in Balochistan have usually a few scattered houses without any order. The site is usually governed by the availability of a water source. Balochistan is known for its hill site shelters, carved in the mountain fronts. These dwellings are no longer fashionable and are yielding place to constructed houses. One of the chief characteristics of Balochistan houses is the absence of windows, admitting air and light only through ventilators in walls, at 3 m above the ground.

Unlike Balochistan, the villages in Khyber Pakhtunkhwa are more compact, often huddled together. Usually located on higher ground, not too far from a source of water, a stream or a spring, most villages in Khyber Pakhtunkhwa are clan villages, often having a common ancestry. Unlike Balochistan where the villages are small, or Sindh where a village is quite large with many satellite settlements, Khyber Pakhtunkhwa has villages in all population size groups. There is no obvious shift towards very large or very small villages.

To summarise, Pakistan has 47,300 villages which are home to 132 million, or roughly 64 percent of the total population of the country. Out of this total number, over 25,000 villages, which constitute 54 percent of the total number in Pakistan, are in Punjab. The average size of a village in Pakistan is 2,791; in Khyber Pakhtunkhwa, it is 3,219; in Punjab, it is 2,742 and in Balochistan, it is only 1,390. The villages are exceptionally large in Sindh and very small in Balochistan.

7.5 Some Aspects of the Village Society in Pakistan

The rural society in Pakistan, in general, is feudal. It is estimated that "*five per cent of the agricultural households in Pakistan own nearly two-thirds of Pakistan's farmland*".[2] Feudalism is very much a part of Pakistan's rural socio-economic landscape. The rural societies in the two major provinces of Punjab and Sindh are dominated by landlords owning large estates. These feudal landlords control the politics, and no less the economy of Pakistan.

The Rural Society in Punjab – The village society in Punjab is very much tethered to the traditional clan or caste, and a large number of villages were populated by members of a specific caste as in Punjab. One may say these were clan or caste dominated villages. An important, yet paradoxical aspect of social organisation in rural Punjab is the prevalence of caste system, a system associated with Hinduism in India. "*Caste System in Rural Punjab exists in the form of indigenous categories of Quom and Zat, which divide the zamindar Quoms, Kammi quoms in distinct and rigid birth-ascribed status groups on the basis of their parental occupation.*[3]

In a contemporary Punjab village, the caste system plays a significant role in the village social structure. The hold of the rich landlords supported by their kinsmen is visible in the affairs of the village local elections and in other political and economic activities. "*Inferior caste status of service providers, in rural Punjab, restricts their right to political participation including right to vote, contest elections, participate in election campaigns and opportunities to acquire leadership roles. Consequently, they are excluded from the system of political patronage that connects villages with politicians through their Biradari representative*".[4] Similarly, southern Punjab still has some dominance of feudal lords. "*With a much less share of the canal colonies, the South was less affected by the greater social mobility and economic dynamism they brought in their train, and also received relatively fewer refugees from East Punjab. It also contains fewer egalitarian Jats than the northern and central parts of Punjab and more Baloch with their traditional deference to their autocratic chieftains. In consequence, Southern Punjab is more feudal than the north, in a way that connects it culturally to Sindh.*"[5]

The Rural Society in Sindh – The most spectacular aspect of the rural society in Sindh is the dominance of traditional landlords, called *waderas* who own much of the land in the countryside. The waderas own thousands of acres of land. The entire system of wadera landlords is almost like the subset of the Government. The tenants depending on the mercy of these landlords are referred to as *Haris* who depend entirely on the mercy of these land-owning waderas. In an opinion note written in 2022 Nasrullah Khadim, Kandhkot, present a severe indictment of the rural feudal system in Sindh[6] observing "*feudalism is one of the intricate and grave issues of*

Pakistan, particularly in Sindh where landlords, called feudalists or ***waderas****, quash the paramount and basic rights of people living there.... People in interior Sindh are still strangled with this pathetic sub-system of government...People are made subordinate to the orders of landlords without any decent wages. They are made helpless and destitute to work below their feet. They barely fill their empty bellies for one time.... The landlords keep people backward and uneducated, economically dependent, and rule them blindly to consolidate their power over the area."*

The agricultural labour is formed by Haris, a labour caste. Talking of Haris the author adds "*the only duty they (the Haris) know is to work like bullocks for their landlords and money lenders, to touch the feet of Zamindars, Masters and Pirs and worship them literally as living Gods, and lastly, to instruct their children to do likewise*".

With just over 5,000 villages with a large number of satellite settlements, the rural landscape of Sindh is dominated by *waderas,* the land owning feudal lords, each owning thousands of acres of land on one end and the vast majority of poverty-stricken working hands who having been toiling on the fields on the other. Unnoticed by the Government, public leaders and left-leaning workers unions, the *waderas* continue the exploitation of the land and the people unabated and with impunity as their divine rights. The lawmakers of Pakistan, amidst a multitude of other economic and political problems of great importance remain unconcerned and unmoved.

Notes

1 Pakistan (2008) *Mouza Statistics*, Government of Pakistan, Statistics Division, Agricultural Census Organization, Pakistan.
2 Kugelman, Michael (2013) Pakistan`s Fixation with Feudalism, *Dawn*, 3 January.
3 Ahmed, Usman (2011) Abstract of the Ph.D, thesis Social Stratification in Punjabi Village in Pakistan: The Dynamics between Caste, Gender and Violence, University of Leeds.
4 Ahmed, Usman (2016) Marginalised Voters and Supporters: Biradari System, Caste Hierarchy and Rights to Political Participation in Rural Punjab, *Journal of Political Studies*, vol. 23 (2), pp. 515–524.
5 Lieven, Anatol (2011) *Pakistan: A Hard Country*, London, Allen Lane, p. 280.
6 Nasrullah, Khadim and Kandhkot (2022) Feuadalism in Sindh, *Pakistan Today*, 22 June, p. 2.

8

URBAN LANDSCAPE OF PAKISTAN

Cities and Towns

Cities are a symbol of civilisation, and it was so even during the pre-historic times. Ancient civilisations were, in fact, defined by the contemporary social structure and organisation of city life. May it be the Indus valley civilisation, the Nile valley or the Babylonian civilisations, cities were the defining characteristic and set a benchmark of the progress of civilisation. Archaeologists always considered city as an inalienable element in all pre-historic civilisations.

In the contemporary world, the cities have attracted many more functions than they traditionally performed. Their multifunctional character, as drivers of economic growth, depositories of knowledge, ideas, technologies, and many innovations, has enabled them to play vital roles in national decision-making. Presiding over the countryside, they are the dispensers of goods and services and purveyors of justice. Economically and politically, the cities have become so important that in diplomatic parlance countries like the UK, the USA, Japan or Pakistan are referred to by the capital as London, Washington, Tokyo or Islamabad. Today, cities also represent points of concentration of wealth and economic and political power.

The urban landscape of Pakistan, as indeed of any other country, is characterised by a large number of urban centres, not only of varying population size but even of varying generations. Some are very old, in fact, of a pre-historic origin, some are of early historic period, while some others grew or established during the medieval period, and as expected some others are young, having originated during the last 150 to 200 years. Thus, the country has towns and cities of different generations, though all have a mantle of some modernity.

DOI: 10.4324/9781003510147-10

8.1 The Legacy of Pakistan's Urban Civilisation

Pakistan inherited the legacy of Indus valley civilisation, marked with an advanced level of urban life, about 4500–5000 years BP. The historic sites of Mohenjo-Daro and Harappa, the former in Larkana district of Sindh, close to the river Indus, and the latter in Sahiwal district of Punjab, not far from the river Ravi, bear testimony to an urban civilisation. The pre-historic urban civilisation ended around 1500 BC leaving a complete hiatus between the pre-historic urban landscape and the evolution of contemporary urban centres.

Most of the cities of Pakistan originated during the medieval period; and only a very few, like Charsadda and Taxila of the early Buddhist period, appear to have existed before the Christian era. During the early historic or medieval period, cities have often emerged as royal seats, capitals of empires, focal points of transcontinental routes, trade centres or important sites of religious congregations. Many of these cities were also the centres of learning and the promotion of knowledge, as today, from where diffusion of ideas and radiation of cultural impulses took place, spreading to the hinterland. Thus, as explained earlier, there have been different phases of urban growth. This historical development of towns is termed in the present context as phases of urbanisation.

Phases of Urbanisation in Pakistan: Cities, regardless of their functional character and population size, have always existed in history. The present urban landscape of Pakistan is partly inherited from history and is partly the result of contemporary development. There have been phases of urbanisation with changing modes of economy and technology. In fact, not only in Pakistan but in most parts of the world, urban evolution corresponds closely with the development in technology, changing phases of economies in the mode of their production and distribution, resulting in complete transformation in the pattern of distribution of urban centres. This has also resulted in a change in urban growth, morphology and functional bases of the towns. Historically, therefore, a number of phases in the evolution of the urban landscape of Pakistan can be identified.

1. Pre-historic phase, 2. Early historic phase, 3. Medieval phase, 4. Modern phase and 5. Contemporary phase

1. *The Pre-historic Phase*: A very early phase of urbanisation in Pakistan relates to a pre-historic phase, exemplified by the pre-historic cities like Mohenjo-Daro in Sindh, Harappa in Punjab and Mehrgarh in Balochistan. There are certainly other towns dating their origin to the pre-historic phase. But these are either not very prominent or occur in adjacent India. Large settlement sites like Dholavira in Kutch, or Lothal, a port city in Saurashtra, which have their origin in the pre-historic

phase of urbanism. Very little is known about the evolution of these settlements, but a great deal of details have emerged after the excavations of their sites. These towns are seen only in ruins, nonfunctional and have only a cultural and historical significance.

2. ***Early Historic Phase of Urbanisation***: This phase includes the towns which developed in the early historical period spanning from a few centuries BC to a few centuries AD. Pakistan has a large number of towns dating back to this period. Indeed, the entire region of Gandhara culture covering a large part of Khyber Pakhtunkhwa and a few ancient towns in the piedmont zone of the Himalayas come in this category. Peshawar, Charsadda, Taxila and Sialkot with several others come in this group.

   Some of these early historic towns have not only survived the ravages of times but even grown with times, while some others exist only in ruins. There are a number of cities in Pakistan which originated in pre-Christian or early Christian era, which have survived devastation and have even accumulated modern urban functions and attained pre-eminence in national affairs. Peshawar, a city controlling an international route and the Khyber Pass, known as Purushapura in ancient times, Multan, with its sun temple or Sialkot, the Sagala of the yore, in the piedmont zone of the Himalayas, a city believably razed by Alexander the Great, are all ancient towns which have survived, and stand out in the urban landscape of Pakistan. Unlike the above-mentioned towns, Charsadda and Taxila, though existing as administrative units, remain as ruins, as a pale shadow of their glorious past.

3. ***The Medieval Phase of Urbanisation***: The medieval phase is characterised more by the establishment of towns as a fort city, a garrison town, a defence post, an administrative headquarters, a trading centre, a place of pilgrimage, or simply the cross roads of long distance caravans. Most of the medieval towns were promoted by Muslim rulers and conquerors after the 10th century, during their route marches, conquest of a territory and its administration, or simply as a capital with royal residence. A few evolved as trading centres for exchange of commodities. The medieval towns, though important, were far and few between. There was no network of towns in Pakistan, as it is today. Hyderabad, Sikarpur, Nawabshah or Mirpur Khas in Sindh, and Lahore, Gujranwala, Sargodha, Pakpattan and Mianwali could be quoted as examples of medieval towns.

4. ***Modern Phase of Urbanisation***: The modern phase of urbanisation began after the arrival of the British which coincided with the exploitation of country's resources. The forests, fuel resources like coal, a variety of minerals like iron, manganese, copper, aluminium and mica, and many other minerals began to be exploited, processed and exported to Europe. This gave rise to mining and industrial towns. As the economy

grew, especially during the colonial period and the administration was streamlined by establishing a hierarchy of administrative areas, like divisions, districts, *tehsils* or *thanas*, a network of towns developed. Equally important was the arrival of modern industries, following the industrial revolution in Europe and a slow and creeping industrialisation in India. Spurred by the railways, developed by the British Colonial rulers for their own interest, mechanised industries, promoted by Indian entrepreneurs began to grow. This led to congregation of people in certain areas giving rise to towns.

The export of raw materials and import of processed goods for the Indian markets necessitated the development of ports, giving rise to cities like Karachi, Bombay and Calcutta. For defence, the British developed cantonments, often adjacent to the existing towns. But above all, they organised a framework of administration, with territorial divisions and sub-divisions, giving rise to a large number of tiny administrative centres which grew into larger towns. The British territorial organisation of administration which remains intact even today with some modifications and additions, produced hundreds of district headquarters which, with the accumulation of some additional secondary functions, were, in due course, classified as towns. The system of modern education, the development of industries, transport network and trading centres led to multiplication of urban centres. By the time, the British left India Pakistan had a well-developed network of transport and towns.

5. ***The Contemporary Phase***: For simplicity, the emergence of Pakistan as a nation state is taken to be the beginning of the contemporary phase. These are the towns which have either grown or developed as a planned city in independent Pakistan or have been considerably improved in Independent Pakistan. After Pakistan came into existence, the country began to develop its resources in a planned way for its own benefit. Though Rawalpindi as a cantonment, Faisalabad (earlier known as Lyallpur), as the headquarters of Chenab colony, and Sahiwal (earlier known as Montgomery), were developed during the British rule, before independence, are classed as contemporary towns, because of the absence of an associated old town and the planned nature of their morphology.

   No town exemplifies the contemporary phase better than the federal capital, Islamabad. The capital is developed in a planned way, keeping in view the needs and aspirations of the people of Pakistan, criticism of the plan of Islamabad and its growth notwithstanding. In fact, politicians, sociologists and economists of the liberal persuasion are the most vocal critics. They talk of the utility and benefit to the last person on the street and disregard the robust functioning of the city as the federal capital of an important nation.

8.1.1 Process of Modern Urbanisation: Europe vs. Pakistan

The process of rapid urbanisation began with industrial revolution in the 19th century in Europe. It saw an enormous expansion of manufacturing and international trade, facilitated with a secure market in the overseas colonies, which had no industrial base of their own. As the secondary and tertiary sectors grew in the economies of the western countries, there was a corresponding growth in the pace of urbanisation. For almost a century, the industrialised countries of the west remained the principal suppliers of manufactured goods, giving rise to a lot of allied activities, witnessing at the same time of emergence of metropolitan cities. With the shrinkage in the manpower need to engage in agriculture, the towns, with their potential to provide increasing employment to surplus labour released from agriculture, grew in number, became more populous and pushed the level of urbanisation to an all-time high.

Urbanisation, in the countries of Europe and the USA, has attained a stage of equilibrium, a situation where the cities don't attract any further migration from the countryside causing displacement from primary sector, nor do they induce any shift in population to the countryside depopulating the cities. The most developed countries of the world have reached a stage of urbanisation where more than 80 percent of the population lives in cities.

8.1.2 Process of Modern Urbanisation in Pakistan

The study of urbanisation in Pakistan and other South Asian countries presents a study in contrast with that of Europe. While by the beginning of the 20th century, much of Europe was not only industrialised but also largely urbanised, urbanisation in Pakistan, based on economic growth, especially industrialisation, was in its infancy. The largely agrarian economy supported by artisans, craftsmen and cottage industry, remained largely rural. A comparison of the level of urbanisation in Pakistan, India and the UK is very revealing. The countries in South Asia had not developed their own industrial base. The colonial administration of these countries had no economic policy to create an industrial base and expand the non-agricultural sector of the economy. Pakistan and other South Asian countries remained, almost exclusively, agricultural countries, and served as market for capital and even consumer goods manufactured in Europe, especially the UK. The towns that developed in Pakistan, as in other South Asian countries, were either administrative centres, cantonments or centres of distribution of goods manufactured overseas, and as centres of collection of some agricultural commodities like tea, jute, indigo and several varieties of oil-seeds, and in case of Pakistan even wheat. A few textile-producing towns like Bombay (now Mumbai), Ahmadabad and Kanpur developed, thanks to the domestic

TABLE 8.1 A Comparison in the Level of Urbanisation in the 20th Century between the UK, Pakistan and India

Year	*Level of Urbanisation in Percentage Terms*		
	England and Wales	*India*	*Pakistan*
1901	77	10.8	9.8
1911	78	10.3	9.5
1931	80	12.0	13.0
1941	N,A.	12.7	15.5
1951	80.8	17.3	17.8

Source: For Pakistan: 1. Census of Pakistan 1951, vol. I, Tables and Reports,, Karachi, pp.11–54; 2. Population Census of Pakistan 1972, Islamabad, and Pakistan Statistical Year Book 1981, p. 6. - For UK, Bruce Wood (1988) Urbanisation and Local Government, in A. H. Halsey (ed. 1988) British Statistical Trends, since 1900. A Guide to Changing Social Structure of Britain, Macmillan, London, pp. 322–356.

entrepreneurship. Notwithstanding this limited industrial development, much of the countryside remained rural with its agricultural landscape (Table 8.1).

The urban landscape of Pakistan remained dormant, as there was no growth of secondary sector; especially manufacturing and a few urban centres functioned largely as administrative and religious towns like Multan, cantonment or garrison towns like Rawalpindi, a large cultural centre with many educational institutions like Lahore or a port city like Karachi. Even these cities of Pakistan grew imperceptibly slowly till 1941, when during the wake of the Second World War, the cities became the focus of war efforts and grew faster than in the previous decades. The slow growth rate of these cities picked up only after 1951, which is considered a watershed in the demographic history of both Pakistan and India, when the urban population and growth of individual towns, both picked up with expanding economic base of towns and the additional of new urban centres, boosting the overall level of urbanisation.

From all accounts, seen either in terms of an overall increase in the level of urbanisation or in the growth of a few urban centres, urbanisation in South Asia in general and Pakistan in particular remained imperceptibly slow: and an accelerated growth of population as well of urbanisation began only after the mid-twentieth century (Table 8.2).

8.2 World Level of Urbanisation and the Place of Pakistan

It has to be emphasised that even today, the countries of the world are at various stages of urbanisation, corresponding to their level of economic development. The economically most developed countries, with their

TABLE 8.2 The Growth of a Few Selected Cities of Pakistan between 1891 and 1941

Year	*Towns*						
	Lahore	*Karachi*	*Rawalpindi*	*Multan*	*Sialkot*	*Peshawar*	*Hyderabad*
1891	176,854	98,195	73,795	74,562	66,087	63,079	54,569
1901	202,964	108,644	87,683	87,394	67,735	73,343	64,790
1921	281,781	201,691	101,142	84,806	71,008	79,427	73,951
1941	671,659	359,492	181,169	142,768	135,283	130,976	128,842

Source: For 1891 to 1941 – M. W. M. Yeats (1941) Census of India vol. I, pt. I table 1943, Govt. of India Press, Simla.

TABLE 8.3 Level of Urbanisation of Groups of Countries at Different Stages of Economic Development

Regions and Countries	*Percentage of Urban Population (2014)*	*Countries*	*% of Urban Population*
A) 1. World	53.6	The USA	81
2. More developed countries	78.0	The UK	82
3. Less developed countries	48.4	France	80
4. Least developed countries	30.9	Germany	75
B) 1. High Income countries	80.2	Japan	93
2. Middle income countries	50.7	China	55
3. Low income countries	30.3	Indonesia	53

expanded secondary and tertiary sectors, producing large volumes of goods and services, with a very high per capita GDP, as a rule, have a higher level of urbanisation. As a broad generalisation, it is assumed that the higher the per capita GDP, the higher will be the level of urbanisation.

Generally, the countries of the world are grouped as the more, less or least developed countries, also as high income, middle income and low income countries. Some examples follow (Table 8.3 and Table 8.4).

8.2.1 *Growth of Urbanisation in Pakistan after 1947* (The year Pakistan came into existence)

It has to be emphasised that Pakistan, before the secession of East Pakistan (now Bangladesh) in 1971, was a much larger country but with an abysmally low level of urbanisation, resulting from a very low urban development in Eastern Pakistan that was largely agricultural with only a few dominant centres like Dacca (Dhaka). The level of urbanisation in Pakistan (including and excluding the eastern wing) was as follows (Table 8.5):

The above table shows a pre-eminent position of the western wing of Pakistan vis-à-vis the eastern wing. The latter lagged considerably behind the western wing of Pakistan. East Pakistan had a measly low percent (10.4 percent) of the population living in urban centres.

8.2.2 *The Growth of Urbanisation in Pakistan from 1951 Onwards*

The growth of urban population as well as the multiplication of urban centres in Pakistan as seen in 1951, against the background of earlier years, marks a clear departure in the urban history of Pakistan. From a low of 17.7 percent of urbanisation in 1951, the country has come a long way, recording a significant growth, showing 32.3 percent urban population in 1988 and 38.6 percent in 2017. The post-independence urban history of Pakistan could be divided into two unequal halves. The first half covered a period of

TABLE 8.4 Urbanisation in the Countries of South Asia

Countries	*Percentage of Urban Population*	*Countries*	*Percentage of Urban Population*
1. Afghanistan	26.3	5. Pakistan	38.2
2. Bangladesh	33.5	6. Maldives	44.5
3. Bhutan	37.9	7. Nepal	18.2
4. India	32.4	8. Sri Lanka	18.3
South Asia including Iran	34.4		

Source: UNO Population Division, Dept of Economic & Social Affairs & World Urbanisation Prospects. The 2014 revision file 1 – Population of rural and urban areas at mid-year (thousands and percentage urban 2014. The actual urban population is omitted here).

TABLE 8.5 Level of Urbanisation 1901–1951

Year	*Percentage of Urban Population Including the Eastern Wing*[1]	*Level of Urbanisation Excluding the Eastern Wing*[2]
1901	5.1	9.8
1911	4.9	9.5
1921	5.4	11.3
1931	6.5	13.0
1941	7.9	15.5
1951	10.4	17.8

Source: 1. Census of Pakistan, Vol. I, Report & tables, Statement 2-J, p. 40.
2. Census of Pakistan 1972, Islamabad, Pakistan

25 years, from the emergence of Pakistan as an independent nation in 1947 to the division of the country and the birth of Bangladesh in 1972. This was a period riddled with unsettled conditions. Though un-conducive to urban growth, the period showed a remarkable urban growth as a result of the immigration of a large number of refugees from India. A large percentage of these refugees settled in cities. The only exceptions were the immigrant farmers of Punjab who settled on the land vacated by migrant landholders from Pakistan to India.

The second half is the period from 1972 to the present. The first half experienced the teething trouble of a newly created state. Besides setting up and activating a National Governance machinery and grappling with the refugee problem, created by the arrival of seven million people from India, Pakistan had to engage in war with its neighbour India causing unsettled conditions. The period closed with the secession of Bangladesh, an event that created disturbed conditions. The second half has not been less turbulent, characterised as it was by periodic domestic convulsions, like the change from democratic to military regimes, creating political instability.

The history of Pakistan including its urban history has been distorted by frequent disruptions in its democratic setup, hampering economic progress and creating unstable conditions that led to the decline of many academic and social institutions and the stagnation of urban life.

8.2.3 The Early Decades and the Impact of Refugees on the Growth of Urban Population

During the two decades (1951–61) and (1961–72), despite all odds, Pakistan recorded an unparalleled increase in its urban population, largely because of the huge influx of refugees from India during the wake of partition of India in the 1947 and even thereafter. In the first decade between 1951 and 1961, the urban population of Pakistan grew 61.3 percent whereas between 1961 and 1972, there was a seventy percent increase in the population of Pakistan (Table 8.6).

The single most important element in the increase in urban population, marked by crowding of cities, was the influx of refugees, the *mohajirs*, who arrived either from India or from East Pakistan, now Bangladesh, following the two historic events, first, the partition of India in 1947, and the second, the emergence of Bangladesh following the 1971 war. Both these events triggered a mass exodus from India post 1947 and from Bangladesh post 1971. The refugees who arrived from India or Bangladesh settled largely in the cities. In 1951, there were 7.2 million *mohajirs* (refugees) in Pakistan, of which five million (5.291 million) were in Punjab and the state of Bahawalpur, forming 25.5 percent of the total population of Punjab. In Karachi, the then federal capital of Pakistan, the number of *mohajirs* exceeded six hundred thousand (617,000) that formed 55 percent of the population of Karachi. The arrival of over seven million refugees rocked not only the economy of the country, diverting resources from economic development, but also had severe impact on city life by inflating urban population and severely

TABLE 8.6 Population Growth and Increase in Urban Population in Pakistan (1951 to 2017)

Year	*Population of Pakistan*	*Urban Population of Pakistan in %*	*Percentage Change in Urban Population*	*Year*
1951	33,740,157	5,985,497 17.7		1951
1961	42,880,378	9,654,572 22.5	+61.3	1961
1972	65,309,340	16,593,631 25.4	+71.8	1972
1981	84,253,644	23,841,171 28.3	+43.6	1981
1998	132,352,279	43,036,404 32.5	+80.5	1998
2017	207,774, 520	75,584,989 36.4	+75.5	2017

Source: Census of Pakistan, 1998 and 2017.

straining the existing infrastructure of the cities, at the same time. Within a decade (1951–1961), the urban population of Punjab increased by 53 percent (53.4 percent) against the backdrop of an increase in the total population of 24 percent. During the same period, the population of Sindh increased from 1.76 million to 3.16 million, an increase of around 60 percent.

8.2.4 Distribution of Refugee Population and Impact on Urban Growth

The impact of refugee population on urban growth can be gauged from the fact that in Sindh 82 percent of the refugees settled in cities, more specifically in Karachi and Hyderabad which formed roughly 54 percent of the total population of Sindh. In 1951, nearly half the population of the large cities in Pakistan was made of refugees. Around 57 percent of the population of Karachi, 43 percent of Lahore, 68 percent of Hyderabad, 37 percent of Rawalpindi, 70 percent of Faisalabad, 50 percent of Gujranwala and 69 percent of the population of Sargodha was composed of Muslim refugees from India.[1] More than a fourth of the population of Punjab was made of *mohajirs*. The most favoured districts were Lyallpur (Faisalabad), Lahore, Montgomery (Sahiwal), Sheikhupura and Multan. The concentration of refugee population was, no doubt, diluted in due course, and replaced by their descendants, who, by virtue of their being born in Pakistan, were no longer refugees. After the secession of Bangladesh, there was another influx of refugees from East Pakistan, of the population, usually termed Bihari Muslims. This influx could be called internal migration, as it was a shift or displacement of people from one part of Pakistan, i.e., East Pakistan to West Pakistan. A far higher percentage of growth of urban population, between 1961 and 1972, as compared to 1951–61, suggests the impact of migration from East Pakistan and the settling of the migrants in cities. While the general population increase during 1961–72 decade was 25.4 percent, the urban population during the same period grew roughly 72 percent.

The most affected region was Sindh province where a growth of 81 percent in the urban population was recorded. In fact, Sindh showed a consistently high growth rate of urban population right since 1951. It must be stated that no province of Pakistan remained unaffected by the influx of refugees. Even Khyber Pakhtunkhwa, then known as North-West Frontier Province, showed an increase in its urban population to the extent of 50.28 percent in 1961 and 57.6 percent in 1972, during the respective decades of 1951–61 and 1961–72.

8.3 Urban Population Growth during the Last Four Decades

The last four decades have proved very tumultuous in the history of Pakistan, affecting all spheres of socio-economic activity. While the adoption of a new constitution in 1973, which still provides the template for the country's government, brought a new enthusiasm and the hope of a fast growing economy, the nationalisation of many industrial units was a temporary dampener that hampered economic development and created a situation of political instability. The military regimes of the eighties and the early 21st century were, to say the least, unwarranted aberrations, and the periodic appearance of democratically elected governments, in the nineties, did not make significant impact on the economy and the urban situation.

The linkage between the growth of the economy and urbanisation is very strong; and as the growth of the economy faltered so did the process of urbanisation. It may be mentioned that in the normal course, the shift of population from the rural to urban areas is gradual, unless there is a rapid economic growth and an increase in the demand of manpower in the cities (Table 8.7).

8.3.1 *Un-uniform Growth of Urbanisation over the Decades*

The growth of urbanisation has not been uniform over decades, nor is it uniform all over Pakistan. What is apparent is that there is an accelerated growth of urban population as a result of both the increasing population

TABLE 8.7 Total Population, Urban Population and Percentage of Urban Population in Pakistan (1951–2017)

Year	*Total Population (in millions)*	*Urban Population (in millions)*	*Percentage of Urban Population*	*Annual Growth Rate of Population during the Inter-census Period*	*Annual Rural Growth Rate*
1951	33.74	5.98	17.7	-----	-----
1961	42.88	9.65	22.5	4.9	1.8
1972	65.30	16.59	25.4	4,8	3.4
1981	84.25	23.84	28.29	4.4	2.6
1998	132.35	43.04	32.5	3.5	2.2
2008	166.40	59.70	35.9	4.08	1.8
2014	188.00	72.50	38.6	3.3	1.32
2017	207.00	75.00	36.36	3.37	

Source: For the years 1951 to 1998, the figures are from the Census of Pakistan 1998. For 2008 & 2014, the figures are estimated by the Planning Commission of Pakistan, taken from Social Development in Pakistan, Annual Review 2014–15, Social Policy and Development Centre Karachi. The annual growth rate of urban and rural population is taken from G. M. Arif & Sabha Ibrahim (1998).The Process of Urbanisation in Pakistan 1951–81, The Pakistan Development Review, vol. 37,4 pp.507–22. The annual urban growth rate for 2008 & 2014 is according to Planning Commission of Pakistan taken from Social Policy and Development Centre, Karachi, Annual Review 2014–15, p. 3.

of existing cities, especially the large ones, till 1981. After 1981, the urban growth rate appears to have declined for a while but appears to have picked up again in 1998. During inter-census period between 1998 and 2008, as estimated by the Planning Commission of Punjab, urban growth reached a level of 4.8 percent.

The impression gains ground that as the growth rate of rural population has declined, the urban population growth rate after 1998 is accelerated. The urban population growth rate appears to have declined again after 2008, recording an annual growth rate of 3.3 percent, though Social Policy and Development Centre, Karachi, disputes the growth and suggests an annual growth rate of 2.8 percent between 2008 and 2014, concluding that urban population of Pakistan in 2014 was 72.5 million. It must be pointed out that the decline in the rural population growth rate is largely because of the decline in fertility rate and not because of rural–urban migration. The importance of the latter is more significant in the increase in the urban growth rate of the country and much less important in the decline of rural population.

For the growth of urban population, the most commonly acknowledged factors are natural growth, rural–urban migration, emergence of new urban centres following the growth of large rural settlements, accumulating a number of central functions, and graduating to be classed urban. There is a general understanding that the cities don't maintain their population by natural growth, and immigration from countryside, termed the rural-urban migration, is the principal cause of an unusually higher growth of cities' population. International immigration in large commercial or capital cities is an added reason for cities' growth. The locational preference of industries or other tertiary sector enterprises for cities is one of the principal reasons for cities' growth. As the economy grows, economic activities multiply, services grow, and so does the population of cities.

8.3.2 Interprovincial Variation in Growth Rate of Urban Population and Towns

There is a stark variation in the level of urbanisation and number of towns, between the provinces, in their growth pattern. As can be seen from Table no. 8.8, all provinces have not grown uniformly either in their percentage of their urban population or in the number of towns.

Sindh has always shown the highest level of urbanisation right since 1951, the first Census of Pakistan. In 1951, when Pakistan had 17.7 percent as an overall level of urbanisation, Sindh had 29.23 percent, higher even than Punjab which had only 17.37 percent of population living in towns, conforming to the national average. During every decade after 1951, Sindh has shown a high growth rate in urban population and also a higher level of

TABLE 8.8 Percentage of Urbanisation and Number of Towns in Provinces during the Different Decades

Provinces		*1951*	*1961*	*1972*	*1981*	*1998*	*2017*
Pakistan	1. % of urban population	17.7	22.5	25.4	28.9	32.51	36.37
	2. No of towns	227	306	375	409	509	620
KP	1. % of urban population	11.07	13.23	14.25	15.05	16.8	16.5
	2. No of towns	26	38	43	42	54	52
FATA	1. % of urban population		1.33	0.53	–	2.69	2.69
	2. No of towns	0	1	2	0	5	xx
Punjab	1. % of urban population	17.37	21.5	24.4	27.6	31.3	36.7
	2. No of towns	148	171	195	210	243	259
Sindh	1. % of urban population	29.4	37.85	40.44	43.31	48.75	52
	2. No of towns	36	66	104	129	160	237
Balochistan	1. % of urban population	12.38	16.8	16.45	15.62	23.84	27.5
	2. No of towns	17	29	30	27	46	61
Islamabad	1. % of urban population	–	–	32.26	60.0	65.7	50.5
	2. No of towns	–	–	1	1	1	1

1. Urban population as percent of the total population of the province
2. Number of urban places (Town Councils, Municipal Councils. Municipal Corporations and Cantonments)

urbanisation. But the most spectacular increase took place between 1951 and 1961 when the urban population of Sindh grew by 79 percent. This trend continued till 1972 when a decadal growth of 80 percent was observed. Till 1972, Sindh showed a higher growth rate of population as well as urban population.

The principal reasons of a higher growth rate and higher level of urbanisation in Sindh are twofold. First, an accelerated growth of Karachi because of its being an industrial as well as a port city added to an increase in the overall urbanisation of Sindh. Second, the concentration of the large population of *mohajirs*, the refugees, in Karachi and other towns like Hyderabad, pushed the level of urbanisation upwards. Though Punjab received a larger number of refugees from India, most of them, being farmers, were settled on land as farmers. A sizeable number, however, also migrated to Karachi, the latter being the federal capital and a cosmopolitan city, with far greater prospects of employment. In 1951, while the number of refugees enumerated in Punjab was 5.28 million, the number of refugees in Sindh (Sindh

TABLE 8.9 Comparison of Population Increase of Karachi, Lahore and the Urban Population of Pakistan, in percentage terms

Percentage Increase in Urban Population	*1951*	*1961*	*1972*	*1981*	*1998*	*2017*
Pakistan's urban population	–	+ 61.3	+ 71.8	+ 43.7	+ 80.5	+ 66.1
Population of Karachi	+ 176	- 79	+ 83.7	+ 48	+ 78	+ 59.6
Population of Lahore	+ 26.3	- 52.6	+ 67.4	+ 36	+ 74.3	+116.3

and Khairpur state and Karachi together) was 1.17 million. It also shows that there must have been a large scale migration of refugees from Punjab to Sindh, especially Karachi during 1951 and 1961. Third, the total population of Sindh being one-third that of Punjab in 1951, in percentage terms, the urban population showed a larger percentage (Table 8.8).

The contribution of Karachi to the overall urbanisation of Sindh is significant. Between 1941 and 1951, the population of Karachi grew from 387,000 in 1941 to 1,068,000 in 1951, a rise of 175 percent. This increase reflects addition of refugees from India. In subsequent decades too, Karachi showed a much higher percentage of increase in urban population, than either the national average, or the rate of increase of population of Lahore, the second largest city of Pakistan (Table 8.9).

In all the decades, except the inter-census period (1981–98), Karachi grew faster than other towns, though its contribution to Sindh's urban population always remained in a narrow band of 60 to 64 percent. Till 1972, the growth of Karachi was much faster than the urban population of Pakistan or any other town. After 1972, the importance of Karachi declined, following the shift in the federal capital to Islamabad. Thereafter, the growth of Karachi levelled to a national average of urban population increase. The cause for a much higher level of urbanisation of Sindh lies in the accelerated growth of Karachi, which forms over 30 percent of the population of Sindh.

Punjab's level of urbanisation, by virtue of it having more than half the total population of Pakistan, conforms to the national average. While the national average in urban population in 1998 was 32.5 percent, for Punjab it was 31.3 percent. Despite the low level of urbanisation of Khyber Pakhtunkhwa (16.8 percent) and of Balochistan (23 percent), the overall urbanisation level of Pakistan is still high. In terms of contribution of individual provinces to the urban population, Sindh's contribution to the total urban population of Pakistan is 35 percent, despite its population being just 23 percent that of Pakistan. Punjab conforms to the average and other provinces are under-urbanised.

8.3.3 Rural–Urban Migration

All over the world, existing towns grow by rural–urban migration. There may be exceptions in some cases where a new city or a capital or a new port is developed, in which case there is a wholesale transfer of offices and bureaucrats from other towns and offices, as has been the case with Islamabad. Though, even in newly established cities, the initial arrival of bureaucrats or technocrats is almost immediately followed by the arrival of skilled and unskilled manpower, moving from the rural areas to organise the informal sector and non-basic functions, like transport, wayside restaurants, laundry and hairdressers and a variety of other services.

The importance of rural–urban migration in the rising level of urbanisation of Pakistan can be judged from the fact that

> one in six persons living in cities and towns of Pakistan is a first generation migrant. Contrary perhaps to perceptions, this ratio is the highest in Punjab. However, in the province, over 80 per cent of the migration is in the form of flows within the province, either in the form of rural-urban or urban-urban migration. The latter movement is referred to as "leapfrogging, whereby a person leaves a small town and comes to a metropolitan city.[2]

The percentage of migrants in the urban population of the provinces varies between 15 and 20 percent, except Balochistan where the strength of the migrants in the urban population is just around 4 percent (Table 8.10).

The rural–urban migration, not so much the urban–urban migration, will continue for many more decades to come, though the percentage of migrants may not substantially change, as there will be generations of migrants, and the successors of earlier migrants, though having their roots in the countryside, would no longer be considered migrants. A disproportionate overload of population in the primary sector, where more than 60 percent of the country's population generates around 15 percent of the national GDP from agriculture, is not a viable proposition. The load of population on the rural economy will be alleviated only with the growth of secondary and service sector in the country, largely in cities, and consequently, an accelerated rural–urban migration will take place. This common phenomenon of rural–urban migration will continue till an equilibrium is reached and the increasing urbanisation reaches a level when the migration from the rural areas ceases altogether. Alternatively, the structure and the functional base of the rural landscape is transformed so much that the villages turn self-sufficient, with no surplus manpower. In such a situation, the push factor ceases to exist and a phase of rural–urban equilibrium is reached.

TABLE 8.10 Number of Migrants in Urban Areas

Provinces	*Number of Migrants (000)*	*Share of Urban Population in %*
Punjab	4,853	18.95
Sindh	2,585	15.42
Khyber Pakhtunkhwa	543	16.33
Balochistan	53	4.18
Pakistan	8,034	17.02

Source: Social Policy and Development Centre, Karachi, Social Development in Pakistan, Annual Review 2014–15, p. 3.

8.3.4 The Effect of Redefining Urban: The Count of Urban Population

Right since the days of partition, any settlement with a population of over 5,000 having a local self-government, like a municipality or a municipal committee, and with 70 percent of its population engaged in non-agricultural occupation, was considered urban. This definition still holds in neighbouring India. In Pakistan, these criteria were applied till 1972. During 1981 Census, only those settlements were identified as urban, which were governed by a local body like a Town Committee, a Municipal Committee, a Cantonment or a Corporation. Each urban centre had to fall under the ambit of one of these categories of local bodies.

In 1998, as calculated by the author, there were 507 towns classed under different categories. These were distributed as follows (Table 8.11):

TABLE 8.11 Categories of Towns in Pakistan as per Local Government Definition, Status 1998

Provinces	*Categories of Towns, as Defined by the Local Government, Status 1998*				
	MC	*TC*	*Cantonments*	*Corporations*	*Total*
Punjab	80	138	18	8	244
Sindh	33	112	8	4	157
K P	30	15	3	1	55
Balochistan	14	28	2	1	45
FATA	0	5	0	0	5
Islamabad	–	–	–	1	1
Pakistan	157	298	37	15	507

Source: Area and Population of Administrative Units by rural/urban -1951–1998 Census, Table I & II

Notes: MC – Municipal Committee, TC – Town Committee, Cantt – Cantonment, Corp – Corporation, K. P. Khyber Pakhtunkhwa (former NWFP), FATA – Federally Administered Tribal Area.

TABLE 8.12 Provincewise Growth in the Number of Towns between 1951 and 2017

Provinces	*1951*	*1961*	*1972*	*1981*	*1998*	*2017*
K P	26	38	43	42	54	62
FATA	0	0	2	0	5	5
Punjab	148	171	195	210	243	259
Sindh	36	66	104	129	160	237
Balochistan	17	29	30	27	46	61
Islamabad	–	–	1	1	1	1
Pakistan	227	304	375	409	509	625

Source: Calculated from Census Organisation Pakistan 1998, table 1: (General), table 2: NWFP, table 3: FATA, table 4: Punjab, table 5: Sindh, table 6: Balochistan, table 7: Islamabad, all under the heading "Area & Population of Administrative Units by Rural/Urban" – 1951–1998 Censuses, Census 2017.

The impact of the introduction of a new set of criteria, solely in terms of the existence of a local self-body, Municipal Council, Town Council, Cantonment or a Corporation, is apparently seen in a decline in the decadal growth rate of the number of towns (Table 8.12).

While in the previous decades starting from 1951–61, there has been a substantial growth in the number of towns, in 1981 only 24 new towns were added to the earliest list, a growth of just 6.2 percent. A very strict application of the new criteria led to minimising the number of settlements that graduated to the category of towns. A large number of villages, though, grew in population, could not develop their own local self-government body and remained in the category of a village. Besides, a few, indeed a very few, settlements that were classed as towns in 1972, were declassified to revert to their status of a village. The introduction of the new set of criteria definitely had a restraining effect on the increase in the number of urban settlements. For example, a number of small towns in sparsely populated districts of Balochistan were declassified. Dalbandin (1972 – P 2862), Duki (1972 – P 1787) in Loralai district, Barkhan (1972 – P 1610) in Barkhan district, Musakhel (1972 – P 917) a mining town in Musakhel district, Qila Saifullah (1972 – P 835) and Muslim Bagh (1972 – P 2694) both in Saifullah district, were declassified. The paradox was that even the district headquarters of Balochistan was declassified. The most significant was the town of Dera Bugti, with a very small population of not even a thousand, though a district headquarters was excluded from the list of towns. Such cases of declassification were rare in Sindh, but not uncommon in Khyber Pakhtunkhwa. To take some examples, Chitral and Drosh in Chitral district and Bat Khela in Malakand protected area were taken off from the list of towns.

An important aspect to be mentioned in this context is the absence of any urban centre in some of the districts of Khyber Pakhtunkhwa. Districts like

Shangla, Buner, Kohistan and Battagram had no urban centres in 1998, the year of the census in Pakistan. The most important settlement in FATA that was declassified in 1981 was Parachinar, which was in the urban category in 1961 and 1972. Parachinar has a history behind it. It was here in the lap of the Spin Range that the alignment of the famous Durand Line was settled in 1893. Parachinar reappeared as a town in 1998.

A positive consequence of the statutory requirement of a local self-body (TC, MD, Corporation or Cantonment) is the reorganisation of the administration of many settlements that acquired the status of a town council (TC) and joined the rank of towns. This happened specifically in Sindh where 32 towns appeared in 1998. Most of these towns had a small population ranging from 5,000 to 10,000 but a few had a population of even less than 5,000 people like Bulri (TC) in Hyderabad district, Kario Ghanwar (TC) in Badin district or Keti Bunder (TC) in Thatta district. Similarly, many towns in Sanghar district emerged newly in 1998 with a population of less than 5,000.

During the last two decades, there has been a great deal of change, in the number of towns as well as the urban population (Table 8.13).

The Census has compiled the population of urban localities according to local self-bodies. These are Metropolitan Corporations, Municipal Corporations, Municipal Councils, Town Councils and Cantonments. In the process, the number of million-cities is disturbed. For example, Quetta is not treated as a Million City, nor is Hyderabad (Sindh). Quetta is split into two parts, Quetta Metropolitan Corporation (907,925) and Quetta Cantonment (91,460). Similarly, Sindh is shown to have only one million city, and that too under the name District Municipal Corporation Korangi (1,071, 514). More striking is the fact that Karachi is split into ten District Municipal Corporations, none of which has a population of over one million. At the other end is Punjab with seven million-cities instead of the usual five. Lahore is split into three million cities with three distinct local self-governments, viz., (1) Lahore Metropolitan Corporation (part of Lahore city Tehsil), (2) Lahore Metropolitan Corporation (part of Model Town Tehsil) and (3) Lahore Metropolitan Corporation (part of Shalimar Tehsil).

This technical definition, though logical, appears to distort the common sense understanding of the population of these cities. To know the actual population of Karachi, as commonly understood, one has to add up the population of all the district Municipal corporations.

Distribution of Towns: Pakistan has over 600 large, medium and small towns. Of these, about 80 percent occur in two of the most urbanised provinces, viz., Punjab and Sindh. Balochistan and Khyber Pakhtunkhwa, on the other hand, have 60 towns each. Punjab has over half (53.45 percent) of the total urban population of Pakistan living in 259 towns. This suggests that Punjab has more of larger towns. In contrast, Sindh with 33 percent of the

TABLE 8.13 Distribution of Urban Localities (Urban Centres) in 2017, in the Provinces in Different Population Range

Provinces	*Group of Towns According in Different Population Range*									
	<1,000.000	*500,000 to 1,000,000*	*200,000 to 500,000*	*100,000 to 200,000*	*50,000 to 100,000*	*25,000 to 50,000*	*10,000 to 25,000*	*5,000 to 10,000*	*5,000*	*Total*
Punjab	07	07	18	34	65	76	50	02	–	259
Sindh	02	14	17	19	27	55	81	19	04	238
Balochistan		01	01	03	08	23	19	04	02	61
Khyber Pakhtunkhwa	01	01	03	06	15	24	07	04	02	62
Islamabad	01	–	–	–	–	–	–	–	–	01
Total	11	23	39	62	115	178	157	29	08	621
% of Total	1.6	3.7	6.2	10	18.5	28.6	25.3	4.66	1.3	99.8

Source: PBS, Pakistan Census Organization; Final Results of 2017 Census, Table-2. Urban Localities by their population size and their population by sex, Annual Growth rate and the Household size. Number of Million- cities is in parentheses.

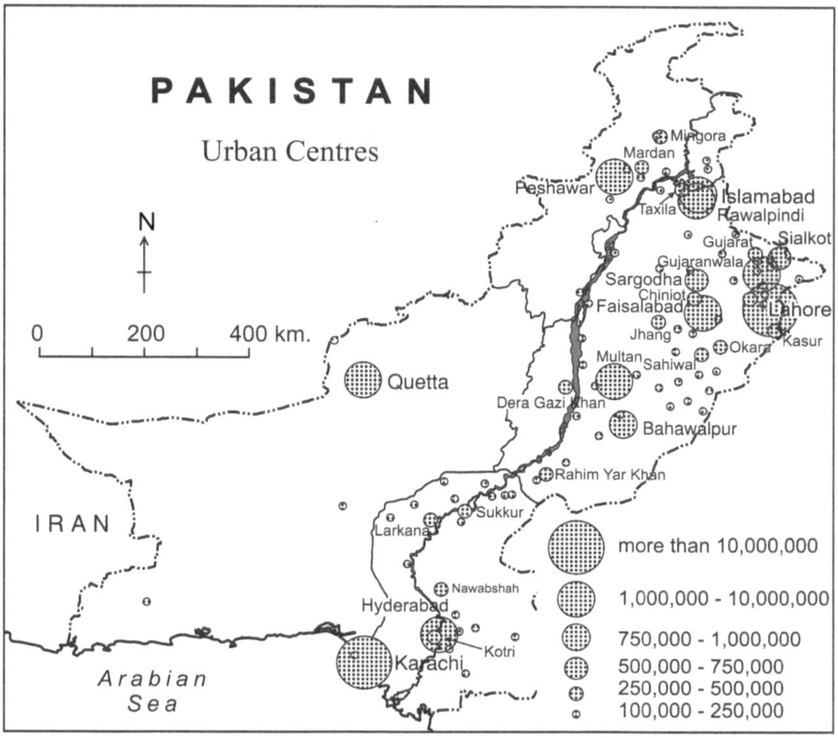

FIGURE 8.1 Urban Centres in Different Population Range (2017)

total urban population of Pakistan and 38 percent of the total number of towns suggests the presence of a large number of small towns.

The largest number of urban centres is in the population range of 25,000 to 50,000. The median population of the towns is around 32,000. Seventy percent of the towns have a population ranging from 10,000 to 100,000, and only 110 cities in Pakistan have a population of over 100,000. More than 50 percent of the towns have a population ranging from 10,000 to 50,000.

It may be noted that more than half the total urban population, (roughly 55 percent) of Pakistan lives in the ten-million cities of the country. And, Karachi and Lahore, the first two-million cities accommodate more than one-third of the total urban population of Pakistan.

8.4 Regional Pattern of Urbanisation

The contemporary spatial pattern of urbanisation in Pakistan is largely inherited from history. The patterns are strengthened in most cases, while new clusters based on more recent economic activities are added. The three or four important focal points that form the urban magnets, Lahore, twin city Rawalpindi–Islamabad, Karachi and Peshawar still have a dominant position in the politico-economic landscape of Pakistan. These were important even during the British rule and their importance has not diminished. The important urban centres ever remained tethered to transport routes and the historically known Grand Trunk Road, from Calcutta to Peshawar, known in Pakistan as the Sher Shah Suri Road.

According to 2017 Census, over 75 million people, more than a third (36.36 percent) of the population of Pakistan, live in towns. This represents a considerable increase from 17.7 percent in 1951, just after the emergence of Pakistan as a state. During the last two decades, from 1998 to 2017, there has again been a significant increase in the urban population of Pakistan rising from over 45 million (45,497,048) in 1998 to over 75 million (75,548,989) in 2017, an increase of 66 percent. This also meant a rise in the level of urbanisation, from 34.37 percent in 1998 to 36.36 percent in 2017.

The urban population of Pakistan, as indeed of any other country in the world, is not uniformly distributed in all regions, and there are vast regional variations. The level of urbanisation in Pakistan varies from 0 percent in certain hilly areas of very low population density of Khyber–Pakhtunkhwa to almost 100 percent, as seen in the case of Karachi district.

Comparing the level of urbanisation of the provinces of Pakistan, it is seen that more than half (53.45 percent) of the urban population of Pakistan is concentrated in Punjab. And, one-third (32.90 percent) of the total urban population occurs in Sindh. Balochistan and Khyber Pakhtunkhwa, together, account for 12 percent, and Islamabad just over 1 percent (1.34 percent) of the total urban population (Table 8.14).

Punjab has its due share of urban population in proportion to total population. It has 53 percent of the country's population, and the percentage of its urban population is also 53 percent of the total urban population of Pakistan. Thus, the level of urbanisation of Punjab closely corresponds to the national average. Both, Pakistan and Punjab, have each 36 percent of their population living in urban centres. In contrast, Sindh has a larger share of urban population, thanks largely to Karachi. It has only 23 percent of the country' population, but 33 percent of the urban population of Pakistan. Sindh is comparatively over-urbanised. The urban population in Khyber Pakhtunkhwa and Balochistan is relatively low. Khyber Pakhtunkhwa with 17 percent of Pakistan's population has only 7.7 percent of the country's urban population. It is comparatively under-urbanised. The most depressing case is that of Balochistan. A province with 43 percent of the area of Pakistan and 6 percent of its population has less than 5 percent (4.5 percent) of the country's urban population, though the level of urbanisation is relatively high, albeit not reaching the national urbanisation level of 36 percent. A 27 percent urbanisation of Balochistan is, in fact, deceptive, and

TABLE 8.14 Distribution of Urban Population in the Provinces of Pakistan, a Comparative View (2017)

Provinces	*Total Population of the Province in 2017*	*% of Total Population of Pakistan*	*Urban Population of the Province*	*Urban Population of the Province as % of the Total Urban Population of Pakistan*	*Level of Urbanisation of the Province*
Khyber Pakhtunkhwa	35,525,047	17.09	5,871,532	7.77	16.52
Punjab	110,012,442	53.00	40,387,298	53.45	36.71
Sindh	47,886,051	23.04	24,910,458	32.90	52.00
Balochistan	12,344,408	5.94	3,400,876	4.50	27.54
Islamabad	2,008,572	0.96	1,014,625	1.34	50.50
Pakistan	207,774,520	100.00	75,548,989	100.00	36.36

Source: PBS, Final results of the Census of 2017.

attributed to the province's low population, and a few towns including the million-city of Quetta make it appear moderately urbanised though below the country's average.

8.4.1 Urbanisation and Urban Centres in Khyber Pakhtunkhwa

Broadly speaking, the dry and mountainous western half of Pakistan, spread over Balochistan and Khyber Pakhtunkhwa, presents a bleak urban landscape. Except a string of cities along the traditional Grand Trunk Road, and some others like Peshawar, a million-city, lying in the fertile Kabul Valley, or Mingora in Swat Valley, or some others like Abbottabad along the Karakoram Highway, there are no major towns. There is, however, a string of towns like Charsadda, Mardan, Nowshera, Swabi and several others in the Kabul River Valley (the Peshawar Valley) that extends eastward, from Peshawar to Attock. This urbanised belt with many historic cities, along the Grand Trunk Road, is the economic and cultural core of Khyber–Pakhtunkhwa. Agriculturally prosperous, moderately industrialised and lying at the northern extremity of the Sher Shah Suri Road, the area is not only well urbanised but strategically the most important region of Pakistan. It controls and dominates the Khyber Pass, the gateway to Pakistan from Central Asia and Afghanistan (Peshawar is discussed separately as one of the million-cities of Pakistan).

The extreme northern part of Khyber Pakhtunkhwa, falling under Chitral district, a mountainous, hilly and glaciated high relief terrain, with an area of around 15,000 km^2 (14,850 km^2) with a population density of just around 20 persons to a square kilometre, has only one town represented by Chitral, the district headquarters, having a population of 49,780 (2017). So is the case of Upper and Lower Dir, Upper Swat, Kohistan, Battagram and Mansehra, all representing the northern and the north-eastern part of Pakistan, just south of Gilgit region. All these hilly districts have only their headquarters that are counted as urban centres. Dir, Timergara, Mingora, Dasu, Battagram and Mansehra are headquarters of different districts. Of these, Mingora, the third largest city of Khyber Pakhtunkhwa, with a population of 331,377 (2017), has an ancient urban base. Presently the headquarters of Swat district, it is an important city in northern part of Khyber Pakhtunkhwa with many Buddhist monuments. With a very pleasant summer season, the presence of fruit orchards and high altitude pastures in the vicinity, Mingora has the potential of turning into tourist paradise.

South of Peshawar, in Khyber Pakhtunkhwa, the land, bordering FATA, is hilly. In this area, a few towns that exist are administrative centres. Of these, Kohat, Bannu and Dera Ismail Khan are the principal towns. Dera Ismail Khan with a population of 212,324 (2017), the fourth largest town of Khyber Pakhtunkhwa, after Peshawar, Mardan and Mingora, caters to the Indus Plain area and to the piedmont zone of the western mountainous region, through a number routes radiating from the city, reaching in its hinterland as far west as Wana and in Waziristan.

The southern part of Khyber Pakhunkhwa is equally bleak, except Kohat and Bannu districts both of which, though located in a piedmont zone, have more than half their areas covered with alluvial plains of the tributaries of the Indus, and enjoy a mountainous climate, good connectivity and some benefit of irrigation, with the result that some small towns have developed besides the district headquarters. Kohat is a cantonment town. During the British period, troops were stationed here to suppress the tribal unrest in the area, or mount a punitive expedition against them. Known largely as a town of the Bangash tribe, the people in the town speak not only Pashto but also Hindko. The city has a functioning airport. Bannu is another city lying in the centre of South Khyber Pakhtunkhwa.

8.4.2 Urbanisation in Balochistan

Balochistan is the least urbanised province of Pakistan. It has 4.6 percent of the total urban population of the country, less than what is warranted by its share of total population of 12.3 million (2017), which is 5.94 percent of the total population of Pakistan. About 30 percent of the urban population of the province is concentrated in Quetta, the capital of Balochistan, having a population of 1,001,205 (2017). In its vast expanse of territory, there are just a couple of towns with a population of more than 100,000. The largest town of Balochistan, next to Quetta is Turbat with a population of 213,557 (2017). The town lying in the western part of Balochistan, on the route connecting Iran, is an important trading centre, known especially for its wholesale market of dates. Khuzdar is another important town, the headquarters of the district of the same name. Lying on the principal Quetta–Karachi route, it has a population of 182,927. It is also an industrial centre. Hub with a population of 175,376 is the terminal point of the route leading to Karachi and lying just around 20 km northwest of Karachi. Chaman, a border town on Pakistan–Afghanistan border, around 125 km northwest of Quetta is the last of the five towns in Balochistan which recorded a population of over 100,000 in 2017. Besides being a transit town, it is known for the large volume of trade it handles.

A large number of urban centres are towns only by classification having a municipal committee or a town committee and have a relatively poor functional base. Some are only district headquarters and have a population ranging between of 20,000 to 50,000. There are a few, which have a population of less than 20,000. One town that is also the headquarters of the district of the same name, with a population of less than even 5,000, is Ziarat with a population of 3,392 (2017).

Balochistan, with 36 districts has 61 towns of which a few like Quetta, Zhob, Loralai and Ormara, have also cantonments, which are counted separately as towns. The most notable example is Gwadar district which occupying a narrow coastal strip, extends for over 300 km, both east and west of Gwadar port. It has four urban centres including Gwadar. The three port cities Ormara, Pasni and Jhiwani, occupying three promontories, each having an air strip, are strategically significant as naval ports. Gwadar, the main port city, and the district headquarters, is an important seaport on the Arabian Sea coast, next to Karachi. Located on a promontory between two natural bays, the Gwadar East and the Gwadar West Bay, it has lately grown in importance, as the proposed terminus of the China–Pakistan Economic Corridor, a trans-Pakistan – trans-Himalayan route connecting Gwadar on the Arabian Sea coast, to Xinjiang Uigur, the autonomous province of western China. The port of Gwadar in Balochistan is leased to China, which is developing the port to its full potential. Under the China–Pakistan–Economic Corridor (CPEC) agreement, the port is to be linked to Karakoram Highway via Rotedoro in Sindh and National Highway No. 5, leading further across the Khunjerab Pass to China.

A high level of urbanisation of Balochistan is very deceptive. It is the low density of population, punctuated by a few towns that give the high figure of urbanisation. Gwadar, with the remaining three small port cities, account for more than half (54 percent) of the total population of the region. So is the case with Quetta, the provincial capital with 75 percent of the population of the district. Quetta alone, accounts for 36 percent of the total urban population of Balochistan. Quetta (1,001,205–2017), Khuzdar (182,927–2017), Turbat (213,557–2017) and Chaman (123,191–2017), have, together, almost half the total urban population of Balochistan. Quetta. Khuzdar, Turbat, Hub, Chaman (a border town between Quetta and Afghanistan), Sibi, Gwadar, Zhob, Usta Muhammad and Dera Murad Jamail, all district headquarters, are the ten important towns of Balochistan, which together account for 63 percent of Balochistan's urban population.

8.4.3 Urbanisation in Punjab and Sindh, the Provinces with a Well-developed Urban Landscape

Together, the two provinces, Punjab and Sindh, have 80 percent of the total population of Pakistan. Of the total urban population of Pakistan, Punjab has a larger share of 48 percent as compared to 32 percent of Sindh. In terms of level of urbanisation, Sindh is more urbanised than Punjab, with more than 50 percent (52 percent) of its population living in urban centres, as compared to 36.7 percent of Punjab.

8.4.3.1 Spatial Pattern of Urbanisation in Punjab

With 55.6 percent of the total population, 53.9 percent of the total urban population and 48 percent of the total number of urban places (259 out of 509) of Pakistan, almost half the total number of urban places is placed in the population range of 25,000–100,000. The province has five million cities and nine having a population between 500,000 and 1,000,000.

Almost all the towns in Punjab have a population of over 10,000. The largest number of towns (29.34 percent) is in the population group 25,000–50,000, and more than half (55.4 percent) of the total number of towns is in the population range of 25,000–1000,000. Out of the ten million-cities in Pakistan, Punjab has five (Lahore, Gujranwala, Rawalpindi, Faisalabad and Multan). Thus, urban population distribution of the towns in Punjab is skewed towards a higher population range where, as in Sindh, Balochistan and Khyber Pakhtunkhwa, it is skewed towards lower population range with a large number of small towns.

Though Punjab has a relatively better developed urban landscape, both, in terms of the distribution of towns in space and the functional character of cities, inside Punjab, however, there are wide contrasts depending on the density of population and economic prosperity, based on the development of industry and trade. Thus, there are regions with lower agricultural and economic productivity and lower population density, with lower level of urbanisation. There are other areas with high concentration of economic activity and better connectivity and more urbanised features. In addition, there are areas which have emerged during the last 60 to 70 years with intensive resource development, attracting industries and promoting growth of towns. One may group the urbanised areas into three classes.

TABLE 8.15 Number and Percentage of Urban Localities in Different Population Groups in Punjab

Population Size Group	*No. of Towns*	*No. of Towns as % of Total no. of Towns in Punjab*	*Population Size Group*	*No. of Towns*	*% of Towns in the Group of the Total no. of Towns*
< 5,000	00	00.00	50,000 – 100,000	65	25.09
5,000–10,000	02	00.77	100,000 – 200,000	34	13.12
10,000–25,000	50	19.30	200,000 – 500,000	18	06.94
25000–50,000	76	29.34	500,000 – 1,000,000	09	03.47
			>1 million	05	99.95
Total No. of Towns				259	99.01

1) Areas with a relatively high level of urbanisation
2) Areas with a relatively low level of urbanisation
3) Areas with recent accelerated economic and urban growth

The Highly Urbanised Belt of Punjab: This is the northern east–west extending belt of Pakistan along the Grand Trunk Road, from Lahore to Peshawar, with a string of cities. This linear belt takes in its fold the districts of Lahore, Sialkot, Gujranwala, Jhelum, Rawalpindi and Sheikhupura, extending as far south as Faisalabad and as far as Attock in the west, with Islamabad, the federal capital, in the vicinity. This highly urbanised belt extends further west, and merges with the cluster of cities (Nowshera, Mardan and Charsadda with Peshawar at the apex) of Khyber Pakhtunkhwa.

This highly urbanised belt, with four-million cities (Lahore, Gujranwala, Rawalpindi and Faisalabad) is the political and strategic region of Pakistan, presided over by Islamabad, the federal capital, attached to the belt like a head to a spine. This belt is also industrialised with textile and cement as the main industries. Of the five-million cities of Punjab, four (Lahore, Faisalabad, Gujranwala and Rawalpindi) are located in this belt. But most important, the region is home to the Army General Headquarters located at Rawalpindi. Lahore, a mega city and the provincial capital of Punjab, is the heart and soul of Punjab, and the literary and cultural capital of Pakistan.

Areas with Low Level of Urbanisation: The western part of Punjab, especially Sind Sagar Doab, lying between the Jhelum and the Indus, also known as the Thal Desert is a poorly developed region of the province. Deprived of early canal irrigation, unlike the Chhaj *Doab*, the area is agriculturally poor, industrially undeveloped, and with a desert-like condition, the least preferred zone of Punjab. It has not only a low density of population but also a very low level of urbanisation. The north-western part of Punjab, appearing like a triangle with three sides formed by the Indus in the west, the Salt Range in the south and the Grand Trunk Road in the north is occupied by the Potwar Plateau, a ravined terrain with relatively poor agriculture. The Chakwal district, with its headquarters at Chakwal, is one of the least urbanised districts of Punjab. The Indus, unlike its tributaries, remained unharnessed for long for irrigation, except in Sindh where the Sukkur barrage provided the lifeline for the entire province of Sindh. The result was a low agricultural development and stagnation in the growth of urban centres. Chakwal, Bhakkar, Layyah, Dera Ghazi Khan, right up to Rajanpur, and Muzaffargarh districts in the south, all bordered by the Indus on their west, have a poor growth of urban centres. The recent construction of a few barrages, viz., Jinnah, Chashma and Taunsa barrage, have infused some agricultural growth. But the region, having suffered a long period of neglect, is still devoid of large towns. The northern edge of south-eastern part of Punjab has no significant urban centres. What is observed, instead, is a large number of textile units that have promoted towns of moderate size like Kasur and Okara.

The south-eastern part of Pakistan, coinciding with the former princely state of Bahawalpur is a desert-like area, though transfer of water across the rivers, in the northern part, has helped ameliorate the situation, making it moderately green. The region has not yet experienced the level of prosperity that the *Rechna Doab* has. Bahawalpur, Lodhran and Vehari are a few towns in the region, though over all urbanisation is low, notwithstanding Bahawalpur, a large town, the capital of the former princely state of the same name.

The Emergence of the Highly Urbanised North Central Punjab: The district of Faisalabad was the earliest beneficiary of Chenab irrigation project. Then, known as Lyallpur, the city itself was the administrative centre and the headquarters of Chenab colony. It was a small town of 9,000 in 1901 and even as late as 1931, it had a population of just around 40,000 occupying 8th rank among the towns of Punjab. The agricultural prosperity, following the Chenab canal irrigation, supplemented subsequently from the Jhelum, brought prosperity to this tract. After Pakistan came into existence, Faisalabad city and region, gained importance; and today, it is the second largest town of Punjab, competing with Lahore, Faisalabad region is a rich agricultural-cum-industrial region, with the dominance of textile industry. The establishment of a textile university at Faisalabad has further enhanced the importance of the region and the city.

8.4.3.2 Urbanisation in Sindh

With 52 percent of the population living in towns and cities, Sindh is the most urbanised province of Pakistan. It must, at once, be pointed out that 58 percent of the urban population of Sindh lives in Karachi and its cantonments. The province has 237 urban places of different categories.

Sindh experienced a rapid increase in the number of towns during a period of five decades, faster than the increase in the number of towns in other provinces, and yet slower than the growth of urban population in Sindh. To take an example, while the number of towns in Punjab grew from 148 in 1951 to 259 in 2017, a growth of 75 percent, the towns in Sindh grew from 36 in 1951 to 237 in 2017, a more than six-fold rise. But urban population of Sindh grew still faster, rising from 1.77 million in 1951 to 15.7 million in 1998, and 25 million in 2017. This enormous growth in urban population can be safely attributed to a very much accelerated growth of Karachi, which has a population of over 14 million in 2017.

The distribution of towns belonging to different size groups is highly skewed and there is a predominance of small towns (Table 8.16).

TABLE 8.16 Number of Towns (leaving Karachi) in Different Population Size Groups in Sindh (2017)

Population size group	*No. of towns*	*No. of towns as % of total no. of towns in Sindh*	*Population size group*	*No. of towns*	*% of towns in the group of the total no. of towns*
< 5,000	04	1.68	50,000 – 100,000	27	11.40
5,000–10,000	19	8.00	100,000 – 500,000	36	15.18
10,000–25,000	81	34.17	500,000 – 1 million	13	05.48
25,000–50,000	55	23.20	>1 million	02	00.84
Total no. of towns				237	99.95

Source: PBS, Census of Pakistan 2017, table 2, PBS.

It can be observed that about two-thirds (67 percent) of the towns in Sindh have a population of less than 50,000, about 44 percent, have a population of less than 25,000. Many of these towns have graduated from rural to urban, during the last several decades. In some large districts, like Sanghar, many settlements away from the district headquarters were given the status of a local self-government like, town committees in successive decades. In 2017, Sindh had 36 towns with a population of over 100,000. Many districts headquarters don't have a city with a population of over 100,000.

8.4.3.3 The Spatial Concentration of Urban Population

Broadly speaking, much of agricultural prosperity, economic activity and even urbanisation are concentrated in the longitudinal core of Sindh, formed by the river Indus. Lying between the dry eastern belt, juxtaposed to the Thar Desert of India, and the equally dry region in the western part under the shadow of the Kirthar Range, is the central longitudinal belt straddling across the Indus. With the East and West Nara Canals, one on each side of the Indus, and running parallel to it, and irrigating land right up to the Indus Delta, Central Sindh is agriculturally the most prosperous region. Parallel to the Indus are transport lines, railways and roads on either side that facilitate smooth movements of people, goods and services all across the region. The efficient flow of goods and people, especially from the city port of Karachi has made the region easily accessible. This has promoted the growth of a string of towns in this central belt. The twin city of Kotri (P-259,358) and Hyderabad (P-1,734,309), Nawabshah (P-279,688) Moro and Dadu (P-171,191) and further north Larkana (P-490,508), together with Khairpur (P-183,181) and Shikarpur (P-195,437) make a highly urbanised linear zone. Jacobabad (P-191,076), further north, functions as a control point between Sindh and Balochistan, and serves as the terminus for Jacobabad–Quetta railway route.

In terms of urban clusters, one may identify a few urban clusters. One is in the extreme north of Sindh, comprising Sukkur, Jacobabad and Shikarpur. At the other end, in the extreme south, is another cluster placed around Karachi and Hyderabad. Between these two extremes, in the centre, is the major city of Nawabshah, the headquarters of the district of the same name, now rechristened as Shaheed Benazirabad.

Most of the large cities and urban centres occupy a linear belt, on the two sides of Indus. The Karachi–Hyderabad Corridor, extending north up to Nawabshah, is the most industrialised, the richest and at the same time, most urbanised part of Sindh. In contrast, the dry eastern part of Sindh consisting much of Khairpur, Sanghar, Umarkot and Tharparkar – all occupying the dry eastern part, facing the Thar Desert, are the least urbanised parts of Sindh. Tharparkar, the dry south-eastern district of Sindh has only eight percent of the people living in the towns, and even Mithi, a district headquarters, has a population of only 47,135 (2017). Thatta, the district occupying the Indus Delta as well as Badin on its east have a low percentage of urban population, former having 11 percent and the latter 16 percent of their population living in towns.

8.4.3.4 The Two Important Urban Corridors of Pakistan

1. The Northern Urban Corridor – It extends from Lahore to Peshawar, with its axis formed by the Grand Trunk Road. And a string of cities, like Lahore, Gujranwala, Jhelum, the twin city of Rawalpindi and Islamabad, and further west Mardan terminating at Peshawar on its north-western extremity.
2. The Second Urban Corridor occupying a belt in the extreme south of Sindh, extending between Nawabshah and Karachi, with Hyderabad, appearing halfway in the middle and forming an important transit point and transport hub.
3. The Nowshera – Mardan – Peshawar Urban Corridor Linked by motorways, along the Grand Trunk road and connected with the main railway line, this is also an important corridor and a vital region of Khyber–Pakhtunkhwa.

The two urban corridors are not quite similar. The Lahore–Rawalpindi–Islamabad–Peshawar corridor is a part of the traditionally well-recognised transcontinental northern route that was the axis of communication of undivided India linking the fertile Gangetic Plain, Delhi, and the plain of Punjab. Peshawar, the gateway to the sub-continent on the one end, and Lahore, for some time the capital of the Mughal Empire, and the present capital of Punjab province, on the other, with the federal capital Islamabad and Rawalpindi, the general headquarter of Pakistan army, thrown in, this corridor represents the political, strategic and cultural core of Pakistan.

The Karachi – Hyderabad – Nawabshah corridor, with Pakistan Steel Mill, has a series of power plants and petroleum refineries, together with international trade passing through Karachi. It has also some premier financial institutions, including the Karachi Stock Exchange and cotton market. This corridor is a reflection of the economic growth of the country. In fact, Karachi contributes believably, a disproportionately large share to the national revenue and the country's GDP.

8.5 The Million-Cities of Pakistan

The first ten cities of Pakistan, also the million-cities, as recorded in 2017 are as follows (Table 8.17).

The ranks have been changing with the rise or decline in the fortunes of specific cities. Before 1947, Lahore had the highest population in Western India and Karachi had a second rank. The first ten cities and their rank are given below (Table 8.18).

While the first four cities viz., Karachi, Lahore, Faisalabad and Rawalpindi have maintained their ranks, Multan and Hyderabad have slid down the ladder. Their places are taken by Gujranwala, a fast growing industrial and commercial city and Peshawar, the capital of Khyber Pakhtunkhwa.

8.5.1 A Bird's Eye View of the Principal Cities of Pakistan

8.5.1.1 Karachi

Located on the Arabian Sea coast (24°51'N and 67°02' E), Karachi, the provincial capital of Sindh, is the largest city of Pakistan with a population of 15 million. The city originally known as Kolachi, a fishing hamlet and a

TABLE 8.17 Million-Cities of Pakistan (2017)

Rank	*Town*	*Population (2017)*	*Rank*	*Town*	*Population*
1	Karachi	14,916,456	6	Peshawar	1,970,042
2	Lahore	11,126,285	7	Multan	1,871,843
3	Faisalabad	3,204,726	8	Hyderabad	1,734,309
4	Rawalpindi	2,098,231	9	Islamabad	1,009,832
5	Gujranwala	2,027,001	10	Quetta	1,001,205

TABLE 8.18 Change in the Ranking of Major Towns of Pakistan between 1951 and 2017

Town	*Before Independence 1947*	*1951*	*1961*	*1972*	*1981*	*1998*	*2017*
Karachi	2	1	1	1	1	1	1
Lahore	1	2	2	2	2	2	2
Faisalabad	9	6	4	3	3	3	3
Rawalpindi	3	4	6	5	4	4	4
Multan	5	5	5	6	6	4	7
Hyderabad	7	3	3	4	5	6	8
Gujranwala	8	9	8	7	7	7	5
Peshawar	4	8	7	8	8	8	6
Quetta	11	10	11	12	11	9	10
Islamabad			–	–	–	10	9

Source: Census of Pakistan 1998, Table 16.11, Ranking of Major Cities.

fortified village, was established in the early 18th century. It was completely transformed and rebuilt as a port town during the British colonial period in the 19th century,

Being a port city, it is the industrial, commercial and financial hub of Pakistan. Developed into a modern metropolis after the British-wrested Sindh from the native rulers, the Talpur Mirs, during the mid-19th century, the city has witnessed unprecedented accretion of business and commercial functions. Karachi is the principal port of Pakistan, though another port, Port Bin Qasim, was developed nearby, to facilitate the international trade of Pakistan Steel Mill, that is located around 40 km north-east of Karachi. As a port, Karachi has a rich and well-connected hinterland. The port city is linked to Punjab by roads as well as railways. In fact, the Indus corridor in Sindh, with railways and roads, running parallel to Indus on either side, is the most heavily loaded transport route of Pakistan. This routes travel through the entire length of Punjab, joining the northern Lahore – Peshawar route. Being the only worthwhile port, Karachi is a point of convergence for goods arriving from all parts of Pakistan, before being shipped abroad, and a point of distribution of bulk commodities and merchandise to inland destinations

The city itself is highly industrialised with a variety of industries like oil refineries, textiles, cement, chemicals, machinery and above all the Pakistan Steel Mills in the neighbourhood. Being the hub of banking and finance, with a large industrial base, and other commercial activities related to international trade, Karachi is considered the financial capital of Pakistan. Though no more the national capital, the city controls most of the trade and international financial transactions. Most of the industrial houses and

TABLE 8.19 Population of Karachi in Different Census Years

1901	136,000	1941	387,000	1981	5,208,000
1911	187,000	1951	1,068,000	1998	9,269,000
1921	244,000	1961	1,913,000	2017	14,916,456
1931	301,000	1971	3,515,000		

Source: Census of Pakistan 1998, table 16.12 p. 337, for the period 1901 to 1998, for 2017, Census 2017, provisional results table-7, population of major cities.

the multinationals have their headquarters in Karachi. Added to this is the Karachi Stock Exchange, the largest in Pakistan. The city is also the nerve centre of the Media and almost all major newspapers bring out their daily editions from here. With a large commercial and economic base, Karachi is aptly called the financial capital of Pakistan. Accounting for a third of the federal excise tax and almost half the total collection of direct taxes, 75 percent of customs duty and 80 percent of the total sales tax on imports, Karachi is indeed a key driver of Pakistan's economy. The city contributes 25–30 percent to Pakistan's GDP.

Demographically, Karachi is one of the fastest growing cities in the world. From 136,000 in 1901, the city grew to 9,239,023 in 1998, and further to 14.9 million in 2017. It is a significant jump in population growth (Table 8.19)

Immediately preceding the formation of Pakistan, Karachi was essentially a port city like Bombay, inhabited by merchants, traders and local population. The population composed of Hindus (51 percent), Muslims, largely Sindhis (42.3 percent) and Christians (2.3 percent).[3] Then came the partition of India with one of the most voluminous cross border movement of people in a short span of time. Between 1947 and 1951, around 180,000 Hindus and a few hundred Christians moved to India, and their place was taken by the newly arriving Urdu-speaking refugees from India, not only to replace the migrants, but a larger number, amounting to several hundreds of thousands. These Urdu-speaking immigrant refugees, called *Mohajirs*, are an important element in Karachi's population today. Besides the refugees from India and later from Bangladesh, Karachi is also a place of refuge for the émigré, the outlawed and the sundry elements seeking financial opportunities. From Balochistan, Sindh's western neighbour, very close to Karachi, arrived in a perpetual stream, workers, traders and livelihood seekers. The Pakhtuns or Pathans of Northwest Frontier Provinces (now Khyber Pakhtunkhwa), though not a large group, were always present in Karachi, known for their hard work and security services.

8.5.1.1.1 Change in the Socio-political Organisation of Karachi

The disturbed conditions in Afghanistan, following the Russian invasion, the Taliban rule or the retaliatory American invasion of Afghanistan to root out the Al-Qaida leadership, engulfed in their wake much of Khyber Pakhtunkhwa, forcing a number of Pakhtuns to move out of the province. They, like the Baloch, preferred to move to Karachi, along the Indus Highway. The numerical strength of Baloch and Pakhtuns immigrants has gradually increased during the last few decades; and today, they are the second largest group after the Urdu-speaking Mohajirs, besides the Punjabi-speaking Muslims. A large number of Mohajirs, Baloch and immigrant Pakhtuns are settled in Orangi, a part of Karachi West, and reportedly the largest slum in Asia.

A congregation of several ethno-linguistic groups has given rise to an urban life that is less than harmonious. The sectarian differences often lead to eruptions of violence and group interest becomes the dominant theme of urban life. The strange fact is that Sindhi-speaking Sindhis, the original inhabitants of Sindh and Karachi, representing today, not even 10 percent of the population of the city, have receded from the political and economic scene of the city, and are relegated to the countryside. The city is largely controlled and administered by domiciled Sindhis, of whom the Mohajirs, the Urdu-speaking immigrants, with their political outfit the Muttahida Quami Movement, form the dominant group.

8.5.1.1.2 Morphology

Spread over an area of 3527 km^2, the city is divided into six administrative districts (1. Karachi South, 2. Karachi East, 3. Karachi Central, 4. Karachi West, 5. Korangi and 6. Malir), each further sub-divided into towns. Thus, the city has six districts divided into 18 towns and six cantonments.

Morphologically, the city can be divided into two distinct units: 1.The Inner City of Karachi, consisting of the four districts of Karachi and Korangi, and Malir, the huge outer part, enveloping the city like a shell. The main Karachi city starts from Jinnah bridge in the south-western part, separating the Karachi port from the Karachi city, and ends at the other end at Quaid-e-Azam Mausoleum. Karachi Central, partly inherited from the British colonial period, like the Fort area of Mumbai (Bombay), is the most important part of the city, with wide roads and stately buildings, built during the British colonial period. The principal artery is the Muhammad Ali Jinnah Road with a number of secondary roads radiating from it, like the Agha Khan road or Faqir Muhd. Khan road and Nishtar road in the north and Shahrah-e-Liaquat and Chundrigar Road in the south.

Karachi Central, confined between Muhd Ali Jinnah Road and Shahrah-e-Liaquat or Plaza Chundrigar, has some of the well-known stately buildings constructed during the British Raj and some built after 1947. The Karachi Port Trust, Frere Hall, Empress Market (now razed to the ground), Hindu Gymkhana, St Patrick's Church, Mohatta Palace and above all the Karachi Port Trust Building are the most magnificent and awe inspiring buildings of the British colonial period. Between Shahrah Liaquat and the Club Road is the administrative segment of Karachi where most of the provincial administrative buildings are clustered. These include the Sindh Legislature, the provincial Secretariat and the Sindh High Court, besides a number of financial institutions like Habib Bank, Karachi Stock Exchange and Karachi Metropolitan Development Authority.

Further south is the SADAR, home to the Governor`s house, Holy Trinity Cathedral, and the PTDC office. In Sadar also lies the Bagh-e-Quaid-e-Azam, spread over a large area. Adjacent to Sadar, and southeast of it, is the Karachi Cantonment with Cantonment railway station. The most noticeable feature in the cantonment area is the Three Sword Memorial, built in 1973 to commemorate the guiding principles of Quaid-e-Azam: UNITY, FAITH and DISCIPLINE. The three swords built in marble remind one of the Father of the Nation, Muhd. Ali Jinnah. To the northeast, on the M.A. Jinnah road, at a tri-junction with Jahangir Road and Quaideen Road is the Mausoleum of Quaid-e-Azam.

The south and south-eastern part of the city is one of the very important the high residential area of Karachi the Clifton, .known for its opulence and the highest class residences of Karachi. Not far off is the Defence Housing Colony, a well-known residential area of Karachi, with Mohatta Palace Museum in the centre. Originally meant for defence service officers, the colony houses largely civilians and is considered a prestigious address in Karachi.

8.5.1.1.3 Land-use of Karachi

A large part of the city area is residential. More than a quarter (26.66 percent) of the area is residential (Akhtar and Sharif, 2007). Unlike the western cities, "*High class residential areas are located close to the city centre while low-class residential areas are generally located in the peripheral areas of the city. The middle class residential are located between the two*". The city periphery is marked by the presence of a number of squatter settlements, known as *katchi abadis*.

> *Inhabited by low income people, the squatter settlements are generally located at the peripheries but a few settlements are located within the high class and middle class residential areas, like Punjab colony, Akhtar*

colony etc. Similarly, Lines area and Sultanabad are located close to the city centre.[4]

Commercial activities, especially retailing, once confined to the city centre, are now more scattered as a number of secondary shopping centres have emerged like Tariq Road shopping centre, Hyden market, and Liaqatabad market. It must be emphasised that exclusive shopping commercial areas are far and few between in Karachi. "*The commercial core of Karachi, Saddar also has intermingled non-residential and residential activities. Shops and workshops often occupy the ground floor while the upper storeys are used by residential purpose*".[5]

It is to be emphasised that Karachi Cantonment occupies 180 km^2 of land.

Recent buildings of importance include the National Museum and the Mausoleum of Quaid-e-Azam Jinnah.

In terms of the planned spaces, residential densities, and buildings and their structural character, Karachi carries two extremes: 1. The high and middle residential areas and 2. The slums, locally known as *Katchi Abadis*. While the high and middle residential areas are described above, the slums of Karachi are difficult to enumerate and describe. The most well-known of the slums are 1. Orangi, 2. Lyari and 3. Machar Colony. *Orangi* is home to roughly 2.4 million people. It has a number of localities like Baloch Goth. Orangi is the largest slum of Pakistan. Originally established as a township, it soon turned into a squatter settlement with a large mass of immigrant population, especially the Mohajirs from East Pakistan. Besides, a large Pashtun group also made Orangi its residence. With an area of 57 km^2, Orangi is not so densely populated as many other slums. The settlement suffers from violence and crimes of different kind, and gang wars are not uncommon.

Lyari is another slum, though of a pre-Karachi existence, it is home to a large number of ethnic groups. Originally a place of Baloch and Iranian immigrants, it received a large number of Mohajirs after 1947 and a number of colonies of immigrants like Bihar colony developed. Lack of employment gave rise to crimes like drug trafficking and gang wars. Things have improved now, but the lot of the suffering mass of humanity in these slums, as much as in any large city of South Asia, will improve only with attaining greater prosperity and employment for the people of these *katchi abadis*.

Transport in Karachi is always in the process of being upgraded. The recent introduction of green Bus line, similar to Orange line of Lahore appears to be making a difference. Karachi Circular Railway is still in the works, and once completed, it may prove a very important means of transport. Among other things, Karachi has the oldest international airport of Pakistan, the Jinnah International airport. This, together with the two seaports, Karachi and port Qasim, the city is well connected internationally. The city has been, and still is, the principal centre of education, art and culture. It boasts of a number of universities and professional institutions including the most prestigious University of Karachi. The city is home to many cultural institutions like the National Academy of Performing Arts.

Once the capital of Pakistan, between 1947 and 1958, Karachi had a leading position in the politics and economy of Pakistan. It was the pre-eminence of Karachi as a capital city and a commercial hub that attracted the Mohajirs to Karachi. The shifting of the capital from Karachi to the newly established capital city of Islamabad has been a serious setback to the importance, status and political standing of Karachi. While the political decision making has shifted from Karachi to Islamabad, the former still remains the commercial and financial capital of Pakistan with its long standing advantage of being the principal port city and the nerve centre of the country's banking and finance. Being a transit point for international trade, a tradition of business and industrial enterprise, having ease of transport and above all the heritage of an established metropolis, the city has made up partially the loss of advantages associated with the shifting of the federal capital to Islamabad. Known in Pakistan as the "city of lights" or the "Quaid-e-Azam's city", Karachi has undergone a temporary convulsion, caused by ethnic violence in the initial decades of this century. With the federal and provincial governments paying special attention to restore a normal law and order situation, Karachi is sure to regain its reputation as a peaceful city. The metropolis has adequate intellectual and cultural energy and economic, commercial and entrepreneurial dynamism, to survive and, in due course, emerge as the premier city of Pakistan, comparable to any of the great cities of the world.

8.5.1.2 Lahore

With a population of over 11 million (11,732,693–2017) Lahore, the capital of Punjab province, is the second largest city of Pakistan. The history of the city dates back to a thousand years, having been the royal seat of different dynasties. Starting from Hindu kings in the 11th century to a succession of subsequent dynasties, like the Ghaznavids and Ghurids in the 12th and 13th century, the Mughals in the 16th to 18th century, and the Sikhs in the first half of the 19th century, Lahore always enjoyed the position of a provincial capital. After the Sikh rule came to an end in 1849, Lahore retained its status as the capital of a large British province of Punjab, incorporating the territory between the Yamuna on the east and the Indus on the west. After the partition of India in 1947, Lahore became the capital of the Pakistani province of Punjab.

A very important nodal point on the trans-continental northern route, the medieval Grand Trunk Road running from Kolkata via Delhi and Lahore to Peshawar, Lahore has the added advantage of lying in the midst of a fertile and productive agricultural region. Situated on the left bank of Ravi, the town and its agricultural hinterland benefitted greatly from the canal irrigation introduced in the early 20th century. It had, since, also developed into an industrial town. The *umland* of Lahore is dotted with textile and sugar mills. The skewed location of the town, in the northeast corner of the province, close to international border with India, has not put the city to any disadvantage.

The city has grown rapidly after 1947. The influx of Mohajirs, besides the rural–urban migration, as a result of rising industrialisation of Lahore and a greater pull exerted by the city, has led to a disproportionate increase in the population of Lahore (Table 8.20).

Though mid-twentieth century is a watershed in the demography of South Asia, showing a sudden acceleration of growth rate in the latter half of the 20th century, Lahore, in contrast, has shown a consistent high growth rate, right after the first quarter of the 20th century. Between 1921 and 1931 it grew by over 50 percent. The growth rate declined during the Second World War ending with the partition of British India. Between 1941 and 1951, the city`s population grew only by 26 percent. The subsequent decades witnessed a great surge in the population growth of Lahore, following the creation of Pakistan, arrival of refugees from India, and an overall rise in urbanisation. With growing economic base of the city and a consequent rural urban migration led to unprecedented growth of Lahore. It has to be noted that between 1955 and 1971, Lahore enjoyed the privilege of being the capital of a much larger province, the West Pakistan, and attracted many industries, and several financial institutions that further boosted the growth of the city. During the decade (1961–72), Lahore showed an exceptional growth of 68 percent, after which the growth rate is moderated. During the last two decades (1998–2017), the city has recorded the fastest ever growth; and its population is more than doubled, from 5 million in 1998 to 11 million in 2017. Lahore has since attained the status of a megacity. Being a capital and industrial city, it may still grow at a rate faster than the average growth rate of urban population in Pakistan.

TABLE 8.20 Population of Lahore in Different Census Years

Year	*Population*	*% Increase*	*Year*	*Population*	*% Increase*
1901	203,000		1951	849,000	26.3
1911	229.000	12.8	1961	1,296,000	52.6
1921	282,000	23.1	1972	2,170,000	67.4
1931	430.000	52.4	1981	2,953,000	36.0
1941	672,000	56.2	1998	5,143,495	74.1
			2017	11,126,285	116.3

Source: Same as table 8.19.

The city has a rich cultural and educational heritage. The Mughal monuments, like Akbar's Fort, the octagonal palace of Jahangir, Badshahi Mosque built by Aurangzeb, the last Moghul king, the Shalimar Garden, the Mausoleum of Jahangir and his queen Nur Jahan, on the other (right) bank of Ravi at Shahdara, are some of the main monuments inherited from the Mughal era. The British further added to the grandeur and importance of Lahore by adding a number of educational institutions like Government College of Lahore, Forman Christian College, and a number of professional colleges like Law and Medical College, Mayo College of Arts, besides a number of technical institutions. Today, besides a number of universities like the University of the Punjab, a reputed addition to the educational landscape of the city is the Lahore University of Management studies (LUMS).

The city has been a witness to several political events of far-reaching consequences, including the 1941 Session of Indian Muslim League where a resolution about the separate homeland for Indian Muslims was passed. Lahore has added two recent monuments to its landscape, the Minar-e-Pakistan and Gadafi stadium, besides the Mausoleum of the poet laureate of Pakistan Allama Muhd. Iqbal. A city known for publication of books, musicals and literary fests, art and architecture and entertainment, Lahore is, indeed, the cultural capital of Pakistan. The history of the city, its cultural tradition and literary achievements are reflected in a number of institutions that proudly preserve their heritage.

8.5.1.3 Faisalabad

A city in central Punjab, known, until 1979, as Lyallpur[6], Faisalabad was a planned irrigation colony with a population of 9,000 in 1901, in the midst of cotton and wheat producing area. The town was known for its cotton and wheat trade, sending these commodities to Karachi for export. It grew at an accelerated rate after 1947. In fact, no town in Pakistan grew as fast as Faisalabad (Table 8.21).

The reasons for its growth are its central location in the midst of the Punjab Plain, irrigated by lower Chenab Canal and the growth of cotton textile industry in the city's hinterland. The city has earned the sobriquet "The Manchester of Pakistan", because of its cotton textile industry. This also justifies the establishment of a National Textile University in the town. Starting with agro-based industries like sugar and textiles, the town has attracted, over the decades, many other industries like chemicals and fertilisers.

The city has one of the most reputed agricultural universities in Pakistan and an agricultural research institute. Centrally located and well connected, it has an international air- port as well as a dry port. Faisalabad is a relatively rich and modern city, gradually developing its own political and cultural base.

TABLE 8.21 Population of Faisalabad in Different Census Years and Rank among Towns in Pakistan

Year	*Population*	*Rank among towns in Pakistan*	*Year*	*Population*	*Rank among towns in Pakistan*
1901	9,000	–	1951	179,000	6th
1911	19,000	–	1961	425,000	4th
1921	28,000	–	1972	823,000	3rd
1931	43,000	14th	1981	1,104,000	3rd
1941	70,000	9th	1998	2,009,000	3rd
			2017	3,204,726	3rd

Source: Same as table 8.19.

8.5.1.4 Peshawar

Peshawar (34° N 71° 34' E), the provincial capital of Khyber–Pakhtunkhwa, is located at a height of 330 m in Kabul Valley (also called Peshawar Valley), close to the international border between Pakistan and Afghanistan. A frontier city, strategically located, commanding the access to Khyber Pass, Peshawar has been the gateway to the Indian sub-continent: and it remains so even today, retaining its strategic importance as a frontier city. Though a Pakistani city, it represents the transition between an Islamic Republic of Pakistan and an Islamic Emirate of Afghanistan, and their cultural ethos and practices. It is a growing city with a population of two million (1,970,042).

No city in Pakistan carries the archaeological traces of its antiquity as much as Peshawar. Known in ancient times as "Puruspura", the city had been the capital of many ruling dynasties from very early times. The existence of the ruins of a large Buddhist *stupa*, dating back to first century AD, suggests the spread and practice of Buddhism in the region in the distant past. It was during the reign of the Kushan King Kanishka's rule (120–150 AD) that a 400 feet high *stupa* was built. The locality known as "*Gorkhatri*" (Warrier`s grave) was the site of a Buddhist monastery. The city, by virtue of controlling the access to Afghanistan, remained an important regional centre all through history, from the Kushan period to the end of the British Empire. Till 1901, a city of Punjab, Peshawar became more important after Northwest Frontier area of Punjab became a separate province named "Northwest Frontier Province", with Peshawar as the capital. The British, to secure the town against invading forces, established a cantonment and the headquarters of their Frontier Corps, utilising the site of the old Bala Hisar Fort. In the 21st century, Peshawar remained a cantonment and a garrison city, the provincial capital, and a symbol of Pashtun culture. Peshawar is a city with layers of history and culture, from the early historic times to the contemporary history of Pakistan.

The Pakistan administration has given Peshawar its modern look, with planned expansion of the city, establishment of a number of higher education and research institutes and several universities, Besides the University of Peshawar, the city has Khyber Medical University, University of Engineering and Technology, Agricultural University, Institute of Management Sciences and Gandhara University. Peshawar Museum, housing valuable artefacts, is a unique archaeological museum of Pakistan.

The city has three distinct periods of growth: a pre-colonial period, a British Colonial period and the modern period as the capital of Khyber Pakhtunkhwa province. The old Peshawar is a walled city with several gates, meant essentially for defence, a strategy no longer relevant in modern times. The Cantonment and the later extensions, spreading beyond the walled town, represent the modern city. Some of the famous localities in the old town are Kotla Mohsin Khan, Chowk Yadgar and Qissa Khabani bazaar. The last one is very famous and has been a witness to many political events of far reaching importance.

Morphologically, the city has a linear alignment, extending south-westward, from the old city. The colonial and the modern extension remained confined between the railway line in the south and the Grand Trunk road in the north. The central road is the *Shahreh*-e-Quaid-e- Azam, the Mall, of the colonial period. The westward extension includes the Cantonment, Saddar, Defence colony, and administrative buildings, like the Civil Secretariat, District Court, Central Jail and several other institutions. The recent extension is home to several consulates and airline offices, lined along the Airport Road.

Enjoying the status of a provincial capital with a growing population reaching a million mark, Peshawar has not been able to attain the status of a liveable city. It has endured the adverse impact of the USA–Afghanistan war that lasted for two decades (2001–2022) as well the recent rise of militancy in some areas of the province. Peshawar has suffered greatly as a result of the influx of Afghan refugees, straining the resources of the city. During the recent decades, the city has experienced several violent events causing death and destruction. The attack on the All Saints Church and the Army Public School, and the massacre of innocent people will remain in the collective memory of Pakistan for a long time to come. These events have hampered the growth, not only of Peshawar, but of the entire province.

The city doesn't have a large industrial base. The main industry in the neighbourhood is making of marble slabs for export. Recently, a large industrial estate on Jamrud Road is developed. Besides, A Special Economic Zone is developed at Hattar. This may promote many industries.

The city has an international airport, called Bacha Khan International Airport, named after Khan Abdul Ghaffar Khan.

8.5.1.4.1 The People and Population

As a city in British India, Peshawar's population had many religious and ethnic groups. The trade and commerce conducted in the city was handled largely by Hindu traders, Sikhs and other communities. The Pashtuns had only a marginal role in trade and shop-keeping. The Sikhs and Hindus, in the city, spoke Hinduki, a melange of Hindustani and local language. After the partition of India in 1947, the non-Muslim population migrated to India. The present population is almost exclusively composed of Muslims, largely Pashtuns. The minority community is represented by a handful of Christians and a few Sikhs.

The city grew from a town of 95,000 in 1901, to a city of 173,000 in 1941, and to a multi-million city with a population of two million, occupying the 6th rank in 2017, among the cities of Pakistan. It had an accelerated growth between 1981 and 1998, when its population doubled, doubling again between 1998 and 2017. The population growth of the town could be seen from the following Table 8.22:

The doubling of population in a period of nine years from 1972 to 1981, and again a steep rise from 566,000 to almost a million in 1998, and then doubling again between 1998 and 2017 can be explained partly by the dynamism of the city, but in large measure to the importance Pakistan attaches to this provincial capital.

It must be stated that no city, with the exception of Karachi, has suffered in the wake of the Soviet–Afghan War or the USA–Afghan War, as Peshawar. Yet, despite all the catastrophic events, the government and the people of Khyber Pakhtunkhwa are struggling to transform Peshawar into a potential ideal city.

8.5.1.5 Islamabad

A planned city, specially designed and built to function as the Federal Capital of Pakistan, became operational as a capital in 1966, though the capital was shifted from Karachi in 1958. In the interim period, between 1958 and 1966, Rawalpindi was temporarily adopted as the capital. Over the last five decades,

TABLE 8.22 Population Growth of Peshawar

Year	*Population*	*Year*	*Population*	*Year*	*Population*
1901	95,000	1941	173,000	1981	566,000
1911	98,000	1951	152,000	1998	968,000
1921	104,000	1961	213,000	2017	1,970,042
1931	122,000	1972	273,000		

Source: Same as table 8.19.

Islamabad has not only grown in size, it has an enlarged functional base and has assumed great importance as a capital, in the affairs of the state.

Situated on a height of over 600 m on at the foot of the Margalla Hills, the city occupies a piedmont zone sandwiched between the Margalla Hills in the north and Rawalpindi in the south. Its latitudinal position (33°43'), comparable to Srinagar (34°), coupled with its altitude (500 to 600 m), and a rainfall about 1,100 mm, spread through the year, make for a salubrious climate, making Islamabad a liveable town. Among all the capitals of South Asia, Islamabad is by far the best in its natural beauty, expansive planning and architectural appeal and is well provided with hotels, parks and gardens, theatres and art houses and clubs, and a first class international airport that was named after a former Prime Minister Shahid Benazir Bhutto International Airport. The city, no doubt, carries the grandeur of a capital city and all the convenience of a modern city.

The decision to shift the capital from Karachi was taken by General Ayub Khan during his martial regime in 1958, to overcome the vulnerabilities of Karachi from the defence point of view. The site of Islamabad was chosen by a committee entrusted with the task of recommending a suitable site for the federal capital. The plan of the city was finalised by a Greek architectural firm, "Konstantinos Apostolos Doxiadus", which produced a gridiron layout.

The layout of the city, aligned northeast–southwest, is parallel to the alignment of the Margalla Hills that overlook the city. It is divided into eight functional zones (administrative, commercial, educational, industrial, diplomatic, residential, green area and a rural sector). These zones are subdivided into seven sectors. The entire city is longitudinally divided into two halves by the most important Jinnah Avenue, also running northeast–southwest. The highest part of the city is occupied by the apex institutions of Governance. Called the Red Zone, the area is home to Pakistani Parliament, consisting of the National Assembly and the Senate, the Central Secretariat, the Supreme Court and the Presidential palace.

Besides the Red Zone, there are sectors which are better developed and there are others which still remain undeveloped. The city has some magnificent buildings and monuments that have sprung up during the last four decades. The Shah Faisal Mosque with four pinnacle-like minarets and the Centaurus, a seven-star hotel are a few buildings that stand out. The capital complex, the Secretariat, the parliament buildings, the Assembly and the Senate and the Presidential Palace, are all aesthetically designed buildings.

The main highway coming out of the city is the Islamabad highway, which ultimately joins the Grand Trunk Road. The city is separated from Rawalpindi in the south by I.J.P. Road. West of the Islamabad highway, and south of the capital city, is the Benazir Bhutto International Airport, which serves the twin city. At a higher level, at the foot of Margalla Hills, is the Rawal dam, creating a reservoir that supplies water to Islamabad.

More than a commercial and industrial centre, Islamabad has attained a pre-eminent position as the apex of an educational pyramid. The Quaid-i-Azam University, National Institute of Science and Technology, COMSATS Institute of Information Technology, Pakistan Institute of Engineering and Applied Sciences, Allama Iqbal Open University and Pakistan Institute of Medical Sciences are leading educational and research Institutes of Pakistan. The capital is planning another airport.

Islamabad, the well-planned capital and the pride of Pakistan, has fulfilled the function for which it was designed. It is a beautiful city with all the grandeur of a capital. Yet, there are a host of problems that crop up every day and it is not easy to find quick and lasting solutions to all the urban problems. Pakistan is a developing country and it will take time before all the rough corners of urban landscape are smoothened and the city functions efficiently. The problem of housing and the emergence of "*kachhi* abadis", in the fringe area, is not only an eye sore but also reflects the general socio-economic condition of the people. One has, however, to look at the brighter side of the capital story, with the hope that in coming years Islamabad will be counted as one of the most liveable capital cities of the world. Islamabad is a city that generates optimism that every country in South Asia needs.

8.5.1.6 Quetta

Quetta, the capital of Balochistan province, grew from a small town of around 25,000 in 1901 to one having a population of 568,000 in 1998 and to a million-city with a population of 1,001,205 in 2017. The city, besides being the capital of Balochistan, is known for two principal reasons: its strategic location and its salubrious climate. Quetta lies on a railroad from Jacobabad in Sindh to Chaman on the Pakistan–Afghanistan Border. It controls the Bolan Pass, about 50 km to the southeast where the railway line and the road pass through a narrow defile. Further, to the northwest, the town, with its cantonment, also supervises the passage to Kandahar town and Kandahar province of Afghanistan via Chaman, a border town of Pakistan. The town is the focal point from where the routes radiate in all directions. The northern route to Khyber Pakhtunkhwa, following the Toba Kakar Range, and the south-western route to Nushki, Chagai, Dalbadin, going further west beyond Balochistan border, linking Zahedan in Iran, are strategically very significant. The southern route running through Mastung, Kalat, Khuzdar, Bela and further south to Hub, links Quetta to Karachi, The eastern route connects Quetta to the Indus Highway at Dera Ghazi Khan, and the south-eastern route connects it to Jacobabad in Sindh via Sibi, running further down to Karachi. Quetta, thus, controls the routes and access to most parts of Balochistan.

Located at a height of 1,680 m Quetta has the semblance of a hill station. It has a healthy climate. It may be mentioned in passing that Quaid-i-Azam Muhammad. Ali Jinnah, the founder of Pakistan, spent his last few months at Ziarat (2,697 m), about 100 km from Quetta. Not far from Quetta is the famous Hazarganji–Chiltan National Park spread over an area of 130 km². Equally important is the Han Lake, in the vicinity of Quetta. Situated at an altitude of around 2,500 m the lake supplies water for Quetta, and also serves as a source of recreation for Quetta city.

Quetta was completely destroyed in a severe earthquake in 1935, but the town was rebuilt. From this commanding height of the city, the British administered the territory held by them, and controlled the princely states of Kalat and Las Bela and several tribal areas.

Today, Quetta is a growing city. It has a university (University of Balochistan), a Medical college and a college of agriculture. It is also the headquarters of the Geological Survey of Pakistan. Being at a considerable height and the least polluted of the cities of Pakistan, it attracts tourists from all over the country. Of late, however, the place has earned the distinction of being a hide-out of several terrorist organisations operating in Afghanistan as well as Pakistan. This has very adversely affected tourism in the region.

In addition to the cities discussed above, there are several other cities with a population of over a million like Multan, Hyderabad (Sindh), Bahawalpur, Gujranwala and several other cities which are economically and culturally important. Yet, these cities could not be discussed because of space constraint, these have been described in the regional discussion of the provinces.

8.6 Urban Economy of Pakistan

Leaving aside the primary sector, including agriculture, fishing, forestry, mining and quarrying, most of the economic activities, in the secondary and tertiary sector, are located in the urban areas or in their vicinity. Urban centres present a congregation of factors that promote economic development. By virtue of the advantage of their location on a transport hub, or of being an administrative or a trading centre, or simply a market place, they enjoy and provide the advantages of agglomeration effect, attracting industries, trade centres, banking and insurance, which, in turn, further promote accumulation of economic activities.

Of the total GDP of approximately US$ 260 billion of Pakistan, about 50 to 55 percent is contributed by cities. While the SPDC[7] places the contribution of urban places to the GDP at 49.1 percent in 2013–14, increasing from 48.1 percent in 2001–02, the Consortium for Development and Policy Research, Pakistan (CDPR[8]) estimates the contribution of Pakistan cities to GDP at 55 percent, out of which Punjab contributes 25 to 33 percent. The different sectors of the economy, however, contribute unequally to

TABLE 8.23 Sector-wise Contribution of Urban Economy in Pakistan (2014–15)

Sector	*Contribution to the GDP generated by urban areas*
Agriculture	02.9 %
Industry	29.8 %
Services	67.3 %

Source: SPDC Karachi, Annual Review 2014–15.

the national economy. According to SPDC (Social Policy and Development Centre, Karachi) their share to the GDP is as follows (Table 8.23).

The economic activities generating high income in the secondary and tertiary sector are also not uniformly distributed among the cities. There is obvious bias towards large cities that develop industrial clusters. Even small-scale industries seek vicinity of large cities to avail of the facilities like capital, raw material, and markets. It is the small-scale industries and the activities in the informal sector that attract even unskilled labour to the cities. The concentration of manufacturing activities of all descriptions occurs at places that offer the best locational advantage. Even those industries that are foot loose look for market and prefer vicinity of towns. Historically, Karachi has been the most suitable place for industries and the reflection of its locational advantage can be seen in a large number of industries like oil refinery, iron and steel, textiles, chemical and fertilisers and several others. The all too important Pakistan Steel Mills is located in its vicinity. Though not as vital as Karachi, Lahore and Faisalabad have become quite important for their textile industries based on the principal raw material, cotton, produced in the region. What is to be emphasised that urban economy, contributing substantially to the GDP, is the economy of half a dozen cities, Karachi, Lahore, Faisalabad, Hyderabad, Rawalpindi and Peshawar where most of the industries are located.

8.7 Urban Living in Pakistan

The living conditions in urban areas can be rated on the basis of spaciousness or congestion in living spaces, adequacy or otherwise of housing, ease of transport, the provision of daily necessities and other urban amenities. It has to be noted that with the exception of Islamabad, the federal capital, and Faisalabad (the old Lyallpur), the former developed in the mid-twentieth century and the latter established in the late 19th century, and a few others like Karachi and Quetta, most of the towns of Pakistan are quite old. In fact, most have grown by accretion to the old core of the city of the pre-automobile age. The original city, often called the old city is invariably crowded, and occupied by lower income group people

Talking in general terms, *"an estimated 47 per cent of the country's urban population - around 35 million people - lives in squatter settlements, where low income residents pay high premium to live in what are often in squalid conditions.*[9] There is a shortage of roughly 10 million housing units.[10] The five largest cities of Punjab (Lahore, Faisalabad, Rawalpindi, Gujranwala and Multan) are predicted to account for 78 percent of the total housing shortage by 2035. According to one estimate, around 30,000 people die every year in Karachi because of unsafe drinking water.[11] The sprawl in many cities and especially in Karachi, often marked by substandard housing, turns into crime dens and affect social life of residents. According to an estimate, 60,000 acres of land are eaten up by sprawl in the cities of Pakistan.[12]

Water and sanitation, besides housing are other areas of concern. In many cities, water supply is limited to four to sixteen hours a day. Karachi suffers immeasurably during summer. According to World Bank estimate, poor sanitation and poor municipal services cost Pakistan four percent of its GDP.

8.7.1 The Slums in Urban Pakistan

Most large cities of Pakistan have their share of slums. It must be noted that the phenomenon of slum is a common feature of most undeveloped countries or even developing countries, of the world. The countries of South Asia – India, Bangladesh and others – are no exceptions.

8.7.1.1 Orangi Town – the Largest Slum of Asia

The largest slum of Pakistan developed in Karachi; the *Orangi Town*, is not only the largest in Pakistan, but also in Asia and the second largest in the world, next only to the slums in Mexico. Parts of *Machar* colony (the Fisherman's colony) and part of Lyari town are also classed as slums. Orangi Town has 13 official neighbourhoods (like Baloch Goth, Bilal colony, Chistinagar, Haryana colony, Iqbal Baloch colony, Madina colony, etc.). Having a population of 1.6 million people, Orangi Town is inhabited by different linguistic and ethnic groups. The residents of Orangi include *Mohajirs* (Urdu-speaking refugees), Punjabis, Sindhis, Pakhtuns and Balochis. The *Machiara* Colony, a fisherman's colony in the neighbourhood of Kiamari town of Karachi, has another 700,000 people, presenting the biggest neighbourhood and inhabited, as the name suggests, by fishermen. Lyari town, one of the 18 towns of Karachi and one of the oldest localities of Karachi, colonised by Sheedi (also written Siddi) community of African origin, is now a stronghold of Baloch group.

The slums of Lahore are more like the crowded localities inside the walled town. The poor and deprived often throng in the old city occupying the north-western part of the city.

Slums have a way of establishing themselves and proliferating. These often represent the residential areas of people engaged in informal sector of the economy, or in the ancillaries of major industries. Many work as watchmen, gardeners, drivers, construction workers, domestics or take odd jobs, earning daily wages. This can be seen in the planned city of Islamabad, the much admired capital of Pakistan. Over 80,000 people live in the capital's *katchi abadis*, as revealed by a report on slum survey, as narrated by Danish Hussain (2014).[13] There are 24 *katchi abadis* (settlement of mud houses) in urban and rural areas of Islamabad with 13,521 families consisting of 84,591 individuals who are living in them. Of these, 21 are located in urban and seven in rural Islamabad. Mera Jaffar slum in Saddar Bazaar is the capital's largest *katchi abadi* and is infamous for crimes.

8.8 Future Prospects

As estimated by the United Nations, the less developed regions of the world are being urbanised at a much faster rate than the developed regions, which are already urbanised to the limit. The urban population of two groups of countries, more developed and less developed, is estimated to grow at a vastly different rate.

By 2030, as given in the above table, half the population of Pakistan will be living in cities. The growth of population is not necessarily a source of optimism. This increase could be more a symptom of rural distress and depopulation of the countryside than the pull of cities born out of the need for additional manpower for their growing economic activities. The need for adequate housing, as seen in most developing countries, is hard to be fulfilled, as the builders and corporate houses are more interested in high end housing that promise better returns. Transport that could facilitate easy and quick intra-urban circulation is another problem. The inter-city transport between Islamabad and Rawalpindi is the showpiece of the federal capital. So are the Lahore's metro-bus and the Bus Rapid Transport (BRT) showing steps in the right direction, but the system should defuse to all other cities in the capital's backyard. The advantage of a good public transport, as observed by Ayesha Siddiqua[14] (2012) is that it dismantles the class divide and brings people together. It may be mentioned that Karachi Circular Railway is still in the works, but once completed, could be a model urban transport system (Table 8.24).

The urban centres, no doubt, need to be equipped with necessary infrastructure and public services. As the population grows, urban sprawl takes places, distances increases, so do the money, time and cost of transport. Decongesting the cities and urban renewal are a far more difficult exercise than designing a new city on the drawing board, as has been done in the case of Islamabad. In an enlightening piece, Ishrat Husain talks of "*messy*

TABLE 8.24 Percentage of Urban Population over Hundred Years (1950–2050) in Different Regions

	1950	*1970*	*1990*	*2014*	*2030*	*2050*
World	29.6	36.6	42.9	53.6	60.0	66.4
More developed regions	54.6	66.7	72.4	78.0	81.5	85.4
Less developed regions	17.6	25.3	34.8	48.4	56.2	63.4
Pakistan	17.7 (1951)	25.4 (1972)	32.5 (1998)	38.6 (2014)	50.3	NA

Source: 1. Satterthwaite, David (2002) Scale of Urban Change Worldwide (1950–2000) and its Underpinnings, Royal Institute of Chartered Surveyors, London.

2. Figure for 2014 is the estimate of Planning Commission of Pakistan taken from Social Policy and Research Centre (2014–15) *The Urban Economy.*

3. *Figure for Urbanisation in Pakistan for 2030 form Consortium for Development and Policy Research, Urban Development in Punjab: A Political Economy Analysis.*

urbanisation,[15] based on the World Bank's report on urbanisation 2016. He is rather too critical in pointing out the *"diseconomies"* in the urban system

> *due to congestion and scarcity, premiums that are paid by migrants for land, housing, water, transport (indicative of dysfunctional institutions of governance); and exploitative mafia backed by political parties. Half the urban population lives in inhuman slums. Productivity gains of urbanisation have eluded us; this is messy urbanisation in practice.*

He hopes that following the increase in urban population and the greater representation of cities in the National Assembly and Senate, the problem of cities will receive greater attention.

Many of the problems faced by urban Pakistan are universal in developing countries, especially in South Asian countries. In terms of urban population growth, Pakistan is the leader in South Asia. In some aspects, it compares favourably with the neighbouring countries and in some it lags behind. To give a comparative view, statistical values of some of these aspects of South Asian countries are given below (Table 8.25).

As can be seen, the urban situation in Pakistan is similar to other South Asian countries. The population in Pakistan is, no doubt, growing at a rate faster than its South Asian neighbours, India and Bangladesh. What is worrying is the population living in slums. It is not that every province has such a large percentage of urban population living in the slums. But, Karachi accounting for around 25 percent of the total urban population of Pakistan with some of the largest slums of the world like Orangi, pushes the overall average of urban population living in the slums in Pakistan.

TABLE 8.25 Actual and Projected Level of Urbanisation in Most and Less Developed Countries

	Level of Urbanisation		*Rate of Change in Urban Population in % 1995–2015*	*Urban Population Living in Slums in % 2014*
	2015	*2025 Projected*		
World	54.0	58.2	2.18	
Most developed countries	78.3	80.4	2.77	
Less developed countries	49.0	54.0	1.55	
Southern Asia	34.8	39.5	2.87	34.4
Afghanistan	26.7	31.4	4.50	62.7
Pakistan	38.8	43.8	2.96	45.5
India	32.7	37.0	2.51	24.0
Bangladesh	34.3	41.8	3.74	56.1
Nepal	18.6	22.7	4.29	54.3
Bhutan	38.8	45.3	5.27	
Sri Lanka	18.4	19.6	0.81	

Source: UNO (2016) World Cities Report, Statistical Annex, A-1, B-1 & B-2.

With a huge population living below the poverty line, the South Asian countries find it difficult to develop a well-organised urban landscape and an efficiently functioning urban system. The cities are economy driven. With some of the lowest per capita GDP, the countries of South Asia cannot create a show piece of urban architecture, urban economy and urban life. Till such time as the economy picks up and the countries of the region, including Pakistan, have the resources to renew and reconstruct the cities, what is needed is "assuring the minimum" to all the city dwellers. Sanitation, regular and uninterrupted water and power supply, a functioning public transport, adequate number of hospitals and health care services, open public places and an honest and efficient administration, if assured, could be very satisfying to the mass of city dwellers. The governments, even in the absence of adequate resources, can ensure law and order and personal safety to citizens, assuring a peaceful, if not a prosperous existence.

People's Republic of China has transformed its urban economy and its landscape in a period of four decades. It is not beyond other nations' capacity to achieve a level of prosperity that could transform the urban system, especially the large cities into enviable living spaces.

Notes

1 Ali, R. (2002) Underestimating Urbanisation, Economic and Political Weekly, Vol. XXXVII (44 & 45), Mumbai.

2 Social Development in Pakistan, Annual Review 2014–15, Social Policy and Development Centre (SPDC), p. 2. The details of the migrant population given here is based on the Annual Review of the SPDC, 2014–15.
3 Census of India (1941) vol. I, Pt. I, Tables, p. 91.
4 Akhtar, Shamshad and Sharif, Mohammad (2007) Population Growth and Land Use Pattern of Karachi City, *Pakistan Geog. Review,* vol. 62, pp. 63–71.
5 Azad, Azra Parveen and Razzaq Ahmed (2006) A Geographical Study of Land Use in the Commercial Heart of Karachi (Saddar), *Pakistan Geog. Review,* vol. 61, pp. 64–82.
6 The city was named in honour of the then Lt. Governor, Sir James Broadwood Lyall, who promoted the colonisation of Lower Chenab valley. In 1979, the Government of Pakistan changed the name to Faisalabad to honour King Faisal of Saudi Arabia.
7 Social Policy and Development Centre, Annual Review (2014–15) Social Development in Pakistan, ch. 1, pp. 3–4.
8 Consortium for Development and Policy Research (CDPR) (2015) Urban Development in Punjab: A Political Economy Analysis.
9 Consortium for Development and Policy Research (2014) Urban Development in Punjab: A Political Economic Analysis, A Note, p. 2.
10 Consortium for Development and Policy Research (2014) Urban Development in Punjab: A Political Economic Analysis, A Note, p. 2.
11 Kugelman, Michael (2013) Urbanisation in Pakistan, Causes and Consequences. Norwegian Peace Building Research Centre, *NOREE Expert Analysis*, June, p. 4.
12 Siddiqua, Ayesha (2012) Urbanisation in Pakistan and its consequences, *Tribune*, 21 November 2012.
13 Hussain Danish (2014) Slum Survey, Over 80,000 People Living in Capital's Katchi Abadi Says Report, *The Express Tribune*, Pakistan, 2 February 2014.
14 Siddiqua, Ayesha (2012) Urbanisation in Pakistan and its consequences, *Tribune*, 21 November 2012.
15 Ishrat Husain (2016) Messy and Hidden Urbanisation, *Dawn,* 16 May 2016.

PART III
Economy

9

WATER RESOURCES OF PAKISTAN

The principal sources of water, in any country, include rainfall, snowfall, glacial meltwater and groundwater stored in the aquifers. These sources, especially the rainfall and meltwater from the glaciers, occurring beyond the borders of a country, also find their way through the allochthonous streams, into another country, from the region of their origin. Pakistan is gifted with a natural drainage system and a stream network that ensures the transfer of water from the neighbouring countries, through its principal river, the Indus, and its tributaries. It has to be noted that Indus runs a course of 3,200 km and has a catchment of over a million square kilometres (1.12 million km^2), of which 520,000 km^2, or roughly 47 percent, lies in Pakistan. The river, however, collects water directly or through its tributaries, from Tibet, India and Afghanistan. This includes, besides rainfall, the meltwater from the snow-covered peaks and glaciers, all reflected in the discharge of the Indus, lower down in the plains.

9.1 Rainfall in Pakistan

There is a large periodic and regional variation in the amount of rainfall in Pakistan. The summer months receive a much higher rainfall than the winter months. In fact, around 70 percent of the rainfall is concentrated during the three months (July–September) of the summer monsoon. The winter (January–March) rains, received largely from western disturbances, account for only 30 percent of the total. Regionally, the annual rainfall varies from less than 100 mm in the lower parts of Sindh to over 1,000 mm in the foothills of the Himalayas, parts of Khyber Pakhtunkhwa and the upper reaches of the Indus basin.

DOI: 10.4324/9781003510147-12

The monsoons originating in the Bay of Bengal reach Pakistan in early July. Though Sindh falls early in the summer monsoon track, the province does not receive much rainfall because of the absence of any prominent relief feature. The north-south–oriented Kirthar Range is ineffective in forcing the winds to ascend and generate orographic rains, as its alignment is parallel to the direction of the monsoon winds. It is only while approaching the northern part of the Punjab Plain, crossing the Salt Range and facing the foothills of the Himalayas that the summer monsoons bring moderate to heavy precipitation. The situation during winter months is very different, as the precipitation in winter is caused largely by the western disturbances, which bring rains to northern parts of Punjab, much of Khyber Pakhtunkhwa and Balochistan.

Estimates of Rainfall – There are various estimates of the rainfall normals in Pakistan; the most reliable is the Pakistan Meteorological Department's estimate. According to PMD, the rainfall normal for summer and winter for the whole of Pakistan is as follows (Table 9.1):

As seen from the table above, Pakistan's mean rainfall for winter (January–March) and summer monsoon (July–September) seasons are 74.3 and 140.9 mm, respectively, which contribute to 23 and 47 percent of the annual rainfall. In contrast, another estimate of the Irrigation Department of Pakistan estimates the average seasonal rainfall of the entire Indus Plains (canal command area) at 212 mm during kharif and 53 mm during *rabi* season, bringing the total to 265 mm for the *kharif* and *rabi* season together.

Here, reliance is placed on a technical note by Faisal and Gaffar[1] (2012) of the Pakistan Meteorological Department, following a rigorous scientific method (Thiessen's polygon) of calculation. Their calculation is based on the monthly rainfall data of 56 climatological stations, spread over the entire country, for a period of 50 years (1961–2010) and weighted against the area of the polygon in which they lie. The following table (Table 9.2) is reproduced from the above paper.

TABLE 9.1 Rainfall Normals for Winter and Summer Monsoons in Pakistan

Rainfall Normals	*Amount in mm*	*Percentage of the Mean Annual Rainfall*
Summer monsoons (July–September)	140.9	47.0
Winter monsoons (January–March)	74.3	23.0
Periods of thunderstorms (April–June and October–November)		
Driest period in Pakistan (October and November)		

Source: Pakistan Meteorological Department, Revised Long Period Normal (1960–2010) taken from Pakistan Economic Survey 2012–2013, p. 26. These normals are only for the two rainy periods but do not strictly represent the entire year.

TABLE 9.2 Mean Seasonal Rainfall of Pakistan (After Faisal and Gaffar)

Month	*Pakistan*	*Azad Kashmir*	*Balochistan*	*Gilgit Baltistan*	*Khyber Pakhtunkhwa*	*Punjab*	*Sindh*
January	18.9	28.6	18.1	11.4	45.0	15.6	2.8
February	24.9	39.0	18.8	15.5	70.6	23.2	5.3
March	30.6	51.0	20.6	24.3	96.7	28.0	4,7
April	22.5	45.2	10.9	38.8	73.9	20.4	3,1
May	14.9	30.4	6.0	32.2	44.3	15.6	2.9
June	18.6	25.9	9.0	13.6	39.6	29.4	10.2
July	63.3	66.7	30.0	16.9	120.7	103.8	60.3
August	56.1	58.9	22.4	20.7	117.3	91.1	53.7
September	21.4	26.6	6.2	11.4	51.3	34.6	19.9
October	8.8	14.8	3.0	9.4	26.6	8.1	4.3
November	5.2	11.8	2.6	4.7	17.5	4.2	1.6
December	13.4	20.0	13.0	9.2	31.2	10.2	3.6
Year	297.6	418.7	160.6	208.0	734.6	386.3	172.4

Source: Faisal, N. & A. Gaffar (2012) op. cit. p. 114.

According to Faisal and Gaffar, the area-weighted mean of rainfall of Pakistan is 297.8 mm, with the lowest mean in Balochistan (160.6 mm) and the highest in Khyber Pakhtunkhwa (734.6 mm). Balochistan (160.6 mm), Sindh (172.4 mm) and Gilgit Baltistan (208 mm) receive rainfall below the mean for the whole of Pakistan, whereas Khyber Pakhtunkhwa, Azad Kashmir and Punjab receive rainfall above the mean for the country. Khyber Pakhtunkhwa receives not only the highest mean among the provinces of Pakistan but also the highest amount of rainfall every month. The advantage of relief is obvious. It receives the mean rainfall of 120.7 and 117.3 mm in July and August, respectively, the rainiest months in Pakistan. It also receives 45, 70.5 and 96.7 mm of rainfall, far above the rainfall average in other provinces, during the three winter months of January, February and March. In winter, Khyber Pakhtunkhwa receives rainfall, three times higher than Punjab, four times higher than Balochistan and 16 times higher than the scanty rains in Sindh. Despite the area of the province being only 10 percent of the area of Pakistan, Khyber Pakhtunkhwa contributes significantly to the total volume of rainfall in the country.

9.1.1 The Total Volume of Rainfall in Pakistan

The total volume of rainfall in Pakistan is 262.45 billion m^3 (or 212.7 million acre feet). This, however, does not quite agree with the figure obtained by adding the volume of annual precipitation of different provinces (Table 9.3).

There is a 2% anomaly in the figures obtained by calculating the volume of rainfall of each province and then summing them up and those obtained from a direct multiplication of the area with the area-weighted mean rainfall for the entire country: that is, 297.6 mm.

The total volume of precipitation of Pakistan is 255–260 $x10^9$ m^3 (255–260 billion m^3) or 206–212 $(10)^6$ acre ft (212 million acre feet). This could be a bit of an overestimate. This, in the view of the present author, may be attributed to large polygons formed in Balochistan, under Thiessen's polygon method, by Faisal and Gaffar (2012). In Balochistan, occupying 47 percent of the area of Pakistan, there are hardly a dozen meteorological stations, each representing a large area. In most cases, the weather stations are in towns that have the advantage of a hill station or any other prominent high point that receives a higher rain. These weather stations have a much higher rain than the surrounding dry areas. Thus, Balochistan appears to contribute more than 20 percent of the rain of Pakistan.

Periodic and Regional Variations – July is the rainiest month, and Khyber Pakhtunkhwa is the rainiest region of Pakistan. All the provinces of Pakistan, with the exception of Gilgit-Baltistan where the rainiest months are March, April and May, receive the highest amount of rainfall in July. The

TABLE 9.3 Area-Weighted Rainfall (mm) of the Provinces of Pakistan (After Faisal & Gaffar-2012)

Province	*Area of the Province (km²)*	*Area-weighted Mean Annual Rainfall (mm)*	*Volume of water*		*Percentage Contribution to the Total Volume of Rainfall*
			In Billion m³	*In Billion Acre Feet*	
Khyber Pakhtunkhwa and FATA[1]	101,741	734.6	74.73	60.60	29.54
Azad Kashmir	13,297	418.7	5.56	4.51	2.17
Punjab	205,345	386.3	79.32	64.30	31.10
Gilgit and Baltistan	72,496	208.0	15.07	12.22	5.90
Sindh	140,914	172.4	24.30	19.70	9.26
Balochistan	347,190	160.6	55.55	45.03	21.17
Pakistan[2]	881,889	—	255.20	206.89	99.14
Pakistan[3]	881,889	297.6	262.45	212.7	

1. The areas of FATA and Islamabad Capital Territory are added in the area of Khyber Pakhtunkhwa, since their mean weighted rainfall is not known, and that figure is assumed to be the same as that of Khyber Pakhtunkhwa.
2. This row is obtained by adding the volume of rainfall of all the provinces.
3. This calculation is based on the area-weighted rainfall of 297.6 mm for entire Pakistan, multiplied by the area of Pakistan.
4. The area-weighted rainfall is taken from Faisal & Gaffar (2012) op. cit.

three summer monsoon months of July, August and September, together, receive around 140 mm of rainfall forming roughly 47 percent of the annual rainfall of Pakistan. The winter months of January, February and March receive around 75 mm of rainfall, which is roughly one-fourth (25 percent) of the mean annual rainfall of the country. October and November are the driest months. An important aspect of the regional distribution of rainfall is a single monthly maxima in Punjab and Sindh, but the hilly areas of Khyber Pakhtunkhwa in the north-west, and to some extent even Balochistan, have a second maxima in March, though its intensity is low.

Variable Estimates of Rainfall – As seen above, area-weighted rainfall for Pakistan as a whole and for all the provinces is calculated by Faisal and Gaffar (2012) taking into account 56 meteorological stations for which 50 years of rainfall data are available. Their finding, that Pakistan as a whole receives 297.6 mm of rainfall annually, does not conform to the estimates of other agencies. For example, the FAO of the United Nations estimates the long-term annual precipitation of Pakistan at 494 mm, 65 percent higher than the estimates of Faisal and Gaffar (2012). Equally important is the volume of water that varies greatly in the two estimates. Working on the basis of 297.6 mm as the area-weighted mean rainfall of Pakistan, the total volume of precipitation within the boundary of Pakistan comes to 262 (10^9) m^3 year or 212.7 (10^6) acre feet/year. The FAO's estimate is 393.3 (10^9) m^3/year, a figure 50 percent higher than the estimate by Faisal and Gaffar.

9.1.2 Stream Flow: Most Important Source of Water

Besides the rainfall, which is the first measure of water availability in a country, another equally significant source of water is the stream flow, a mode of collection and transport of water, regionally or internationally. Besides collecting rainfall water after evaporation and infiltration, the rivers transport water from extraterritorial regions, i.e., the areas beyond the borders of a country. In the case of Pakistan, it is the Indus and its tributaries, which bring large volume of water from their source regions and other parts of their catchments. The Indus Basin area is shared by four countries: Pakistan 520,000 km^2 (47%), India 440,000 km^2 (39%), China 88,000 km^2 (8%) and Afghanistan 72,000 km^2 (6%). Thus, a large volume of water is brought by Indus and its tributaries from the neighbouring countries.

9.1.3 Contribution of Glaciers and Snow-Melt to Stream Flow in Indus Basin

A large part of the extreme northern part of the country bordering India, China and Afghanistan, is mountainous with high ranges. The Indus catchment area is unique in the sense that it contains seven of the world's highest peaks after Mt Everest. Among these are K2 (8,600 m), Nanga Parbat (8,100 m), Rakaposhi (7,800 m) and Haramosh Massif on the east and the Chitral on the west. The glacial field of Gilgit Baltistan, by far the largest after the polar ice cap, covers an area of about 22,000 km², where on average 3 to 4 m of snow falls every year. Kazmi and Qasim Jan[2] (1997) mention 30 glaciers in Northern Pakistan, 7 in the Himalayas, 19 in Karakoram and 4 in Hindukush. Some of these glaciers, like Siachin, are 70 km long. It is estimated that the volume of water stored in the glaciated area of Indus is about 340 MAF, whereas the glaciers in the Kabul, Chenab and Jhelum Rivers, all tributaries of Indus, store 300 MAF of water. It appears that there is a storage of 650 MAF water in the glaciated area of the Indus Basin. In fact, the glaciated areas of the Indus Basin, confined to the northern part of Pakistan, India and Afghanistan, represent a veritable storehouse, a water tower of Pakistan. It is also estimated that every year, meltwater from snow and glaciers contributes 110 MAF of water to the flow of streams at the rim stations in Pakistan.

Besides the Indus, Kabul, Chenab and Jhelum, the principal tributaries of the Indus also receive glacial meltwater in summer.

According to one estimate,[3] 80 percent of Indus flow in summer is derived from snow and ice melt. The amount of meltwater brought by Kabul is about a third of the total meltwater received in the Indus. Snow melt accounts for more than 50 percent of the flow of Jhelum. What is more significant is the contribution of snow and glacial meltwater during the lean seasons April,

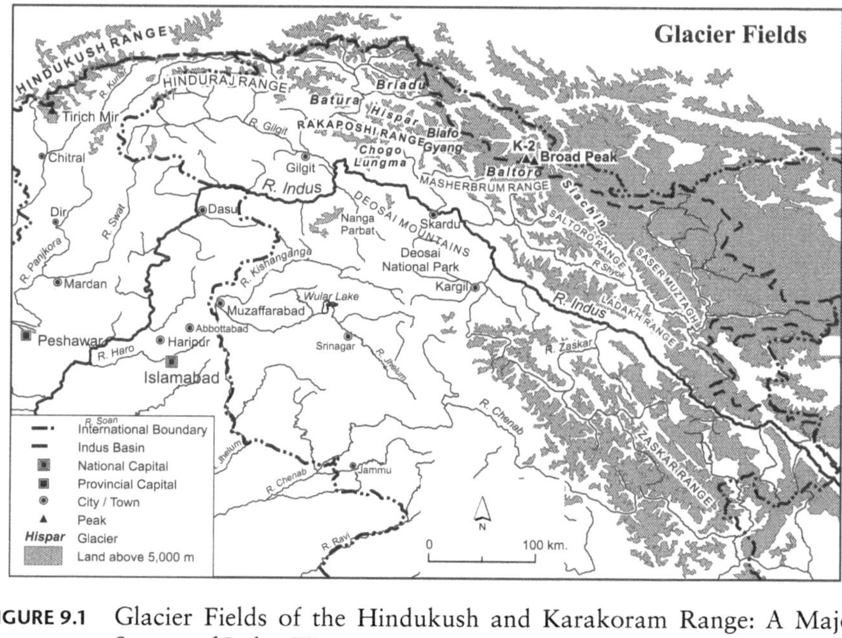

FIGURE 9.1 Glacier Fields of the Hindukush and Karakoram Range: A Major Source of Indus Water

May and June, before the onset of the summer monsoon. In fact, it is the snow melt that keeps the Indus flowing during summer.

9.2 Groundwater

The Indus alluvial basin, occupying much of Punjab and Sindh, extending over a length of 1600 km and an area of 20 million hectares (over 200,000 km^2) is extensively underlain by aquifers. These aquifers have stored water as a result of direct recharge from natural precipitation, stream flows and their sub-surface lateral transfer of water in the flood plain, especially during floods, and the "continued seepage from the conveyance system of canals, distributaries, watercourses and application losses in the irrigated lands, during the last 90 years" (Mohammad Akram et al.[4] The Pakistan Development Review Vol. 46:4 pp. 997–1009.). The total potential of groundwater in Pakistan is 55 MAF. About 79 percent of the area of Punjab and about 28 percent of the area of Sindh are underlain by freshwater aquifers (Pakistan Water Strategy Report). The exploitation of groundwater running to the extent of 38 MAF is done by over half a million tube wells. The quality of groundwater is not uniform and ranges from freshwater (salinity less than 1000 mg/l TDS) to saline water (salinity more than 3000 mg/l TDS). About 80 percent area of Punjab and 30 percent of Sindh have freshwater aquifers.

Of all the provinces, Balochistan is the one that depends almost entirely on groundwater. Groundwater is extracted here through dug wells, tube wells, spring wells and *karezas*. The streams are seasonal and often ephemeral. The potential of groundwater in Balochistan is 0.9 MAF, of which 0.5 MAF is already being exploited.

9.3 Measurement of Stream Flow

The availability of surface water is assessed by measuring the flow of rivers at rim stations. A rim station is defined as a control structure (reservoir, barrage, etc.) on the river just when the river enters into Pakistan territory or upstream of the canal-irrigated Indus Plains of Punjab and Sindh. The rim station for Indus is the Kalabagh Barrage (or sometimes Tarbela reservoir); for Jhelum, it is the Mangla reservoir; for the Chenab, it is the Marala Barrage and for Ravi and Sutlej, the rim stations are at Balloki and Sulemanki reservoirs.

The stream flow of the major rivers has varied from pre-partition days to the contemporary situation when the waters of Sutlej, Beas and Ravi were denied to Pakistan following the Indus Water Treaty (1960). The average annual flows of the pre-1960 Indus Water Treaty and the contemporary period are given below (Table 9.4).

TABLE 9.4 Average Annual Flow for the Period 1922–1961 and 1985–1995 and during the Drought Year 2001–2002 (in MAF)

Rivers	*Average Annual Flow (1922–1961)*		*Average Annual Flow (1985–1995)*		*Annual Flow during the Drought Year of 2001–2002 in MAF*
	Million Acre Feet	*% of Total Average Annual Flow*	*Million Acre Feet*	*% Contribution of Each River*	
Indus	93	49.20	62.7	42.13	48.00
Jhelum	23	12.16	26.6	17.87	11.85
Chenab	26	13.75	27.5	18.50	12.38
Ravi	7	3.70	5.0	3.36	1.47
Sutlej	14	7.40	3.6	2.42	0.02
Kabul	26	13.75	23.4	15.72	18.9
Total	189	99.96	148.8	100	92.62

Source: Govt. of Pakistan (2013) Programme Monitoring and Implementation Unit, Punjab Irrigation and Power Department. Water Resources of Pakistan.

9.3.1 Long-Term Average Supply of Stream Water in Pakistan

The above table, which shows the 40 years mean (1922–1961), including 25 years of pre-partition and 15 years of post-partition period, and the 10 years mean (1985–1995), as well as the annual flow of a single drought year of 2001–2002, presents three sets of contrasting conditions and does not represent a long-term mean of the post-partition period. According to the Irrigation and Power Department of the Government of Pakistan, the Indus River and its tributaries, on average, bring 154 million acre feet (MAF) of water, of which the four western rivers Indus, Jhelum, Chenab and Kabul contribute around 145 MAF, while the eastern rivers Ravi and Sutlej contribute roughly 9 percent of the annual flow of stream water in Pakistan.

Some officials[5] of the Pakistan Government leave out the contribution of Ravi and Sutlej, that is, around 8 to 10 MAF annually, and consider Indus, Jhelum, Chenab and Kabul as the main contributors of surface water, thus bringing down the annual flow to 140 MAF, divided into the contribution of Indus, Jhelum and Chenab as 65, 17 and 18 percent, respectively.

9.3.2 Reservoir Storage

Besides the surface flow and groundwater, there are a few storage reservoirs in Pakistan, one on Indus at Tarbela and the other on Jhelum at Mangla, besides a small reservoir on Indus further down at Chashma. The live storage capacity of these reservoirs has declined because of silting, as given below (Table 9.5 and Table 9.6).

TABLE 9.5 Life Storage Capacity of a Few Reservoirs in Pakistan and the Decrease in Percent

Reservoir	*Year of Commissioning*	*Life Storage Capacity (MAF)*			*Decrease in Percent*	
		Starting	*2000*	*2010*	*2000*	*2010*
Mangla on Jhelum	1967	5.3	4.5	4.2	15	21
Tarbela on Indus	1974	9.7	8.8	7.3	9	25
Chashma	1971	0.7	0.3	0.2	57	71
Total		15.7	13.6	11.7	13	25

Source: Three Years Development Programme (2001–2004) Planning Commission Govt. of Pakistan, quoted from Muhammad Akram Kahloon and Abdul Majeed , *Water Resources Situation in Pakistan; Challenges and Future Strategies*. Pakistan Development Review, Vol.46-4pp.

TABLE 9.6 Water Availability in Pakistan from All Sources and Its Utilisation

Water Availability Source	*In Million Acre Feet (MAF)*	*Utilisation Water Utilised*	*Million Acre Feet (MAF)*
		A. Water utilised	
1. Long-term stream flow average including glacial melt water	154	→1. Surface water irrigation	105 (at 68 percent utilisation of surface water)
2. Groundwater in Punjab and Sindh	50	→2. Tube well irrigation	40
3. Storage (Tarbela) on Indus, (Mangla) on Jhelum and (Chashma) on Indus	14	→3. Reserved for drought irrigation	
		Sub-total	**145**
		B. Water utilised or not available for irrigation	
		1. System loss	9
		2. Flowing to sea	40
		3. Reservoir storage reserved for drought	14
		4. Utilised ground water	10
		Sub-total	73
Total	**218**		**218**

Many other individuals and authorities have developed different models of water availability and utilisation. One such water utilisation budget attributed to Mushtaq Ahmad Gill (T. I.) of the South Asian Conservation Agriculture Network (SACAN – http://www.sacanasia.org) talks of Pakistan Water Budget – Indus Irrigation System. According to him, the total amount of water from stream flow is 142 MAF, and after the addition of groundwater and many losses, only 64 MAF is actually utilised for irrigation. His budget is as follows:

1). Available surface water from stream flow	+ 142 MAF
2). Groundwater abstraction	+ 44 MAF
Total available water	+ 186 MAF
Losses	
a. To sea	- 36 MAF
b. Losses in canals and distributaries	- 23 MAF
c. Losses in water courses	- 28 MAF
Total losses	- 122 MAF
Actual volume of water utilised for crops	64 MAF

Of the above two budgets, which is closer to reality is difficult to say; what is clear is that in all cases, 35 to 40 MAF of water is lost to the sea which, in any case, is essential to keep the mouth of the stream open and save the land from the unseen incursion of the sea inland.

In another estimate, Alizeh Kohari (The Indus Republic - *Dawn* Sept. 1, 2014) gives a summarised account of irrigation water in the Indus Basin as follows:

1. Average canal water diversion in MAF	104.7 MAF
2. Groundwater extraction in MAF	41.6 MAF
	146.3 MAF
Amount of water lost to seepage and transmission losses	76.0 MAF
	222.3 MAF

N. B. – The water lost to the sea is not mentioned.

9.3.3 *Possibility of Enhancing the Available Water Resources*

There is scarcely any possibility of increasing the supply of water. The Water and Power Development Authority (WAPDA) of Pakistan is struggling to enhance the storage capacity by building a few additional reservoirs. Second, not enough attention appears to have been paid to the hill torrents of Khyber Pakhtunkhwa, a series of small reservoirs may enhance water supply and

TABLE 9.7 Estimates of Water Availability by FAO

	FAO[1]	*As Given in the Text*
1. Long-term annual precipitation in depth (mm/year)	494	297.6[2]
2. Long-term annual precipitation in volume (10^9 m³/year)	293.3	262.45[3]
3. Groundwater produced internally (10^9m³/year)	55	50
4. Snow and glacial meltwater	N. A.	N. A.

1. *FAO, UNO Aquastat Data base (http://*www.fao.org/nr/water/aquastat/data/query/results – Accessed on 1.3.2016.

2. *& 3. Calculated on the basis of mean rainfall as given in Faisal, N. & A. Gaffar (2012) Development of Pakistan's New Area-Weighted Rainfall using Thiessen Polygon Method, Pakistan Journal of Meteorology, Vol. 9, June 17:*107–116.

agricultural productivity in the region. There is a tremendous loss of water in irrigation because of faulty irrigation practice, the losses could be minimised. Yet, the scope of increasing the supply remains limited. Conservation and planned utilisation, when practised, could enhance the prospect of additional availability of water.

The above statement presents a much-generalised picture based on averages, a condition that does not always occur. In the case of Pakistan, droughts are frequent, and the occasional floods are quite devastating. While the above water budget is the product of WAPDA of Pakistan and the Irrigation Department of the country, the estimates given by the Food and Agriculture Department of the UNO don't conform to domestic estimate of the available water and its utilisation.

For record, the estimates of water availability as estimated by the FAO of the UNO are given below (Table 9.7):

9.4 Indus Water Treaty (1960) and Resolution of the Pakistan–India River Water Dispute

It may be emphasised that the present estimate of the available water from the rivers of Pakistan is based on the Indus Water Treaty, signed in 1960 between India and Pakistan, a very significant agreement between the two countries on the distribution of water of the five international rivers (Sutlej, Beas, Ravi, Chenab, Jhelum and Indus). All these rivers have their origin outside Pakistan. Some of these rivers like the Indus and Sutlej originate in the Trans-Himalayan region, while the others have their origin in the Himalayas. But all of these rivers either originate in, or pass through, the Indian Territory. The water dispute between the upper and the lower riparian countries is quite common, and the sharing of water is invariably decided by agreements between the states claiming access to the waters of specific rivers.

Genesis of the Dispute – India being the upper riparian state had the option of using any amount of water, sometimes even disregarding the interests of the lower riparian state, in this case, Pakistan.

Before 1947, the year of the creation of Pakistan, all parts of Punjab, including present-day Haryana State of India, were under one administration, that is, the provincial Government of undivided Punjab, that regulated the supply of water to different regions. After the partition of Punjab, the upper parts of most of the rivers, including some of their head-works, fell in India, giving India an undue advantage. In order that the river waters of these rivers are distributed and utilised rationally by the two countries, a formal treaty, known as the Indus Water Treaty, mediated by the World Bank, was signed by the two countries in 1960 (September 19, 1960) at Karachi, effective retrospectively from April 1, 1960.

Under the provisions of this Treaty, the drainage system of Punjab was divided into (a) Eastern rivers (The Sutlej, The Beas and The Ravi taken together) and (b) Western rivers (The Indus, The Jhelum and The Chenab taken together). The Treaty, having 12 articles and many appendices, has well-defined terms and jurisdictions, with a provision for mediation and arbitration. This historic Treaty has been held for six decades, and many points of disagreement, including those arising out of the building of hydroelectric projects, in the category of non-consumptive use, have been amicably resolved. The eastern rivers are not in the Pakistani domain, and they are used only as channels for water transferred from western rivers, as the Treaty provided for the transfer of 20 MAF of water for irrigation of the area that was cutoff from the irrigation systems of the Ravi, the Sutlej and the Beas.

Utilisation and Demand of Water in Pakistan – Irrigation claims the largest share of water that is available. Of the 154 MAF of water available annually from the Indus and its tributaries (144.91 MAF from the three western and 9.14 MAF from the eastern rivers), about **104.73 MAF or 67 percent** is diverted for irrigation. This is done through a series of canals and associated water structures. Besides the three major reservoirs (Tarbela on the Indus, Mangla on the Jhelum and Chashma on the Indus), the Indus irrigation system has 16 barrages, 2 head-works, 2 siphons across major rivers, 12 inter-river link canals, 44 canal systems (23 in Punjab, 14 in Sindh, 5 in Khyber-Pakhtunkhwa and 2 in Balochistan) and more than 107,000 water courses. The aggregate length of the canals is about 56,073 km. In addition, the water courses, farm channels and field ditches cover another 1.6 million km. The irrigation thus uses 67 percent of the surface water available from the stream flow.

Of the total groundwater potential of 55 MAF, 41.6 MAF is utilised for irrigation through half a million (500,000) tube wells, including the deep tube wells developed under the Salinity Control and Reclamation Projects (SCARPS). A major part of the groundwater abstraction for irrigation is within the canal command area or the flood plains of the rivers. These tube wells have compensated for the loss of pumping capacity of the SCARP tube wells and also helped in keeping the water table low.

Utilisation for Domestic Needs – Kahlown and Majeed (2007)[6] have estimated the domestic requirement of water based on per capita demand of 46 m^3. Their estimate of the domestic requirement of water in 2025, with an estimated population of 260 million, would be 9.7 MAF. In the overall demand for water, the domestic consumption being insignificant is often ignored, and the attention is shifted to the per capita availability of water. The per capita availability of water is a global term, expressing the total volume of water in a country distributed over the entire population.

9.5 Per Capita Availability of Water in Pakistan

According to the FAO, the per capita total renewable water resource in Pakistan has varied over the decades with the increase in population. While around 1990 (1988–1992), it was 2,104 m^3/year per capita, the figure dipped to 1,333 m^3/year per capita in 2014. There is no doubt that the availability of water is declining at an alarming rate. Quoting the *State of Environment report of Pakistan,* Ahmed, Iftikhar and Chaudhry (2007)[7] have demonstrated that per capita availability of water has declined from 5,300 m^2 in 1951 to 850 m^2 in 2013 and a projected per capita availability of 659 m^3 in 2025, as shown in the following table (Table 9.8).

According to the UNO, an annual water supply of 1,700 m^3 per person is normally required for a balanced use of water. A drop below 1,700 m^3 causes water stress, and the areas having a supply of less than 1,700 m^3, per person per annum are considered water stressed. Similarly, if the supply

TABLE 9.8 Year Wise per Capita Available Water (1951–2025)

Year	*Population (in million)*	*Per Capita Availability (m^3/year)*
1951	34	5,300
1961	46	3,950
1971	65	2,700
1981	84	2,100
1991	115	1,600
2000	148	1,200
2013	207	850
2025	221	659

drops below 1,000 m^3 per person per annum, a water-scarce situation is generated. And if this goes below 500 m^3, it is considered absolute scarcity.

Pakistan, on this scale, is one of the water-stressed countries of the world, and the situation calls for remedial measures,

Measures for Enhancing the Quantity and Quality of Water – Measures proposed by the Government of Pakistan, Ministry of Water Resources are numerous and well thought out. This can be visualised from the NATIONAL WATER POLICY, April 2018. The preamble of the National Water Policy makes the following statement that shows the inadequacy of water in Pakistan.

> With rapidly growing population, Pakistan is heading towards a situation of water shortage and by corollary, a threat of food insecurity. Per capita surface water availability has declined from 5,260 m^3 per year in 1951, to around 1,000 m^3 in 2016. This quantity is likely to further drop to around 860 m^3 by 2025 marking our transition from a water-stressed to a water- scarce country. (The minimum water requirement to avoid food and health implications of water scarcity is 1,000 m^3 per capita per year.) The situation calls for rapid development and management of the country's water resources on a war footing.[8]

Notes

1 Faisal, N. and A. Gaffar (2012) Development of Pakistan's New Area-weighted Rainfall using Thiessen Polygon Method, *Pakistan Journal of Meteorology*, vol. 9(17), pp. 107–116.
2 Kazmi, A. H. and M. Qasim Jan (1997) *Geology and Tectonics of Pakistan*, Graphic Publishers, Karachi.
3 The estimate of glacial melt is from the report of "Programme Monitoring and Implementation Unit – Punjab Irrigation and Power Department", Government of Punjab (Pakistan) in "Water Resources of Pakistan" (2013).
4 Muhammad, Akram Kahlown and Abdul Majeed (2007) Water Resources Situation in Pakistan; Challenges and Future Strategies, *Pakistan Development Review*, vol. 46, p. 4.
5 Mohammad Hanif – Management of Water Resources in South Asia – a Note Agricultural Development Commissioner.
6 Mohammad Hanif – Management of Water Resources in South Asia – a Note Agricultural Development Commissioner, pp. 997–1009.
7 Ahmed, Ayaz, Henna Ifthikar and G. M. Chaudhry (2007) Water Resources and Conservation, Strategy of Pakistan, *The Pakistan Development Review*, vol. 46 (4) pt. II (Winter), p. 1003.
8 National Water Policy, Ministry of Water Resources, Government of Pakistan, April 2018, Preamble, p. 1.

10

IRRIGATION IN PAKISTAN

Irrigation is an important activity allied to agriculture. It concerns with watering the crops to maintain an adequate level of moisture in the soil to promote the growth of crops. Irrigation, as a part of agricultural practices, is most common in the tropics, especially in monsoon lands, where the rainfall is seasonal and a large part of the year is dry. During the drier part of the year, excessive evaporation occurs from the water surfaces, soil and even plants, the latter in the form of evapotranspiration. This depletes the soil of their moisture. Consequently, crop growth, during the non-rainy season, needs moisture from an external source, which is provided by irrigation.

Volume of Water utilised for Irrigation in the world and in some selected countries: Out of the annual global withdrawal of renewable water resource, which stands at 3,800 km^3, 2,500 km^3, roughly 66 percent, is utilised for agriculture, largely irrigation. That is around two-thirds of the total withdrawal. In South Asian and adjacent countries, the utilisation for agriculture is much higher, as can be seen in the following table (Table 10.1).

According to the World Bank Development Indicators Environment (2015), out of the total annual withdrawal of 3,906.7 billion m^3, the share of South Asia is 1,023.5 billion m^3, which means around 26 percent, that is, more than one-fourth of the total withdrawal of freshwater from the renewable freshwater resource of the world, is in the countries of South Asia.

10.1 The State of Rainfall Situation and Moisture Availability in Pakistan

Pakistan is a relatively dry country. Its area-weighted annual rainfall is 291 mm. Balochistan, the largest of the provinces occupying over 43 percent of

DOI: 10.4324/9781003510147-13

TABLE 10.1 Percentage of Water Used for Irrigation in Different South Asian Countries

Countries	*Total Withdrawal Billion m³*	*PerCent Used for Agriculture*
World	3,900.0	65.78
Pakistan	183.5	94.00
India	761.0	90.00
Bangladesh	35.9	88.00
Nepal	9.5	98.00
Bhutan	0.3	94.00
Sri Lanka	13.0	87.00
Maldives	0.0	0.00
Afghanistan	20.3	99.00
South Asia	1,023.5	91.00

Source: World Bank Development Indicators (2015).

the total area of Pakistan, has an annual mean of 158 mm, and in no month except July, the rainfall in the province exceeds 25 mm. Similarly, in Sindh, except for the two rainy months of July and August when the rainfall varies between 50 and 60 mm, no other month receives more than 10 mm of rain. Khyber Pakhtunkhwa is the exception where not only the annual rainfall is 625 mm, but it is also quite uniformly distributed except in October and November, when it shrinks to around 20 mm. The situation in Punjab, with 390 mm of area-weighted annual rainfall, is slightly better. Yet, leaving aside July and August, the months receiving an average rainfall of around 100 mm each, the rest of the months don't receive more than 25–30 mm of rainfall. The two dry months of October and November, the period of minimum rainfall, with their October maxima temperatures, are enough to deplete the soil moisture accumulated during the monsoon months.

To overcome this moisture-deficient situation, Pakistan has developed a network of irrigation system. The country withdraws 183.5 billion m^3 of freshwater every year from its annual renewable water resources, of which 94 percent is utilised for agriculture. In other words, Pakistan utilises 172.5 billion m^3 for agriculture, which is largely for irrigation. To take this water to farms, the country has one of the most elaborate and well-managed system of irrigation utilising both surface as well as groundwater.

10.1.1 Water Availability for Irrigation in Pakistan

The total availability of surface water at rim stations of the western and eastern rivers, on average, amounts to over 130 million acre feet (MAF). This figure varies annually depending on the amount of rainfall each year, in the upper parts of the river basins as well as the amount of snow melt in the higher source regions of the Indus and some of its tributaries. For example, during 2010–2011, the year of a severe flood in the Indus, the inflow at all the rim stations (Jhelum at Mangla, Chenab at Marala, Indus at Kalabagh, Ravi at Balloki and Sutlej at Sulemanki) was exceptionally high, amounting to 161.52 MAF.

Of the total inflow, not all the water reaches the farm gate for irrigation. There is considerable loss through infiltration and evaporation during the process of transport, from the canal heads to the fields. The canal water, however, is supplemented heavily by groundwater, pumped out at the field level. It has to be emphasised that the groundwater in Pakistan is largely from the infiltration from the canals and is tapped by tube wells, in the flood plains of the rivers. The overall availability of water, varying in a narrow band of 130–135 MAF, consists of surface water and groundwater tapped through different devices. The following table (Table 10.2) gives the availability of water during the *rabi* and *kharif* seasons of the year 2017–2018.

As can be seen, 133.4 MAF is the total volume of water that is really utilised through different means. Out of this, 81.4 MAF is the surface water, brought by canals to the farm gate, while the remaining 52 MAF is the contribution of groundwater, withdrawn through public tube wells, private tube wells, SCRAP tube wells and other means. The groundwater storage is largely derived from infiltration from the canals. In fact, it is a safe storage, free from evaporation, to be utilised as and when irrigation is needed.

10.1.2 Irrigation System of Pakistan

The irrigation network of Pakistan includes

> three major reservoirs, sixteen barrages, two head works, two siphons across major rivers, 112 inter-water link canals, 44 canal systems (23 in Punjab, 14 in Sindh, 5 in Khyber Pakhtunkhwa, and 2 in Balochistan) and more than 107,000 water courses. The aggregate length of the canals is about 56,073 km. In addition, the water courses, farm channels and field ditches cover another 1.5 million km.

Indeed, Pakistan has one of the largest canal irrigation network in the world.

Together, this entire network of canals, tube wells and wells irrigate 18.86 million hectares of cropped land (Table 10.3).

Undoubtedly, canals are the most significant part of the irrigation system of Pakistan, accounting for roughly 97 percent of the total irrigated land. The canal wells, tube wells and the canal tube wells derive their water from the canals, only the means of withdrawal differ. Non-canal sources like wells, tanks and other minor sources of irrigation play only a very marginal role in the irrigation system of Pakistan, irrigating not more than 3 percent

TABLE 10.2 Water Availability for Irrigation in MAF During 2017–2018

Year	*Season*	*Surface water*		*Groundwater*				*Total Water Availability*
		At Canal Head	*At Farm Head*	*Public T. Wells*	*Private T. Wells*	*SCRAP T. Wells*	*Others*	
2017–2018	*Kharif*	72.54	56.08	1.50	23.00	3.00	NA	83.55
	Rabi	25.07	25.35	1.00	20.50	3.00	NA	49.85
	Total	98.51	81.40	2.50	43.50	6.00	NA	133.40

Source: Water Resources Section, Planning and Development Division, Indus Water System Authority. Surface water figures at Farm Gate, according to the above source, are estimated after deducting conveyance losses, and the addition of losses saved due to the completion of projects (water courses improvement, lining of distributaries, rehabilitation/modernisation of existing irrigation system and construction of small dams to save hill torrents water) (taken from Pakistan Statistical Year Book – 2018).

TABLE 10.3 Area and Percentage of Irrigated Land by Different Sources in Pakistan (2017–2018)

Sources of Irrigation	*In Million Hectares*	*Percentage of Each Source*
Canals	5.96	32.06
Wells	0.44	2.36
Canal wells	0.27	1.45
Tube wells	3.60	19.36
Canal tube wells	8.15	43.84
Others	0.17	0.91
Total	18.59	99.98

Source: Pakistan Economic Survey 2012–2013, Govt. of Pakistan, Agriculture table 2–11 7 Statistical Yearbook 2018, p.67.

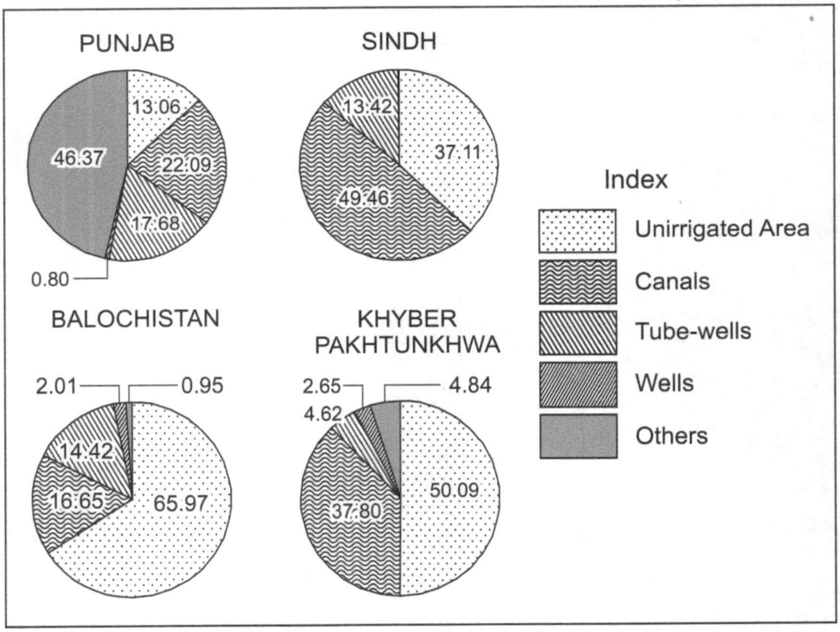

FIGURE 10.1 Province-Wise Percentage of Irrigated Land by Different Sources

of the total irrigated area of the country. In the canal system, however, the actual delivery of water is affected either directly by canal distributaries or by drawing water from the canal through tube wells or canal tube wells. Tube wells either in the canals or in independent places account for more than 60 percent of the total irrigated area of Pakistan. Direct irrigation from canals is only one-third of the total irrigated area.

10.1.3 Distribution of Irrigation Among the Provinces of Pakistan

The irrigated area in Pakistan is not distributed according to the cropped area in each province. While Punjab has 90 percent of its cropped area under irrigation, Sindh has only 63 percent of the cropped area that is irrigated. Khyber Pakhtunkhwa has still less with just over 45 percent of cropped area under irrigation.

In terms of the share of each province in the total irrigated acreage, Punjab, with roughly 15 million hectares of the cropped area under irrigation, has 77 percent of the total irrigated area of the country, and this is 91 percent of its own cropped land. By contrast, other provinces especially Sindh and Khyber Pakhtunkhwa have a much smaller share of total irrigated land, besides the fact they are able to irrigate only around 63 and 46 percent, respectively, of their cropped area. Balochistan has a higher area under irrigation than the area under crops. Obviously, the few cultivated patches receiving irrigation may require irrigation several times during the period of the crop growth.

It may appear paradoxical that the total irrigated area in Pakistan is more than the Net Sown Area standing just over 15 million hectares. Out of this, 7.48 million hectares of Net Sown Area is cropped more than once, bringing the total cropped area to 23.40 million hectares. The total irrigated area of Pakistan is thus 80 percent of the total cropped area, distributed among the provinces as follows (Table 10.4):

In total, 84 percent of the cropped area of Pakistan is irrigated. This is fairly high and compensates for the non-availability of water from rainfall or the loss of soil moisture by evaporation. One has to recall that Pakistan doesn't have adequate agriculturally suitable land. It has only 23 million hectares of land under cultivation that includes the Net Sown Area in a

TABLE 10.4 Percentage Share of Irrigated Area of the Province to the Total Cropped Area of Pakistan (Area in Million Hectares)

Provinces	*Net Sown Area*	*Area Sown More Than Once*	*Total Cropped Area*	*Total Irrigated Area*	*Percentage of Irrigated Area of Total Cropped Area of the Province*
Punjab	10.62	5.87	16.49	14.94	90.60
Sindh	2.68	1.00	3.68	2.32	63.04
KPK	1.29	0.60	1.89	0.88	46.56
Balochistan	1.05	0.01	1.06	1.22	115.09
Total Pakistan	15.64	7.48	23.12	19.42	83.99

Source: Land Utilization, Table 1.1, p.2, Pakistan Statistical Year Book 2019, Irrigation-Table 1.15, p.66, Pakistan Statistical Yearbook 2018.

year (15.64 million hectares) and the current fallow in a year (7.24 million hectares). This is roughly 29 percent of the country's geographical area. The actual cropped area in a year, however, increases considerably because of multiple cropping, as a result of widely practised irrigation. This adds another 7 million hectares to the Net Sown Area bringing the total cropped area in a year to just over 23.12 million hectares of cropped area, 19.42 million hectares, or 84 percent of the cropped area is irrigated.

10.1.4 Impact of Irrigation on Tenure System, Cropping Pattern and Agricultural Productivity of Pakistan

Though the most significant impact of irrigation, as widely understood, is the increase in the area under cultivation, especially in the rainfall-deficient dry regions, an increase in agricultural productivity and diversification of cropping pattern, an equally significant impact of canal irrigation in many areas of Pakistan, not so commonly perceived, is a drastic change in the tenure system and agricultural holdings of the farmers in Punjab and Sindh.

Before the network of canals was laid in the late 19th and the early 20th centuries during British colonial rule, much of the ***doab*** land of Punjab was a grazing wasteland. After the construction of the network of canals, large chunks of barren land in Punjab were brought under cultivation and irrigation at the same time. To cultivate the newly reclaimed land, farmers from many districts like Amritsar, Gurdaspur and other neighbouring districts were invited to migrate to this newly reclaimed land, and huge acreage of land was allotted to these migrant farmers. To distribute the land, a proper survey was conducted, and land records were maintained. These farmers were settled into newly formed settlements called *chaks,* named after a pre-existing village, if one existed, or after the name of a specific canal or its branch. The newly allotted individual holdings ranged from tens of hectares to hundreds of hectares and thus emerged a class of capitalist farmers with huge agricultural holdings. These were the *lumberdars,* owning huge agricultural holdings, who formed the support base of the British Government in Punjab. To manage the canal system of irrigation, revenue collection or resolve internal disputes, the then colonial Government developed a few colonies, like Chenab colony with its administrative centre at Faisalabad, the erstwhile Llyallpur. These colonies, in due course, became economically far more important than the districts. The large holdings and weighty capitalist farmers existed for quite some time. Some distortion, in the system, took place after the partition of Punjab in 1947. A large number of Hindu and Sikh landowners migrated to India, and the lands vacated by them were allotted to immigrant Muslim farmers from India. The capitalist farmers still exist in Punjab though in a smaller number. They still yield considerable influence on the politics of Pakistan.

A similar situation developed in Sindh after the construction of Sukkur Barrage built in the 1930s and two other barrages, Guddu and Kotri Barrage, built in the 1950s. Large chunks of irrigated land, owned by *waderas,* have much to do with the land grab by powerful local chieftain connived at by the British rulers interested largely in revenue collection. These waderas in Sindh, the most powerful section of Peasantry in Pakistan, developed following the development of large-scale irrigation in Sindh. Thus, while irrigation in Punjab and Sindh increased cropped land agricultural productivity, it gave rise to the class of landowners, called *lumbardars* and *waderas* with huge land holding, on the one hand, and the landless class of tenants, the *haris*, on the other.

10.1.5 Increase in Area Under Cultivation and Agricultural Productivity

Pakistan, with a long-term area-weighted mean annual rainfall of 241 mm, can hardly grow crops, especially the winter crops (Rabi crops) without irrigation. Thus, the role of irrigation in Pakistan's agriculture is pivotal and more important than any other inputs. In fact, in the absence of the existing system of irrigation, so ingeniously developed, the country would have depended on rain-fed cultivation, with productivity far below the present level. Irrigation in Pakistan has not only enlarged the area under cultivation and increased agricultural productivity, it has played a significant role in the nation's economy.

Irrigation has enabled the country to develop a cropping combination that includes in its composition some of the tropical crops like rice and sugarcane requiring substantial irrigation during their growth. The dry land areas of Pakistan presently plant 1.5 million hectares of sugarcane and almost 3 million hectares of rice. This was impossible without irrigation in the Southern Punjab receiving 125 mm of rain annually. The most spectacular impact has been on Sindh, the province receiving, on average, 125 mm of rainfall annually.

The impact and importance of irrigation on Pakistan can be judged from the fact that it has made the country self-sufficient in food, besides enabling the country to produce commercial crops like cotton that sustain the most important industry of Pakistan, the textile industry.

10.2 The Canal System of Pakistan

10.2.1 Situation Before the Arrival of the British Rule in 1849 and the Introduction of Canals

The general landscape and the state of agriculture before canal irrigation was introduced in Punjab by the British colonialists were very different from today, often depressing. The *doabs,* the granaries of present-day Pakistan,

> bounded by the great rivers presented formerly a succession of alkaline wastes, often covered with low bushes of the saltwort tribe, or untilled expanses dotted with a scrub of thorny bushes of acacia family and of van (salvadora, a desert representative of the olive) with an occasional row of tamarisks near a creek or a work hole, relieved in the autumn by a short-lived flush of climbing plants, and in a good season by an abundant crop of grasses, which afforded coarse but invaluable pasture to the cattle of the nomad population. Canal extension and systematic state colonisation are now changing all this rapidly and the flora is approximating to the general spring and autumn series of agrestal species of Northern India.

Before the arrival of the British, whatever land was irrigated, it was from wells using Persian wheels or simply leather buckets, but very prominently from inundation canals during the peak flood period of the rivers. Shallow irrigation channels were constructed by the farmers, in which water was diverted during floods. It was, no doubt, a great help, but depended entirely on the flood level of the rivers. Parts of Punjab and Sindh, the two agriculturally most productive lands, had large dry areas, covered with scrubs and remained uncultivated for want of water. The inundation canals irrigated the land close to the river, below the flood level, in the flood plain, but did not reach the interfluves, which remained dry.

This bleak scenario of the agricultural landscape began to change with the opening of irrigation canals, after the British arrived in Punjab 1849. Over a period of half a century, the picture was completely transformed. Beginning with the Bari Doab Canal in 1859 till 1947, a series of canals, practically from all the rivers, were taken out, and these ensured a fair level of agricultural prosperity of the region.

The partition of India and the creation of two states, India and Pakistan, did cause some disruption in the existing arrangement in the canal network and irrigation, but the situation was remedied by the 1960 Indus Water Treaty between Pakistan and India, dividing the rivers, headworks and the canals and granting compensation to Pakistan for rearrangement of its irrigation works.

10.2.2 History of the Canal Network and Extension of Agriculture

Punjab presented a poor prospect of agriculture in the late 19th century. The only way out was to harness the perennial rivers of Punjab, lay down canals and improve agricultural productivity by irrigation canals and other modern methods of irrigation. There were, no doubt, a few, indeed a very few small canals, almost dysfunctional existing in Eastern Punjab. A canal dug by Firoz Shah Tughlaq as a monsoon channel was revived by Shahajhan's Persian Engineer Mardan Khan who brought water to Delhi. The same engineer dug a canal from Ravi near Madhopur to water the royal gardens at Lahore. The canals developed during the British period were functionally well-designed and meant almost exclusively for irrigation.

10.2.3 The Canals Developed Before 1947

The present network of canals in Pakistan includes the canals developed during the British period as well as those developed and added to the pre-existing network during the post-1947 period. The canals developed before 1947 are given below.

1. **Upper Bari Doab Canal** – Bari Doab is the acronym of Beas-Ravi Doab. The canal was completed in 1859. It has its headwaters at Madhopur (now in India). The project was irrigating more than 1.2 million acres of land. After 1947, the headworks as well as the irrigated land fell in the Indian territory. In 1881, another canal named Sidhnai Canal was opened from Sidhnai Barrage on Ravi.
2. **Lower Chenab Canal** – Most revolutionary project for the period was the Lower Chenab Canal, one of the greatest irrigation works of the late 19th century. Completed in 1892, it had a command area of over 3 million acres, irrigating roughly 2.25 million acres of land. The headworks consist of a weir across Chenab at Khanki, in Gujranwala District, and the irrigation project benefitted the districts of Gujranwala, Lyallpur (now Faisalabad) and Jhang.
3. **Lower Jhelum Canal** – The Lower Jhelum Canal takes its water from Jhelum at Rasul. The project was completed in 1901. It has a command area of about a million acres.
4. **The Triple Project** – Upper Jhelum and Upper Chenab and Lower Bari Doab Canal – To compensate for the water deficiency in Ravi, the river that is now assigned to India under the 1960 treaty, surplus water from Jhelum was diverted linking Jhelum, Chenab and Ravi. The linking of the three rivers is the reason the project is known as the Triple project. It is a unique project of inter-basin transfer of water and water management. The project was completed in 1915. The project was designed to irrigate a large expanse of Rechna Doab (Ravi-Chenab Doab), lying west of Gujranwala District. For this, Upper Jhelum, Upper Chenab and Upper Bari Doab Canals were combined. Under this scheme, the Upper Jhelum Canal with its headworks at Mangla, now transformed into a big reservoir, was to transport a large volume of water to Jan Chenab at Khanki, from where the Lower Chenab Canal takes off. This freed a large volume of water from the Upper Chenab Canal taking off from Marla, the headworks of the Upper Chenab Canal. The Upper Chenab Canal could irrigate 640,000 acres in Gujranwala District, and at the

same time, the canal carried water across the river Ravi by an aqueduct at Balloki, south of Lahore. The lower part of the canal, in undivided India, was known as the Lower Bari Doab Canal.

5. **Balloki-Sulemanki Link Canal** – This canal is meant to compensate the loss of water from the Sutlej Canal, following the partition of the country. The Sulemanki headworks was a part of the original Sutlej Valley Project completed in 1922–1927 in undivided India, built at the behest of the Nawab of Bahawalpur. The headworks are located close to the Indian border on the river Sutlej in Okara district. About 16 km upstream from Sulemanki headworks, the Balloki-Sulemanki Link Canal has its outfall connecting Ravi to the Sutlej River and thus offsetting the loss of water to India, as agreed upon in the Indus Basin Water Treaty of 1960.
6. **The Canals of Khyber Pakhtunkhwa** (Former North West Frontier Province) – The canal systems of Khyber Pakhtunkhwa are linked to the river Kabul and its tributary Swat. The Kabul River canal irrigates largely the area south of Peshawar and the land along the lower course of the river Bara, another tributary of Kabul. A much larger area is irrigated by the Swat Canal that was opened in 1885. The Upper Swat

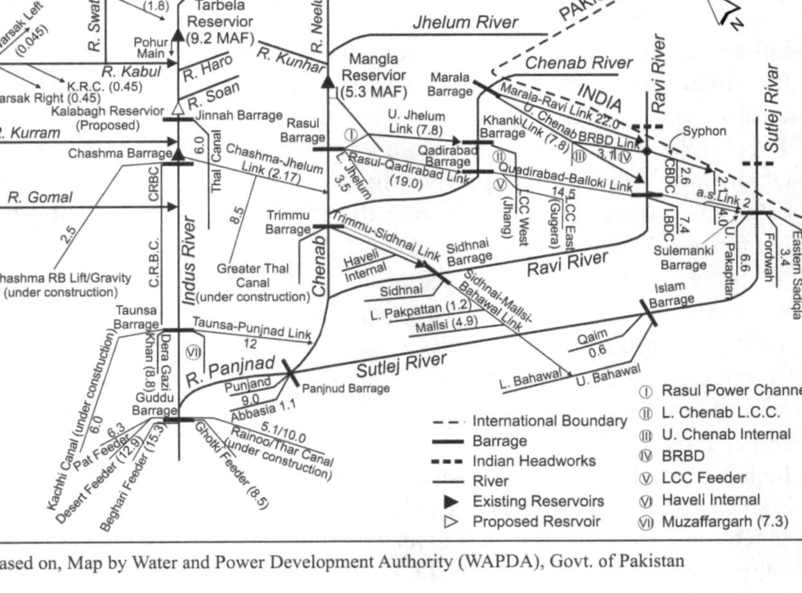

Based on, Map by Water and Power Development Authority (WAPDA), Govt. of Pakistan

FIGURE 10.2 Schematic Diagram of Links Between Canals Based on Map by Water and Power Development Authority (WAPDA), Govt. of Pakistan

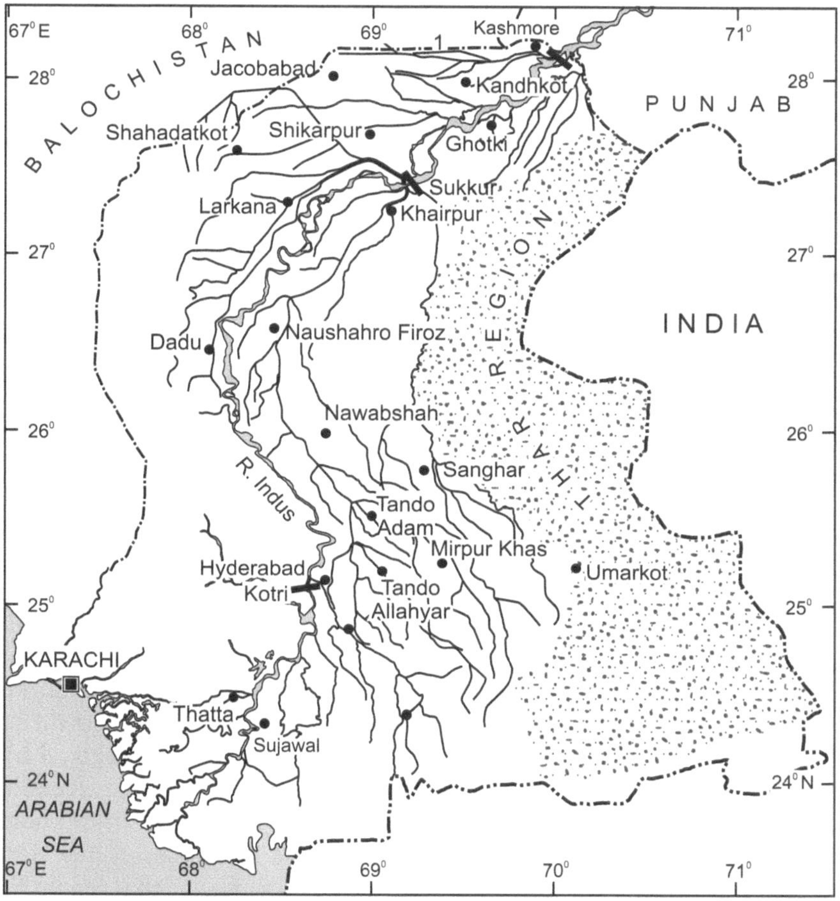

Based on Interanational Million Sheets - NF 42 and NG 42 (1943)

FIGURE 10.3 Network of Irrigation Canals and Major Barrages in Sindh Based on Million Sheets: NF 42 and NG 42 (1942)

Canal takes its water from near Chakdara, much to the north, and is taken through a tunnel about 3 km long and irrigates around 300,000 hectares of land through Machai and Pehur branches.

10.2.4 Other Irrigation Works Existing Before 1947

Timmu barrage – Completed between 1937 and 1938, it was a barrage on Chenab in the Jhang District of Punjab, meant to control flood to protect Jhang city. It is situated downstream from the confluence of Jhelum and Chenab. The barrage, causing a rise in water, is also used for irrigation

Sukkur Barrage – This was, and is even today, the most important irrigation work that was executed in the 1930s, on the Indus in Sindh. The earlier barrages on other rivers were minuscule. Sukkur Barrage, on the Indus in Sindh province, a part of the Bombay Presidency till 1937, was the most significant project launched by the British to irrigate the dry lands of Sindh. The barrage on Indus, located close to the city of Sukkur, built during the British Raj took nine years to complete, from 1923 to 1932. Initially known as Lloyd Barrage after the governor of Bombay, Sir George A. Lloyd, it was later named Sukkur Barrage after the district in which it is located. The canals taking off from the barrage irrigate 7.63 million acres (over 3 million hectares) of land in Sindh, which forms roughly 17 percent of the total irrigated area of the country. The barrage was rehabilitated and improved by the Government of Pakistan in 2004–2005, increasing not only its efficiency but also its longevity by another six to seven decades.

Believed to be the largest single project of its kind in the world, carrying a network of seven canals, width a length of 10,000 km, the water from Sukkur Barrage irrigates practically the entire Sindh, though lately a couple of more barrages on the Indus are added in Sindh province. Of the seven canals taken out from the Sukkur Barrage, three take off from the right bank and four from the left bank

The ***Nara Canal***, one of the seven canals, is the largest in the country. It was developed by excavating the old course of the river Nara, off the left bank of the Indus. It traverses virtually the entire length of Sindh and irrigates the districts of Sukkur, Khairpur, Sanghar and Tharpakar. Running for 365 km, the canal has developed wetlands in its course, the most known of them is the Nara Game Reserve developed in 1972. The second largest canal taking off from the left bank is the Rohri Canal that brings 2.6 million acres of cultivated land under irrigation. Another important canal, taking off from the right bank is the North-West Canal, providing irrigation to 965,000 acres of cultivated land, of which 184,000 acres are in Balochistan.

Jinnah Barrage – Like the Sukkur Barrage, the Jinnah Barrage is also on the Indus, near Kalabagh (Mianwali District). Completed in 1946, it irrigates the dry area of Sindh Sagar Doab. The water impounded by the barrage is diverted into a 50 km long Thal Canal that irrigates the dry area of Bhakkar, Khushab, Layyah, Mianwali and Muzzaffargarh Districts. With 3,362 km of additional canal branches and distributaries, with a discharge capacity of 27,000 m^3/s. Jinnah Barrage is a boon for Sindh Sagar Doab and part of the Thal desert. With a small hydraulic head of around 5 m, the barrage has eight small turbines, each with a capacity of 12 MW, thus having an installed capacity to generate 90 MW of electricity.

10.3 Building of Barrages on the Indus During the Post-1947 Period

Kotri Barrage – Unlike Sukkur Barrage, which is in Sukkur District of the northern part of Sindh, Kotri Barrage on the Indus is in the southern part of Sindh, just south of Hyderabad city and around 100 km north-east of Karachi. Completed in 1955, the barrage is used to irrigate land, in the area south of Hyderabad, in the deltaic region of Indus. With a maximum discharge capacity of 875,000 ft^3/s, the barrage is used for irrigation and flood control in southern Sindh.

Taunsa Barrage – The barrage was completed in 1958 on the river Indus, 90 km north of Dera Ghazi Khan town in Punjab province. Besides irrigating about one million hectares of land (951,400 ha), the barrage is also used to divert the water of the Indus to the Chenab River, through Taunsa-Panjnad Link Canal (TP Link Canal). The barrage also serves as a bridge, facilitating the laying of oil and gas pipeline. The waters from the project benefit Dera Ghazi Khan district, on the right bank of the Indus, and Muzaffargarh district on the left bank of the river. Taunsa-Panjnad Link Canal (TP Link Canal) was added to the system in 1970, extending the command area of the barrage. The barrage has undergone a rehabilitation exercise to ensure uninterrupted irrigation of Muzaffargarh District, to supply water to the Taunsa-Panjnad Link Canal and to ensure the supply of water to larger areas.

The rehabilitated barrage system was blamed for the destruction of Muzaffargarh District during severe floods on the Indus in 2010. There could, perhaps, be some deficiency in both the design and functioning of the barrage, which needs to be addressed.

Guddu Barrage – That is another barrage in Sindh on Indus, built in 1962 just south of the border between Sindh and Punjab, close to the border town of Kashmore. Completed in 1962, the barrage is designed for flood control and irrigation. With a length of 1.4 km, the barrage has a discharge capacity of 34,000 m^3/s. It provides irrigation to 12,000 km^2 of cultivated land in Kashmore, Jacobabad, Larkana and Sukkur Districts in Sindh and Naseerabad District of Balochistan. It also supports a number of feeder canals in Sindh, like Ghotki Feeder, Begari Feeder and Desert and Pat Feeder Canals.

Chashma Barrage – Fifty-six km south of Jinnah Barrage, another barrage on the Indus, in Mianwali District, was built in 1971. It is a power project with an installed capacity of 184 MW. The barrage is used for power generation, irrigation and flood control.

10.3.1 Harnessing Indus and Its Tributaries

After 1870, the year British colonial rulers started building the canal network in Punjab, there has been a complete transformation in the irrigation

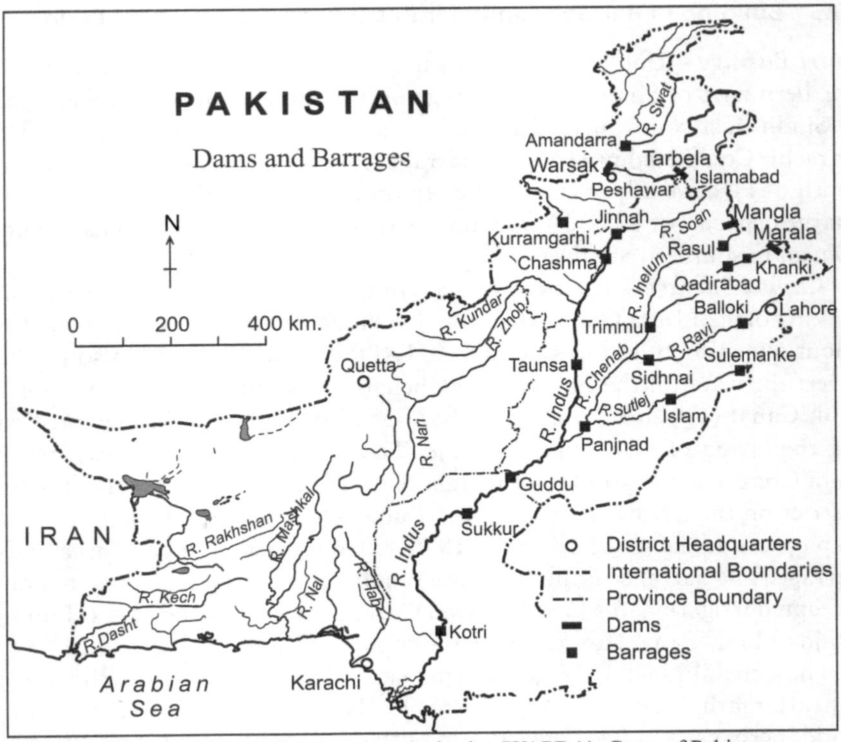

Based on, Water and Power Development Authority (WAPDA), Govt. of Pakistan.

FIGURE 10.4 Irrigation: Dams and Barrages in Pakistan Based on Water and Power Development Authority (WAPDA), Govt. of Pakistan

system, from the inundation canals to well-laid canal network. This resulted in an assured regulated supply of water, leading to the reclamation and colonisation of many parts of Punjab. The most remarkable was the Chenab Colony, following the completion of the Lower Chenab Canal, followed by other projects.

The Indus proper was harnessed by half a dozen barrages, some built during the British period and some by the Pakistani administration. The barrages on the Indus, as discussed earlier were the following in chronological order:

1. Sukkur Barrage	Completed in 1932, Rehabilitated
2. Jinnah Barrage	Completed in 1946
3. Kotri Barrage	Completed in 1955
4. Taunsa Barrage	Completed in 1958
5. Guddu Barrage	Completed in 1962
6. Chashma Barrage	Completed in 1955

Besides these, some additional links were added to carry water to all parts of Punjab and Sindh.

10.4 Dams and Reservoirs

Besides the existing barrages on the Indus, a number of dams have been constructed for irrigation and generation of hydroelectric power. Pakistan has completed a number of *multi-purpose projects*, some with large dams and reservoirs to store enough water to meet any scarcity condition.

1. ***Warsak Dam*** – Built on the river Kabul, about 20 km northeast of Peshawar, under the Colombo Plan, the project is largely a power-generating project. Completed in phases, the first phase was completed in 1960. The power station was commissioned in 1981. It had an installed capacity of 243 MW, raised to 525 MW by the Pakistan Water and Power Development Authority (WAPDA). The reservoir has a capacity of over 30 million m^3 (31,207,090 m^3) of water.

2. ***Tarbela Dam*** – Tarbela Dam, on the Indus, occupies the pride of a place in the irrigation and power system of Pakistan. Conceived and designed during the presidency of Ayub Khan, the construction on the project started in 1968 and completed in 1976. It is one of the largest earth-filled dams in the world. Located in the Haripur District of Khyber Pakhtunkhwa, about 50 km north of Islamabad, the 148 m high dam, above the riverbed, has produced a reservoir with a length of 80 km and a surface area of 250 km^2, holding 13.7 km^3 of water, with a storage capacity of 12.0 km^3. The catchment area of the reservoir is 168,000 km^2. Being a multipurpose project, the project has an installed capacity to generate 3478 MW of power. The irrigation potential of the dam lies more in an assured supply of water downstream during the dry season. The huge storage of water is also a guarantee for drought years, as the reservoir is fed continuously by the snowmelt in the higher reaches of the Indus Basin. The reservoir could thus regulate the flow of water downstream. Tarbela Dam project remains largely a power project, with a huge storage capacity of 13 km^3 (13 billion m^3).

Ghazi-Barotha Project - A run-of-the-river project south of Tarbela, and linked to it is the Ghazi Barotha Hydroelectric power project. The project took a long time to complete. Started in 1995, the project was completed in 2003. Located about 10 km west of Attock, it has a generating capacity of 1,450 MW. The active capacity of the reservoir is 25.5 million m^3. The powerhouse consists of five generating units, each with a capacity of 290 MW.

3. ***Mangla Dam*** – The second most important dam and reservoir in Pakistan, next to Tarbela Dam, Mangla Dam is built on the Jhelum River,

at Mangla in the Mirpur District of Azad Kashmir region of Pakistan. The construction of the dam started immediately after the signing of the Indus Water Treaty in 1960 and completed in 1967 with the support of the World Bank. The dam with a height of 138 m (about 10 m less than Tarbela) and a length of 3 km (3140 m) encloses a reservoir with a capacity of 7.25 km^3 (5.88 MAF), which has since declined because of sedimentation. The power-generating station of the project has a generating capacity of 1,000 MW.

The project, primarily designed for irrigation has suffered enormously because of sedimentation. The present storage capacity of 4.75 MAF (5.86 km^3) has declined from the actual designed capacity of 5.88 MAF (7.25 km^3). The government may restore the capacity or even increase it by further raising the height of the dam.

Neelum-Jhelum Hydroelectric Project – A run-of-the-river Hydroelectric project, Neelum-Jhelum project consists of a dam on the Neelum River, a tributary of Jhelum, at a site around 20 km south of Muzaffarabad. The project has a 26 km long tunnel to divert the water to a power station on the Jhelum River. The project with a projected installed capacity of 969 MW is considerably delayed, despite the fact that the work on the project began in 2008, after a Chinese consortium was awarded the contract. The project, supervised on the Pakistani side by the Water and Power Development Authority (WAPDA), was commissioned in 2018.

The dam on Neelum has a height of 60 m and a length of 125 m. The reservoir has a total capacity of 8 million m^3 (6,486 AF) of water. The project is a little complicated as the tunnel passes below the Jhelum. The power-generating units get a hydraulic head of 420 m, which represents the drop in elevation from the dam level to the power station.

Gomal Zam Dam – This dam on the river Gomal, a tributary of the Indus, is located in the South Waziristan District of the erstwhile Federally Administered Territory of Pakistan, now a part of Khyber Pakhtunkhwa. The dam across the river is located at Khajuri Kach, where the Gomal occupies a narrow gorge. Designed for irrigation, as well as flood control and power generation, the reservoir with an active capacity of 1.1 billion m^3 can irrigate 665 km^2 of land. Completed in 2012, this is an important irrigation project of Khyber Pakhtunkhwa.

10.4.1 Some of the Major Power cum Irrigation Projects not Yet Completed

There are many small and large projects designed by the Water and Power Development Authority of Pakistan. Among these, there are a few large ones, which have not yet taken off, but have undergone feasibility studies. The most important of these are:

1. Diamer-Bhasa Dam (on the Indus)	35°-31' N & 73° 47'-31" E	- 4,500 MW
2. Kalabagh Dam (on the Indus)	32°-57'-21' N & 71°-36'-49"E	- 3,600 MW
3. Thakot Dam (on the Indus)	34°-45' N & 72°-53'E	- 2,800 MW
4. Pattan (on the Indus North. of Thakot)		- 2,800 MW

All these four projects are designed to be executed on the Indus. The northernmost in the mountainous region is the Diamer-Bhasha project, followed by Pattan in the south, and further south by one at Thakot – all three lying in the mountainous region. The most contentious one is the Kalabagh Dam.

Kalabagh Dam – Lying in the Mianwali District of Punjab, hardly 15 km from the border of Khyber-Pakhtunkhwa, the project right from its inception has been controversial, especially opposed by two provinces viz. Khyber Pakhtunkhwa and Sindh. The dam designed to impound 7.52 km^3 of active and 9.7 km^3 of inactive storage is 3.35 km long and 79 m high.

Providing a hydraulic head of 52 m, the project is designed to have a power generating capacity of 3,600 MW, with 12 turbines each capable of generating 300 MW of power. The irrigation capacity of the project is not certain, though a claim is made that the project would irrigate 7 million acres of land, an irrigation potential comparable to Sukkur Barrage.

There is a dearth of consensus among the provinces about building the Kalabagh Dam. The main fear of Sindh is a perceived decline in the volume of water available for its three barrages, Guddu, Sukkur and Kotri, besides a general depletion of water in the Indus. It is not a run-of-the-river project in which the water flows back after being released from the powerhouse.

Diamer-Bhasa Hydroelectric Project – This hydroelectric project on the Indus is designed to have an installed capacity of 4,800 MW to generate 19,028 GWH of electricity annually. The project is named after its location at Bhasha in the Diamer District of Gilgit-Baltistan. The foundation stone of the project was laid in October 2011, the progress on the project is, however, extremely slow.

The project consists of a 272 m high dam on the Indus that will be the highest dam in the world after its completion. The reservoir will have a capacity to store 8.5 MAF of water, capable of having 6.4 MAF (roughly 8 billion m^3) of live storage. The project, originally expected to be completed by 2020, is considerably delayed. Once completed, the project will be an added source of water to the regions downstream and will enhance the capacity and longevity of Tarbela on the Indus, about 300 m downstream. It is proposed to build two underground powerhouses, one on each side of the river, each powerhouse having six turbines, each capable of generating 375

MW of electricity. The two powerhouses will be connected by two tunnels. The surface level of the reservoir, when full, is 1,160 m ASL. The minimum operation level is 1,060 m ASL. This means the water to a depth of 100 m can be used for generating electricity.

The Importance of Diamer-Bhasha Project – Though in terms of water storage, Diamer-Bhasha is 30 percent smaller than Tarbela Dam on the Indus, its importance lies in its location in a much higher protected area with an assured supply of water and low level of evaporation because of its height. It is located at a height of around 1,100 m ASL far above that of Tarbela at 500 m ASL. Besides being an assured source of power, the storage in Diamer-Bhasha of about 10 km^3 (10 billion m^3) is an insurance against drought in the lower and middle Indus Valley. This water could be released in an emergency to be used for irrigation and other purposes.

An infrastructure of this dimension will unleash a series of other economic activities, like tourism, and encourage the growth of power-based foot-loose industries. It will also promote the growth of towns on the Karakoram Highway that is the axis of the Pakistan-China Economic Corridor. Despite all the benefits that Pakistan will have after the project becomes operational, the cost of the project estimated to be around US $14 billion (estimated in 2013) is the most restrictive factor and proving to be a stumbling block in the implementation of the project.

Recent Developments: After considerable delay, the Pakistan Government has signed a PKR 442 billion- contract with a joint venture of China Power and Frontier Works Organization (FWO) for the construction of the dam. With enormous experience in building extra-large dams in difficult terrains and situations, the Chinese Power and FWO should be able to complete the work in the foreseeable future.

Besides the abovementioned projects, The Water and Power Development Authority (WAPDA) and Indus River Development Authority (IRDA) have planned a number of projects of different magnitude in Pakistan.

10.5 Geopolitics of Water Sharing and Indus Water Treaty

Sharing of river water has been a crucial issue between Pakistan and India, as some of the rivers flowing in Pakistan have their sources in the Indian Territory. This problem was realised early enough, and Pakistan approached the International Court of Justice in 1949 to find a solution to the distribution of water of six rivers (Sutlej, Beas, Ravi, Chenab, Jhelum and Indus), all of which either originate in Indian Territory or run part of their upper course through the Indian Territory. The existing headwaters especially of Sutlej, Beas and Ravi, supplying part of their waters to Pakistan were under the control of India, a situation that created the problem of river-water sharing between the two states, India and Pakistan.

To resolve the water-sharing problem, India and Pakistan negotiated the Indus Water Treaty through the intermediation of the World Bank. The Treaty was signed on September 19, 1960, with 12 articles, and many annexures. According to the Treaty, the three eastern rivers, Sutlej, Beas and Ravi, were given to India, while the three western rivers, Chenab, Jhelum and Indus, were to be controlled by Pakistan. To enable Pakistan to develop its own river water resources, India had to pay a sum of 62 million pounds sterling. The amount granted to Pakistan could enable it to transfer water from the western rivers (Chenab, Jhelum and Indus) to such areas in south-east Pakistan that were receiving water from the eastern rivers given to India. Following the Treaty, Pakistan built five barrages, one siphon and eight inter-river link canals between 1960 and 1971. The two major dams, Mangla with a storage capacity of 4.82 MAF and Tarbela on the Indus with a storage capacity of 9.3 MAF, were developed by Pakistan to compensate for the losses of water from the eastern rivers and also to add to its irrigation and power-generation potential. The Treaty, with all the bickering of either country, has stood the test of time and still holds good. Complaints of either side are looked into by an arbitration provided for, in the Treaty.

Another riparian river involving Afghanistan and Pakistan is the river Kabul, a tributary of the Indus. The river brings around 25 MAF of water, derived not only from rains but especially from the glacial melt on the Hindu Kush Heights. Currently, there is no problem. Pakistan has built not only the Warsak Dam on the river but also exploits its tributaries like Swat, irrigating a substantial area. A dispute over the waters of Kabul cannot be ruled out, as and when Afghanistan decides to harness the river Kabul in its upper reaches, affecting the discharge downstream, and especially the supply of water to Warsak reservoir.

10.6 Problems of Domestic Water Distribution Among the Provinces and the Inter-Provincial Water Distribution Accord 1991

The distribution of canal water in Pakistan before 1991-Accord was made on an ad-hoc basis by the Federal Government every year on the season-to-season basis in *kharif* as well as *rabi* season. This was the arrangement since 1970. Before 1970, water was distributed between Punjab and Sindh under the Sindh-Punjab Draft Agreement 1945. After 1970, the distribution was on an ad-hoc basis.

To remedy the ad-hoc arrangement in the distribution of water, the Government of Pakistan appointed a number of commissions and committees that finally brought out a framework of the distribution of water that was accepted by all the provinces. An agreement known as the "Inter-Provincial Water Distribution Accord 1991" was signed by all the chief ministers of all the provinces.

TABLE 10.5 Allocation of Canal Water to Each Province in Kharif and Rabi Season Under the IPWA 1991

Provinces	*Allocation of Water in MAF*		
	Kharif	*Rabi*	*Total*
Punjab	34.07	18.87	55.94
Sindh	33.94	14.82	48.76
Khyber Pakhtunkhwa	a) 3.48	2.30	5.78
	b) 1.8	1.2	3.0
Balochistan	2.85	1.02	3.87
Total Pakistan	78.14	38.21	117.35

Source: Abdul Majid Kazi (2014) Overview of Water Resources in Pakistan: Pakissan.com, April 7, 2014

Under the Accord, 117.35 MAF have been allocated to the provinces from the existing canal system. The agreement also included a provision for sharing the storages and surpluses of water, in years of drought, and more than average availability of water in years of heavy rainfall. The allocation, in broad outlines, is as follows: Table 10.5

This allocation of water to each province is subject to change following a drought or a flood period. The accord has made provision for outflow to the sea to the tune of 10 MAF and also for a situation of reduced supply of water in the river Kabul. Equally, the award is alive to the decline in the volume of available water following an increase in the height of Mangla Dam or even enhanced need of water following the added area brought under cultivation. Despite some flaws here and there, this is an accord that guides the distribution of water among the provinces.

The accord may be reworked following the impounding of roughly 8 MAF by the Diamer-Bhasha project.

Water and Irrigation Management in Pakistan – The water resources of Pakistan, their conservation and distribution, are one of the best managed in the world. The country has added substantial capacity to the inherited structures and capacity after 1947. New reservoirs have been built, more canals and link canals have been added, several barrages have been constructed, groundwater utilisation is regulated and the main river Indus is kept flowing by allowing around 10 MAF of water to flow to the sea.

The Water and Power Development Authority (WAPDA) and Indus River Development Authority (IRDA) are ever watchful and keep devising new structures and mechanisms for enhancing the supply of water as much as to avoid wastage of water.

11

AGRICULTURE IN PAKISTAN

Agriculture in Pakistan can be viewed as a traditional occupation of its people as much as an important segment of the nation's economy. Accounting for over 23 percent of the national Gross Domestic Product (GDP) and employing over two-thirds of Pakistan's population, the importance of agriculture, the traditional occupation of the people, has remained undiminished in the life and culture of the people of Pakistan. Of all the segments of Pakistan's economy, agriculture is one that is buoyant, demonstrating persistent growth in production and yield. With only 27 percent of its geographical area under cultivation, the country has been able, over the decades, to produce adequate food for its people, raw material, especially cotton for its textile mills and even an exportable surplus of a special variety of rice known as "basmati". Today, with just over 22 million hectares of land under cultivation, Pakistan produces over 38 million tonnes of food grains, including 25 million tonnes of wheat and 7 million tonnes of rice, 13 to 14 million bales of cotton, 60 million tonnes of sugarcane and 4 million tonnes of oil seeds. Agricultural production has kept with the growth of population, stalling any catastrophic situation of famine or food scarcity. Lately, however, with an annual growth rate of 2.4 percent of the population of the country, a shortage of wheat, though marginally, is beginning to be experienced, and the country has imported wheat to meet the domestic demand.

Contributing 23 percent of the GDP of Pakistan and employing 45 percent of its labour, agriculture is central to the growth and development of Pakistan. Its importance far outweighs its monetary value, and its contribution to the nation's food security insured by a prosperous agriculture is vital to the sociopolitical stability of the society. Any social unrest following

DOI: 10.4324/9781003510147-14

food insecurity is far more unsettling than a political movement based on ideology. Pakistan has not witnessed any social unrest, and despite political instability, there have been no large-scale peasant movements to complicate the situation. The peasantry, in Pakistan, has by far, remained tethered to its farms, fields and orchards and engaged in agriculture, minimally influenced by any political controversy and squabbles of the political class.

Factors determining the distribution of agricultural land in Pakistan – Agriculture being a land-based activity, the availability of flat lands covered with fertile soil is the basic necessity for any profitable agriculture. Climate, including temperature conditions over the year and the periodic contribution of rainfall, is no less important. Moisture retention in the soil, sunshine for photosynthesis and temperature appropriate for the ripening of crops are vital factors that determine the health of the crops. Deficiency in soil moisture is met by irrigation, insuring optimal crop growth. Of all the factors, however, the availability of suitable land is the basic and the most decisive factor that influences agriculture.

11.1 Land Availability: a Limiting Factor in the Agricultural Growth of Pakistan

A glance at the physical map of Pakistan would reveal that a large part of Pakistan is occupied by high and low mountains, the Hindukush and the western extension of the Himalayas in the north, and the Sulaiman and Kirthar Ranges in the west. The Pakistan Forest Institute has given the following "geographical Break-up of Land Area in Pakistan' (including Azad Kashmir, Gilgit and Baltistan) (Table 11.1).[1]

Much of the flat area of Pakistan is confined largely to the Indus Plain that covers the Punjab and Sindh Plains which, together, account for more than 80 percent of the agricultural land of Pakistan. A large part of Khyber Pakhtunkhwa, including the erstwhile Federally Administered Tribal Area (FATA) is hilly. There are, however, some fertile basins like those of the rivers Kabul, and Swat, and the Indus in the Sub-Himalayan region. Not quite similar, but agriculturally very unfavourable is the Balochistan province with its west-east oriented Coastal and Central Makran Ranges in the southern half

TABLE 11.1 Relief Categories and Land Area (in Hectares and Percent)

Relief Categories	*Area in Thousand Hectares*	*Percentage of Area in Pakistan*
Plains	20,580	23.40
Mountains	66,487	75.57
Water Bodies	913	1.03
Total for Pakistan	87,980[2] or 8,79,800 km ^{2}	100.00

of the province, a dry desert occupying the mid-north area and Ras Koh and Chagai Hills occupying the northern part of the province. The intervening space occupied by mid-altitude plateau ranging from 500 to 1000 m ASL is dry with only ephemeral streams, not leaving much land suitable for cultivation.

The country extending from the lofty peaks in the Himalayas in the north to the mouth of the Indus joining the Arabian Sea is a beautiful mosaic of awe-inspiring Himalayan landscape with snow-covered peaks, splendid river valleys like Swat and Chitral, the crumbling, dry mountain relief of Sulaiman–Kirthar Ranges with deep ravines devoid of much greenery. This agriculturally unproductive terrain is compensated by the endless flat land of the Indus Plain. Starting from 34°N latitude, widening gradually as one proceeds southward, it embraces in its sweep the Punjab Plain, drained by Sutlej, Ravi, Chenab and Jhelum Rivers, all coming together to form Panchnad, and almost the entire province of Sindh further south, leaving only a narrow hilly fringe on the west, formed by north-south aligned Kirthar Ranges.

Thus, out of the total cultivated area of 21–25 million hectares of Pakistan, 17.35 million hectares (81.64 percent) lie in Punjab and Sindh. In fact, Punjab alone accounts for 58 percent (58.64 percent) of the total cultivated area of Pakistan.

The climate, besides relief, is the second most significant factor in limiting the area suitable for agriculture. The year-long low temperatures in the High Himalayas make the region unsuitable for cultivation, except in the narrow valleys of Gilgit and its tributaries, where some crops grow on the lower river terraces in summer. Elsewhere, as in the southern parts, aridity, caused by scant rainfall and non-availability of moisture to the crops, has a restrictive effect on agriculture. The low rainfall areas of Punjab and Sindh are compensated by irrigation. Punjab and Sindh enjoy a high level of canal and tube well irrigation. The canal network that was developed by the colonial British rulers has been further expanded by Pakistan after the country came into existence in 1947. Though most of the rivers, tapped for canal irrigation, originate in the Himalayas, beyond the borders of Pakistan, an adequate and uninterrupted supply of water is assured from these rivers by an international agreement between Pakistan and India, known as the Indus Water Treaty, signed between the two countries in 1960.

Not much grows in Pakistan without irrigation. Rain-fed agriculture exists, on a very small scale, confined largely to the hilly areas of Khyber Pakhtunkhwa and northern Balochistan, which enjoy some amount of summer as well as winter rains.

11.2 Land Utilisation and Agricultural Land Use in Pakistan

Agricultural statistics of Pakistan, though available for the latest years, a universal flaw in the land utilisation statistics is a category called "Reported Area", which is limited to just over 70 percent (72.55 percent) of the total geographical area of the country. Of the total geographical area of 79.61 million hectares, land utilisation categorisation is available for only 57.76 million hectares, that is, 72 percent of the total area. This implies that for the remaining 28 percent of the area amounting to over 22 million hectares, land use statistics remains undefined. The official clarification given in the footnote explains "Total Area Reported" as the "total physical area of the village / *deh, tehsils* or districts etc". Inaccessible snow-covered regions of glaciated areas seem to have been excluded from the records. For all practical purposes, one assumes that these are unproductive areas and no agriculture is likely to be practised there. Thus, it appears fair to calculate the percentage of agricultural land as the percentage of the total geographical areas of the country and not of the "Reported Area" (Table 11.2) (Figure 11.1).

Cultivated Area – In a synoptic view, only 20 percent of the area of Pakistan would appear under crops, which is the actual Net Sown Area. Including the current fallow, the total cultivated area of Pakistan is 21.88 million hectares, which is around 29 percent of the geographical area of the country. In a year, however, part of the cultivated area is cropped more than once, thus increasing the total cropped area in a specific year. For the year 2018–2019, Pakistan had 23.2 million hectares of cropped area, which is 29 percent of the geographical area of Pakistan. Thus, in a year, Pakistan has the following division of cultivated land (Table 11.3).

Despite over 9 percent of land left fallow every year, the total cropped area exceeds the total Net Sown Area by 10 percent. This is because of multiple cropping in over 30 percent of the cultivated land.

Provincial Variation of Land Use – Physical and climatic conditions, favourable for one tract but adversely affecting the other, contribute to a wide variation in the land use of different provinces. Seen in relative terms, as expressed in percentages, a much larger proportion of land is available for cultivation in Punjab and Sindh than in Khyber Pakhtunkhwa (KPK) and Balochistan.

If one looks holistically at Pakistan, a bleak feature of agricultural land use emerges. The two provinces of Balochistan and Khyber Pakhtunkhwa (KPK) with dry mountainous terrain pull the averages down. For Pakistan as a whole, the Net Sown Area in any average year is about 16 million (15.92 million) hectares, which is just about 20 percent of the geographical area of the country. Adding the area sown more than once, the figure of total cropped area in a year comes to 30 percent (29.39 percent) of the geographical area of Pakistan.

The picture turns encouraging when one looks at the agriculturally rich provinces of Punjab and Sindh, especially the former. Around 70 percent of the total Net Sown Area of Pakistan lies in Punjab. If one takes the total cropped area, Punjab accounts for 73.6 percent of the total cropped area. Sindh is a poor second, and the other two provinces pale into insignificance

TABLE 11.2 Land Utilisation in Pakistan (2018–2019) in Million Hectares

Provinces	*Geog. Area*	*Geog. Area Reported (4 to 7)*	*Forest Area*	*Not Available for Cultivation*	*Culturable Waste*	*Cultivated Area (8+9)*	*Current Fallow*	*Net Sown Area*	*Area Sown More Than Once*	*Total Cropped Area (9+10)*
1	*2*	*3*	*4*	*5*	*6*	*7*	*8*	*9*	*10*	*11*
Punjab	20.63	17.52	0.48	2.89	1.55	12.6	1.98	10.62	5.87	16.49
Sindh	14.09	14.08	1.03	6.85	1.61	5.19	2.51	2.68	1.00	3.68
KPK	10.17	8.36	1.24	3.92	1.34	1.86	0.66	1.29	0.60	1.89
Balochistan	34.72	18.05	1.16	9.83	3.83	3.23	0.57	1.05	0.01	1.06
Total	79.61	58.01	3,91	22.89	8.33	22.88	7.24	15.64	7.48	23.12
Percentage in each category	100.00	72.86	4.91	28.75	10.46	28.74	9.09	19.64	9.40	29.04

Source: Pakistan Bureau of Statistics, Pakistan Statistical Year Book 2018, Table 1.1 p.2.

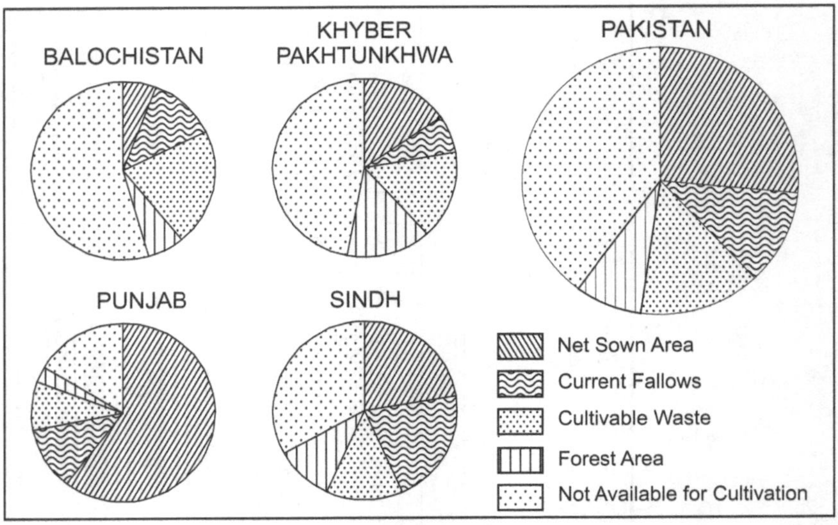

FIGURE 11.1 Land Utilisation in Different Provinces of Pakistan (2018–19)

TABLE 11.3 Principal Land Uses of Pakistan (Area in Million Hectares)

Category	*Area*	*Percentage of Each Category to the Geographical Area of Pakistan*
1. Net Sown Area	15.64	19.64
2. Current fallow	7.24	9.09
3. Area sown more than once	7.48	9.39
4. Cultivated area in a year (1+2)	22.88	28.74
5. Total cropped area in a year (1+3)	23.12	29.04

Source: Extracted from table .2.

as far as agriculture is concerned. This dominance of Punjab is also reflected in the field of irrigation and the level of agricultural production (Table 11.4).

Distribution of Cropped Land – The distribution of cropped area, among the provinces, is highly skewed when compared with the population of each province. Of the total cropped area of 23.12 million hectares of Pakistan, 16.5 million hectares, that is more than 70 percent, lies in Punjab. As against this, Sindh, Khyber Pakhtunkhwa and Balochistan have a relatively small share of cropped area of the country. This skewed distribution is also reflected in the cropped area per hundred of the population. Juxtaposed against the population of different provinces, it is observed that Punjab has more than their share, while the others have less. Punjab with 52.9 percent of the population of Pakistan has 71 percent of the cropped area of the country. It has 15 hectares of cropped area per hundred of its population.

TABLE 11.4 Distribution of the Total Cropped Area of Pakistan Among the Provinces (2018–2019) (Million Hectares)

Provincial Units	*Cropped Area (in million hectares)*	*Percentage Share of the Total Cropped Area of Pakistan*	*Cropped Land (in ha) Per 100 Population*
Pakistan	23.12	100.00	11.12
Punjab	16.49	71.13	14.98
Sindh	3.68	15.96	7.68
KPK	1.89	8.17	5.26
Balochistan	1.06	4.58	8.58

Source: Calculated on the basis of population of Pakistan as per 2017 census.

In contrast, Sindh, Khyber Pakhtunkhwa and Balochistan with 23 17 and 6 percent of the population of Pakistan, respectively, have 16, 8 and 4.5 percent of the country's cropped area. In each of these three provinces, the share of cropped area is less than what is warranted by the size of their population. In each of the three provinces – Sindh, Khyber Pakhtunkhwa and Balochistan – the cropped area per 100 persons is below the national average of 11.12 hectares per hundred population.

It must be emphasised that the hectarage of cropped area depends on the physical conditions of relief, soil and climate. Punjab is undoubtedly better placed with the plains – the different *doabs* – of the tributaries of the Indus, receiving canal irrigation and making the province agriculturally very productive.

Man-made factors determining the productivity of agricultural Land – While the physical factors, like terrain conditions, soils and climatic conditions largely determine the availability of agricultural land, it is the man-made factors that determine agricultural productivity. The use of irrigation, application of fertilisers and use of improved variety of seeds, improved methods of cultivation and other plant protection methods determine the yield and overall agricultural productivity of the land. The two provinces, Punjab and Sindh, have greatly benefitted from the canal and tube-well irrigation from the Indus and its tributaries.

Irrigation though a man-made facility, by far the most important factor to boost agricultural production, is not universally available and requires, besides human enterprise, a permanent source of water like rivers or underground water stored in the aquifers. In this respect, Punjab has a natural advantage over other provinces. Sindh has been able to tap water from the Indus, as the latter flows through the centre of Sindh. In contrast, Balochistan is devoid of any significant source of water, and Khyber Pakhtunkhwa benefits in a limited measure from the river Kabul and a few northern tributaries of the Indus, especially Swat and the sub-tributary Panjkhora.

11.2.1 The Impact of Irrigation on Cropped Area of the Provinces

Almost half the Net Sown Area of Pakistan has the benefit of multiple cropping, supported by irrigation. Even the hilly province of KPK has more than one-third of its Net Sown Area under multiple cropping. This has been made possible by irrigation projects like the Warsak Dam on the Kabul River on the one hand and irrigation from the hill streams in the inter-montane valleys on the other. A lower level of evaporation at higher altitude also helps in restraining the speedy depletion of moisture from the soil in the mid-altitude mountain valleys of Khyber Pakhtunkhwa (Figure 11.2) (Table 11.5).

In most of the provinces, irrigation is received largely from either canals and to some extent from tube wells. Punjab is, however, an exception. It receives large-scale irrigation from what is known as canal tube well – the tube wells sunk in the canal where the groundwater has risen because of seepage. Besides, there are canal wells, which are also used for irrigation.

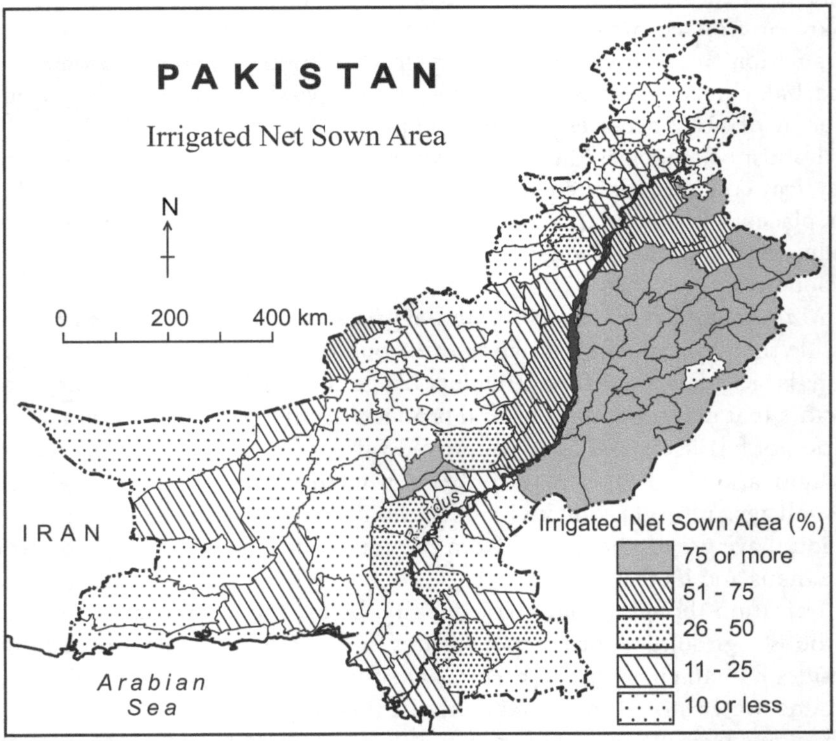

Sources: Directorate of Crops Reporting Services, Balochistan, and Khyber Pakhtunkhwa (Peshawar), Bureau of Statistics, Punjab, Lahore
Crop Reporting Service centre (C. R .S .C) Govt. of Sindh, Hyderabad

FIGURE 11.2 Intensity of Irrigation in Pakistan Expressed as Percentage of Net Sown Area

TABLE 11.5 The Irrigated Areas Against the Total Cropped Areas in Different Provinces (in Million Hectares)

Areal Units	*Total Cropped Area*	*Total Irrigated Area*	*Irrigated Area in Percentage of Cropped Area*
Punjab	17.23	14.41	83.6
Sindh	3.45	2.09	60.6
Khyber Pakhtunkhwa	1.59	0.93	50.5
Balochistan	1.13	1.2*	106
Pakistan	23.4	18.63	79.6

Source: Agricultural Statistics, Pakistan Bureau of Statistics 2011, Govt. of Pakistan, p. 115.
*Several crops receive irrigation more than once.

TABLE 11.6 Fertiliser Consumption in the Provinces of Pakistan in Thousand Tonnes (2016–2017)

Province	*Annual Consumption*	*Cropped Area in Million Hectares*	*Amount of Fertiliser Consumed per 100 ha (in Thousand Tonnes)*
Punjab	3,498.0	16.49	21
Sindh	1,185.5	3.68	32
K. P.	225.4	1.89	12
Balochistan	131.0	1.07	3
Pakistan	5,039.9	23.13	20

Source: Pakistan Statistical Year Book, 2018, Pakistan Bureau of Statistics, Government of Pakistan, p.62.

A very high percentage of irrigated area is explained by the fact that virtually nothing grows without irrigation in Balochistan. (Irrigation being an important aspect of agriculture especially in Pakistan, a separate chapter is devoted to irrigation, especially the canals and canal irrigation).

Application of Fertilisers: The application of fertiliser is a common feature of modern agriculture, though organic farming involving organic manure is being emphasised today. But, it has limited uses, and giving up the application of nutrients directly through chemical fertiliser involves the risk of a sudden fall in productivity, as has happened in certain countries. Pakistan consumes approximately five million tonnes of fertiliser annually, distributed among the provinces as follows (Table 11.6):

Of the total consumption of 5 million tons of fertiliser, roughly 70 percent is consumed by the farmers of Punjab. About 24 percent is consumed in Sindh, and Khyber Pakhtunkhwa consumes a very insignificant amount of fertiliser. It may be noted that Sindh, with a year-long benefit of irrigation from the three barrages (Guddu, Sukkur and Kotri) consumes more fertiliser per hectare than all the other provinces of Pakistan. Even in terms of the amount of fertiliser consumed, the yields per hectare of both crops are higher in Sindh than in Punjab. This is reflected in the yields of wheat and rice.

It may be noted that despite 7 to 8 million tonnes of fertiliser in Pakistan, the country still imports fertiliser worth millions of rupees.

11.3 Structure of Holdings

The size of holdings and nature of tenure are other factors that influence the agriculture of a country. Pakistan is much better placed, compared to its two South Asian neighbours in terms of size of holdings and the nature of tenure. In a comparative view, Pakistan shows a much higher average size of holdings and a much higher proportion of larger holdings as compared to India and Bangladesh (Table 11.7).

What is evident is the excessive fragmentations of holdings in India and Bangladesh as compared to Pakistan. The average size of holdings in Pakistan is six times higher than in India and ten times higher than the average size of holdings in Bangladesh. If we drop the deceptive average figures, and take the median size, the latter will be in the 1–5 ha category in Pakistan, possibly 3 or 4 ha, whereas in the case of India and Bangladesh, even the median size of holdings would be less than 1 ha. This is also reflected in the number of farms. Pakistan has over 5 million farms, a number half of that of Bangladesh and roughly 5 percent of the number of India.

The large size of farms in Pakistan is an advantage to the country in modernising its agriculture, especially in mechanising traditional farming practices. The productivity, measured by yield, however, does not always depend on the size of farms. Often in a situation of high population density in agricultural areas, intensive farming gives a higher yield. The crop yields in Pakistan are still low; the average is always pulled down by very low productivity in poor tracts of low productivity in Balochistan and Khyber Pakhtunkhwa.

11.4 Cropping Pattern

The agricultural crops in Pakistan are divided into five categories by the Department of Agriculture, Government of Pakistan.

1. Food crops – These are largely cereals, like wheat, rice, jowar, maize, bajra and barley.
2. Cash crops – A category that includes sugarcane, cotton, tobacco, jute, sugar beet, guar seed and sun-hemp.

TABLE 11.7 A Comparison of the Structure of Holdings in India, Pakistan and Bangladesh

Country	*No. of Farms in thousands*	*Average Size of Holdings (ha)*	*Percentage of Holdings in Each Size Class*					*Land in Farms (thousand ha)*
			< 1ha	*1-5*	*5-10*	*10-20*	*>20 ha*	
Pakistan (1990)	5,071	9.3	2.7	54.0	12.2	4.7	2.1	22,150
India (1990)	105,300	1.6	59.0	32.2	7.2	1.6		165,600
Bangladesh (1983)	10,045	0.9	70.3	27.6	2.5	0.2		9,137

Source: Britannica Book of the Year 1998:796–801.

3. Pulses – These include gram, *mung, masoor, mash, mattar* and other kharif and rabi crops.
4. Edible oil seeds – These include rape and mustard seeds, sesamum, groundnuts, soya bean, safflower and cottonseeds.
5. Condiments

For the country as a whole, the acreage under each of the crops, besides vegetables and fruits, is given below (Table 11.8).

As can be seen, food crops dominate the agricultural landscape of Pakistan, followed by cash crops, pulses and oil seeds. The most striking feature of the production of cereals in Pakistan is a very spectacular growth in yield and production. From 25 million tonnes during the year 2001–2002, the production of food crops has risen to 40 million tonnes in 2019–2020, a 60 percent increase in a span of two decades. Cereals occupy more than half the cultivated land, followed by cash crops and pulses. The production of cereals in different provinces is neither in proportion to the area nor to the total production but highly skewed in favour of Punjab, as seen from the following table (Table 11.9).

Taken individually, the following food crops, some occupying more than a million hectares, are ranked in the order of area under each of them.

11.4.1 Food Crops

Wheat, rice, maize, jowar, bajra and barley are the principal food crops of Pakistan. Of the total area of over 13 million hectares under food crops, over 8.6 million hectares or around two-thirds of the total food crop area are covered by wheat followed by maize and rice. Wheat and rice together occupy 86 percent of the total food crop area under cereals. The area under

TABLE 11.8 Pakistan: Area Under Different Categories of Crops (2010–2011) in Thousand Hectares

Crop groups	*Area in Thousand ha*	*Percentage of Area under Each Category*
1. Food crops	13,097	56*
2. Cash crops	4,808	21
3. Pulses	1,329	6
4. Oil seeds	664	3
5. Condiments	288	1
6. Vegetables	411	2
7. Fruits	836	4
8. Others	1,967	7
Total	23,034	100

*Decimal points are left out of percentages.

TABLE 11.9 Production of Cereals in Different Provinces (2018–19) in Thousand Tonnes

Provinces	*Production of Cereals in Thousand Tonnes*	*Percentage of the Total Production of Pakistan*
Punjab	24,349.0	75.20
Sindh	5,608.1	16.30
KPK	2,006.0	5.84
Balochistan	902.8	2.63
Total	34,302.5	99.97

TABLE 11.10 Rank and Area Under Cultivation for Each Food Crop in Thousand Hectares and in Percentages (2018–2019)

Rank	*Crop*	*Area under Cultivation in Thousand ha in 2018-2019*	*Percentage of Total Food Crop Area*
1.	Wheat	8,678	63.72
2.	Rice	2,810	20.63
3.	Maize	1,374	10.09
4.	Bajra	456	3.35
5.	Jowar	241	1.77
6.	Barley	57	0.41
Total		13,618	99.97

Source: Pakistan Economic Survey 2019–20, Finance Division, Government of Pakistan, Table 2.6 to 2.9, p. 21–23.

food crops has increased from 11.6 million hectares in 1991–1992 to over 13 million hectares in 2011–2012, showing the increasing concern for greater production of food.

Wheat – Being the staple food of the people, wheat is the most important crop of Pakistan, with over 36 percent of the total cultivated land, and over 24 million tonnes of annual production, wheat occupies a central position in the agricultural economy of Pakistan. Wheat contributes 2.2 percent to the GDP and over 10 percent to the value added in agriculture in Pakistan.[3] It has a modest average yield of 2,787 kg/ha.

Wheat is a ubiquitous crop in Pakistan growing in all parts of the country with a variable share of the cultivated land. The distribution of area under wheat and production in the four provinces of Pakistan (2017–18) were as follows (Table 11.11):

Much of the area under wheat and its production lies in Punjab, the agricultural core of Pakistan. The province accounts for over 75 percent of the acreage under wheat and produces over 19 million tonnes of wheat, which is over three-fourths of the total production of Pakistan. There is a disproportionate distribution of area under wheat in the four provinces. All three

TABLE 11.11 Area Under Wheat in Thousand Hectares and Production in Thousand Tonnes (2017–2018)

Provinces / Pakistan	*Area in Thousand ha*	*Percentage*	*Production Thousand Tonnes*	*Percentage*
Punjab	6,555.8	74.52	19,178.5.	76.48
Sindh	1,089.6	12.38	3,639.5	14.51
Khyber Pakhtunkhwa	753.4	8.56	1,322.7	5.27
Balochistan	394.4	4.48	935.4	3.73
Pakistan	8,797.2	99.94	25,076.1	99.99

Source: Pakistan Statistical Year Book 2018, Government of Pakistan.

provinces, Sindh, KPK and Balochistan, cultivate far less wheat in proportion to their area or population. Even Sindh with the fertile Indus delta with 17 percent of the area and over 23 percent of the population of Pakistan produces only 14.73 percent of the total production of wheat. The two provinces Balochistan and KPK with over half the area of Pakistan produce less than 10 percent of the production of wheat, the staple food of Pakistan.

In Punjab, central belt along Chenab and its tributaries, comprising the districts of Jhang, Faisalabad, Muzaffargarh and further south the region Bahawalpur and Rahim Yar Khan in the irrigated tract of Central Punjab, have a high concentration of wheat production. In Sindh, the central and northern parts produce wheat though not as intensively as in Punjab. In KPK, the crop is confined to the Peshawar Basin with the adjacent districts of Charsadda and Mardan. There is a stretch of land where wheat cultivation was practised a few decades earlier, but the head-ward extension of the Indus Plain along its two tributaries, the Gomal and the Badra, forming a conical plain with its apex at Sibi has produced a low land bordered by below 100 m ASL, known as Kachhi Desert or even Kachhi Plain. The lower part of the plain enjoys irrigation in the form of a desert canal, taken out from the Indus near Kashmore, the tri-junction of three provinces, Punjab, Sindh and Balochistan. Thus, the districts of Nasirabad, Jhal Magsi and Jaffrabad have started to produce enough wheat. Thus, Balochistan can boast of producing over 800,000 tonnes of wheat.

In the agricultural landscape of Pakistan, Punjab stands out in terms of cultivated area and agricultural production. The province accounts for 57 percent of the total cultivated land of Pakistan and produces 75 percent of the total tonnage of cereals.

Rice – Rice is the second important food crop of Pakistan. The area and production of rice in the four provinces is as follows (Table 11.12):

The apparently higher yield of rice in all the provinces as compared to Punjab is not an indication of poor rice culture in the province, but a very

TABLE 11.12 Area Under Rice and Its Production in 2017–2018 in Pakistan and the Provinces

Province / Pakistan	*Area in Thousand ha*	*Production in Thousand Tonnes*	*Area of the Province under the Crop as Percentage of the Total Area under Crop in Pakistan of Total Area*	*Percentage of Total Production in Each Province*	*Yield kg/ha*
Punjab	1840.9	3,698.0	66.75	52.40	1,842
Sindh	828.3	2,850.5	24.75	36.49	3,458
K.P.K.	61.6	147.5	2.07	1.84	2,087
Balochistan	169.6	653.8	6.41	9.20	3,386
Pakistan	2,900.6	7,449.8	99.90	99.93	

Source: Pakistan Statistical Year Book, 2018, Pakistan Bureau of Statistics, Government of Pakistan.

high proportion of "*basmati*" variety of rice is harvested in Punjab. Whereas Punjab has 75 percent rice acreage under *basmati* variety, Sindh, Khyber Pakhtunkhwa and Balochistan have 21, 23 and 24 percent, under *basmati* rice. Rice, the finest variety of rice is usually exported, and this is sourced from the rice fields of Punjab. Over 50 percent of the rice production of Pakistan comes from Punjab, another one-third from Sindh and the remaining 15 percent comes from K.P.K. and Balochistan.

In Punjab, rice, unlike wheat, is highly concentrated in the northern part, taking advantage of higher rains and traditional rice farming areas of Sialkot, in the Himalayan sub-montane region, including the districts of Gujranwala and Sheikhupura, Nankana Saheb and Jhang. In the case of Sindh, irrigated rice of coarse variety grows, largely, in the western part starting from Jacobabad, Kashmore and Larkana in the north to the deltaic districts of Badin in the south. The eastern arid districts of Tharpakar, Umarkot and Saughor are devoid of any rice cultivation. Some rice grows in the Dir and Malakand Districts of Khyber Pakhtunkhwa. But these don't count for much. Balochistan is a new entrant in the field of rice production, and like wheat, rice is produced with irrigation by the Desert Canal in Nasirabad District.

The story of rice in Pakistan is not very old. Even after the Chenab Canal system appeared in Punjab, wheat was the principal crop. Rice was grown in the sub-montane region like Dir and Chitral of erstwhile North-West Frontier Province in very inferior soils with very poor yield. The experiment of growing rice with irrigation was initiated after the mid-20th century, after the emergence of *basmati* rice, an important commodity for export.

It started with Indian Punjab and rapidly spread to Pakistan, initially in Punjab and then down to Sindh reaching the Arabian Sea coast. The widespread increase in the cultivation of rice can be gauged from the fact that the united Punjab before 1947, extending from Yamuna to Indus, the total area under rice in 1903–04 was 1074 sq. miles (2,65,400 ha), while the present acreage of rice in Pakistan Punjab is around 2 million hectares (1,977,600 ha). There is no doubt; there has been a greater expansion of rice farming than any other food crop. Today, rice forms 20 percent of the total food grain production in Pakistan.

The *basmati* rice produced in the country is now around 40 percent of the total production of rice, the remaining 60 percent of the rice production is of a coarse variety. The cultivation of *basmati* rice is officially encouraged by the government for its export value and earner of foreign exchange. The reduction in the area and production of rice, seldom exceeding 10 percent, is caused by the late arrival of monsoons and less rainfall. Besides, the late recession of the floodwaters of the Indus from the rice fields delays the sowing and causes a fall in production.

Maize – Maize is the third important food crop of Pakistan, though historically introduced in the early 20th century, it was the staple cereal ahead of millet and rice. Maize was the principal food crop of the montane and sub-montane regions. The total area of over a million hectares under maize is almost equally divided between Punjab and Khyber Pakhtunkhwa, with insignificant acreage in Sindh and Balochistan. Punjab, no doubt, shows a very high yield of 5,000 kg/ha in a favourable year in contrast to Khyber Pakhtunkhwa, where the yield of maize, seldom, if ever, rises above 2,000 kg/ha. Maize in Pakistan is used essentially as a cereal for human consumption, besides being used as poultry feed and in commercial starch production.

In Punjab, the central and east central districts of Jhang, Faisalabad, Okara, Pakpattan and Sahiwal have a high concentration of maize cultivation. In Khyber Pakhtunkhwa, maize is widely grown, with a heavy concentration in Mardan and Manshera in the central part and Swat and Kohistan Districts in the mountainous north. The introduction of hybrid maize has increased the yield of the crop. Unlike other crops, there is a spectacular increase in the yield and production of the crop. The distribution of maize in Pakistan is as follows (Table 11.13):

Maize accounts for roughly 10 percent of the total production of food grains. In recent years, its production has further risen in recent years. A decade ago in 2008–2009, the area under maize in Pakistan was 1,052,000 hectares that rose to 1,250,000 hectares in 2018, an increase of roughly 20 percent. Similarly, the production of maize reached almost 6 million (5,901,600) tonnes from 3.6 million tonnes in 2008, an increase of over 60 percent over a period of ten years.

TABLE 11.13 Area Under Maize in Hectares and Production in Tonnes (2017–2018)

Province/State	*Area under Maize in Thousand ha (2008–2009)*	*Production in Thousand Tonnes (2017–2018)*
Punjab	770.3	5,028.0
Sindh	3.8	3.6
Khyber Pakhtunkhwa	474.0	867.0
Balochistan	2.9	3.0
Pakistan	1,250.8	5,901.6

Source: Pakistan Statistical Year Book, 2018, Pakistan Bureau of Statistics, Government of Pakistan.

Other Crops – Other food crops include gram, *bajra* (Pearl millet) and *jowar* (sorghum) in that order. Gram (*Cicer arietinum*) occupies as much area as maize and is the principal pulse crop, though its production and yield are relatively low, yet, 2018, over one million hectares of land were under gram, next only to cotton in the non-cereal category, and the production of gram was 325,000 tonnes. The area and production of grams have shown a consistent decline, most likely because of changes in food habits of the people.

12.4.2 Non-Food Crops

Cotton and *sugarcane* are the two principal non-food crops. The most important commercial crop of Pakistan, cotton ranks second only to wheat in terms of area. Grown in an area of roughly 2.7 million hectares, Pakistan produces 10 to 13 million bales of cotton annually, each bale weighing 375 lbs. Unlike rice and some other crops, the acreage under cotton has remained more or less constant, hovering around 2.7 million hectares.

Cotton is the principal raw material for the textile industry of Pakistan. Pakistan has a flourishing textile industry with 453 textile mills and 50 integrated units and 403 spinning units with over 9 million spindles and 140,000 rotors. The industry feeds on the locally produced cotton.

Besides, cottonseed is an important source of edible oil. With more than 60 percent of the edible oil extracted from cottonseed, the overwhelming importance of cotton in the national economy is obvious. The crop accounts for 7.0 percent of the value added in agriculture and 1.5 percent of the GDP. The production of cotton is almost exclusively confined to Punjab and Sindh with just about 1 percent of the area under cotton lying in Balochistan. Khyber Pakhtunkhwa does not produce any cotton because cultivated land is scarce and exclusively devoted to food crops. The distribution of area under cotton and its production in the provinces in 2008–2009 is as follows (Table 11.14):

TABLE 11.14 Distribution of Area Under Cotton, Production in Thousand Bales, and Yield (2008–2009)

Province/ State	*Area in Thousand ha*	*Percentage of Total Area under Cotton*	*Production in Thousand Bales*	*Percentage of Total Production*	*Yield Bales/ha*
Punjab	2,223.6	78.82	8,751.3	74.04	3.93
Sindh	561.5	19.90	2,978.4	25.19	5.30
K. P. K.	0.2	–	0.5	–	–
Balochistan	34.7	1.23	89.2	0.75	2.57
Pakistan	2,820.0	99.95	11,819.4	99.98	4.19

Source: Pakistan Statistical Year Book, 2018, Pakistan Bureau of Statistics, Government of Pakistan, p.13.

Pakistan with a production of 2.12 million tonnes of cotton has a share of 9–10 percent of the world production of cotton. As in most crops, Punjab has the highest share of hectarage as well as production of cotton among the provinces of Pakistan. In fact, more than 99 percent of the cotton is produced in the two provinces of Punjab and Sindh, the former producing 75 percent and Sindh about 25 percent of the total production of cotton in Pakistan. In Punjab, the cultivation of cotton is largely confined to the southern half of the province. The hilly Sub-Himalayan region, or the northern region, comprising the districts along the Grand Trunk Road don't produce any cotton, and there is a concentration of cotton in the southernmost districts of the province, starting from the mid-point of Khanewal District. The eight or nine southern districts in the province account for over 85 percent of the total produce of cotton in the province.

It is miraculous the dry districts of Punjab with an annual rainfall not exceeding 250 mm, once a dry grass–covered wasteland, now form agriculturally the most productive region of the provinces. The districts of Bahawalpur, Multan and Rahim Yar Khan were once desolate wasteland, but the canal network from the Indus and its tributaries, right into Sindh, have rendered this dry land into one of the most productive regions of the subcontinent.

The yield of cotton is higher in Sindh than in Punjab. The average yield of Pakistan is 700 kg/ha, but it is highly variable depending on the weather and the plant disease. The yields are also regionally highly variable. Punjab has a yield of 667 kg/ha, whereas Sindh has a much higher yield of 836 kg/ha, about 25 percent higher than in Punjab. But, there are some years as in 2009–2010, Punjab had a yield of 597 kg/ha against 1144 kg/ha for Sindh. It appears as one moves south, the yield increases. The sunshine and high summer temperatures produce a burst of cotton bolls, setting the entire field aglow with white cotton, not very dissimilar to the situation in Egypt.

To improve the quality of cotton and, in turn, the quality of fabrics, the Government of Pakistan has established a supervisory Institution, the "Pakistan Cotton Standard Standardization Institute", which is charged with regulation and improvement of the quality of cotton in the country. Being prone to attack by pests, the yield and production are highly variable. When attacked by thrips (*Thrips tabaci*), whitefly (*Bemisia tabaci*) or cotton leaf curl virus (CLCV), the production is very severely affected. Despite the best efforts of farmers and official agencies, adverse weather like heavy localised monsoon rains also affects the yield and production of the crop. Despite all the odds, the area under cotton and its production have consistently increased during the last two decades, from 1.637 million tonnes in 1990–1991 to 2.125 million tonnes in 2012–2013.

Sugarcane – Grown over an area of over a million hectares (1.058 million ha in 2011–2012), Pakistan produces around 60 million tonnes of sugarcane annually. The actual production in 2011–2012 was over 58 million tonnes. The area under sugarcane and its production in Pakistan for the year 2008–2009 is as follows. The annual variation in sugarcane production, unlike cotton, is minimal, and the acreage and production figure of any one year are a fair representation of the mean (Table 11.15).

The climatic requirement for sugarcane is high temperature as well as enough high rainfall. In Pakistan, irrigation from canals as well as tube wells is used all over to produce sugarcane. Much of the sugarcane produced in the country finds its way to sugar mills. It has to be emphasised that these sugar mills are located in the midst of or in the vicinity of the sugarcane growing area. In Punjab, sugarcane production is heavily concentrated in the central parts to the extent that over 30 percent of the sugarcane is produced in the four or five districts of Faisalabad, Jhang, Sargodha and Kasur and Toba Tek Singh. An important addition is Rahim Yar Khan, the

TABLE 11.15 Area and Production of Sugarcane in Pakistan (2008–09), Area in Thousand hectares, Production in Thousand tonnes

Provinces / State	*Area under Sugarcane Thousand Hectares*	*Percentage of Area*	*Production of Sugarcane in Thousand Tonnes*	*Percentage of Production*
Punjab	666.5	64.74	32,294,7	64.50
Sindh	263.9	25.63	15,304.3	26.58
K. P. K.	98.3	9.54	4,408.7	8.80
Balochistan	0.8	–0.08	37.9	0.07
Pakistan	1029.5	99.99	50,045.6	99.95

Source: Pakistan Statistical Year Book, 2018, Pakistan Bureau of Statistics, Government of Pakistan, p.12

southernmost district of Punjab. Most of these districts have several sugar mills. In fact, the five districts of Jhang, Faisalabad, Sargodha, Rahim Yar Khan and Muzaffargarh, together have around 25 sugar mills, more than half the total number in Punjab. In Sindh, sugarcane cultivation as much as the sugar mills is concentrated in the Indus delta region of the province. Hyderabad, Thatta, Badin, Nawabshah and Tando Allahyar are some of the districts in the delta region that together cover two-thirds of the total sugarcane grown in Sindh. Naushahro Feroze and Khairpur are important pioneers of cane in the northern part of Sindh.

In recent years, the production of sugarcane has increased, and the country seems to have produced 62 million tonnes of sugarcane. From 2008 to 2012, there has been a steady increase in the area and production of sugarcane as well as in the yield, which has spurted from 48,634 kg/ha in 2008–2009 to 55,580 kg/ha in 2012–2013.

Cultivation of sugarcane in Khyber Pakhtunkhwa is confined to the Peshawar Basin, incorporating Peshawar, Charsadda, Nowshera and Mardan districts and additionally Dera Ismail Khan. The above districts account for 80 percent of the area under sugarcane and its production. Charsadda and Mardan are especially known for their sugarcane. The sugar mills are concentrated at Dera Ismail Khan and Mardan. Leaving aside these principal crops, there is a considerable acreage given to coarse millets and oilseeds.

Oilseeds – Among the oilseeds, besides cotton seeds, which are used as an important source of edible oil, the principal oilseeds include sunflower, rapeseed and mustard, groundnut, sesame and soybean. Pakistan produces more than 400,000 tonnes of edible oil from cottonseeds. Cottonseeds contribute more than 50 percent to the total oil production of Pakistan. In the year 2009–2010, the area under each of the oilseeds and production is as follows (Table 11.16):

While mustard, groundnut and sesame are predominantly grown in Punjab, sunflower is the most favoured oilseed in Sindh. There are some other minor oilseeds, but they are not very significant.

Fruits and Vegetables – Among the fruits, citrus fruits, mangoes and apples are important. Citrus and mango, both are important export crops.

TABLE 11.16 Area and Production of Oilseeds in Pakistan (2009–2010)

Crops	*Area in ha*	*Production in Tonnes*	*Yield kg/ha*
Sunflower	256,121	325,478	1271
Rapeseed and mustard	190,400	162,478	852
Groundnut	87,400	53,200	609
Sesame	79,800	33,400	419

Source: Pakistan Statistical Year Book, 2018, Pakistan Bureau of Statistics, Government of Pakistan, p.25–28.

Over 200,000 hectares of land are under citrus fruits in Pakistan, producing over 2 million tonnes of fruits. Mango, a traditional fruit crop of Pakistan, is grown in Punjab and Sindh. An area of 173,800 hectares was under mango orchards in 2009–2010. In recent years, the production of apples has increased largely in Balochistan, the sole province producing apples in Pakistan. Many other fruits of temperate latitude like apricot, peaches, pears and plums are predominantly produced in the hilly areas of Balochistan and Khyber Pakhtunkhwa.

11.5 Agricultural Progress in Pakistan

In a period of two and a half decades, by whatever measure, one evaluates, Pakistan has made remarkable progress in agriculture. There has been an overall increase in the area and production of the principal food crops, though some of the less important food crops like *jowar* (sorghum), *bajra* and barley have declined in area of production. The change is more marked in Punjab and Balochistan, though the latter's contribution to the national agricultural product is marginal. Paradoxically, the area under food crops has declined by more than 10 percent, but not the production suggesting a clear and substantial increase in yield. The following is a comparable table of area and production of different food crops, as they have changed from 1981–1982 to 2008–2009 (Table 11.17).

Over a period of 27 years, 1981–2008, Pakistan has increased its area under food crops by just 26 percent but has more than doubled the production of food crops from a lowly 16 million tonnes in 1981 to over 35 million tonnes in 2008–2009. The increase in area is most striking in the case of Balochistan because of the introduction of canal irrigation in the Kachhi Plain. There is also an increase of more than one-third in the acreage under food crops in Pakistan, but the most remarkable case is that of Sindh, where despite a decline of over 10 percent in the area under food crops, there has been an increase of 60 percent in the production. The increase in cultivated area in Balochistan followed the arrival of canal irrigation taken out from the Begari and the north-western canals, taken out from the Indus at Sukkur. These have benefitted the districts of Nasirabad and Jhal Magsi in Balochistan.

Low Yields – A Persistent Phenomenon of Pakistan Agriculture – Though the progress in agriculture is undeniably impressive and there is an overall increase in the production of cereals, which has more than doubled in the last two decades, yet even the increased yields of crops are relatively low when compared with other countries of the world. That there has been no doubt an increase in yields during the last 25 years, yet these remain far below the yields of some of the Asian countries. A comparative yield of wheat, rice and sugarcane for the following six countries are given below (Table 11.18).

TABLE 11.17 Change in the Area and Production of Food Crops from 1981–1982 to 2008-2009

Provinces/State	*Area in 000 ha*		*Change in %*	*Production in 000 Tons*		*Change in %*
	1981-1982	*2008-2009*		*1981-1982*	*2008-2009*	
Punjab	7,224	9,951	37.7	10,184	25,070	146.2
Sindh	2,129	1,893	–10.08	3,832	6,161	60.7
K.P.K.	1,285	1,381	7.4	1,582	2,324	46.9
Balochistan	364	654	79.6	638	1,565	145.4
Pakistan	11,002	13,879	26.14	16,236	35,120	116.3

Compiled by Author

TABLE 11.18 Comparative Yields of Wheat, Rice and Sugarcane of Some Asian Countries and Pakistan

Countries	*Yield in kg/ha for 2013*		
	Wheat	*Rice*	*Sugarcane*
Pakistan	2,784.4	3,500.0	56,475.8
Bangladesh	3,013.8	4,375.5	40,580.6
China	5,050.5	6,724.9	69,028.6
India	3,153.8	3,659.8	67,430.8
Indonesia		5,152.0	74,888.9
Sri Lanka		3,888.8	53,846.2

Source: F.A.O. Crop Yields 2013.

As can be seen, the yields of these crops in Pakistan are much below the yield in the rice- or wheat-producing countries. Pakistan is essentially a wheat-producing country, and even in the case of wheat, the country shows a poor yield. The yields in China are incomparably higher, but yields in Pakistan are lower than even the neighbouring countries India and Bangladesh.

Domestically, Pakistan shows noticeable inter-provincial variations. The wheat yields are much higher in Sindh with 3,600–3,700 kg/ha in northern Sindh, the region compares favourably with other countries, but in Punjab, the yields are lower, bringing down the average. The reason appears to be the pressure of multiple cropping, in which case growing more than one crop in a field gets priority over getting a higher yield. Whereas in Sindh, almost half of the cultivated area is left fallow, in Punjab, it is around 12 percent. Proper aeration and permitting the fields to recuperate in their fertility have resulted in higher yields in Sindh. Yet, it is likely that by multiple cropping, Punjab compensates for the yield.

11.6 Livestock

Animal husbandry is as important an occupation in Pakistan as agriculture and is especially important as a source of foreign earnings. Traditionally, a family-based enterprise to meet the requirements of milk and other dairy products; livestock rearing is becoming an organised large-scale enterprise. It contributes roughly 12 percent of the national GDP. A source of additional income, besides agriculture, as well as an avenue of employment livestock has arrived centre stage in national economic planning. The gross value addition of the livestock sector at constant factor cost to the national income, accounts to roughly 756 billion Pakistani rupees every year (Table 11.19).

TABLE 11.19 Livestock Population in 2012–2013 as Estimated by the Ministry of National Food Security and Research*

Species	*Number in Millions*
Cattle	38.3
Buffalo	33.7
Sheep	28.8
Goat	64.9
Camels	1.0
Horses	0.4
Asses	4.9
Mules	0.2
Total	172.2

Source: Pakistan Statistical Year Book, 2018, Pakistan Bureau of Statistics, Government of Pakistan, p.73.

*The estimates are based on inter-Census growth rates.

TABLE 11.20 Production of Milk, Meat, Wool and Hides in Pakistan (2012–2013)

	Production in Thousand Tonnes
1. Milk and milk products	46,440
2. Meat	3,095
3. Wool	43.6
4. Skin and hides	50,712

Source: Pakistan Statistical Year Book, 2018, Pakistan Bureau of Statistics, Government of Pakistan, p.74.

The livestock is a source of dairy products, meat, wool and hides, and the estimated amount for these products for 2012–2013 is as follows (Table 11.20):

The production of milk is gradually increasing by 2–3 percent every year. Besides livestock, there is a flourishing poultry industry that employs over 1.5 million people. The poultry industry produces 26 percent of the total meat production in the country, and the industry is registering robust growth.

11.7 Future of Agriculture

Agriculture contributes around 24 percent to the GDP of Pakistan and employs roughly 43 percent of the total labour force of the country. As the manufacturing and the service sector grow, the overall share of agriculture in the GDP of a country comes down. Here is a comparison of the share of agriculture in the GDP of a few countries (Table 11.21).

TABLE 11.21 Comparison of the Share of Agriculture in the GDP of Different Countries

Country	*Share of Agriculture in the Gross Domestic Product*	*Percentage of Labour Force Employed in Agriculture*
Pakistan	23.9 (2012)	42.3 (2008)
India	17.5 (2008–09)	52.6 (2000)
Indonesia	14.7 (2011)	33.5 (2011)
Bangladesh	15.4 (2010–11)	43.8 (2004–05)
China	10.3 (2009)	38.7 (2009)*
USA	1.2 (2011)	1.5 (2011)
Germany	0.8 (2011)	2.1 (2008)

Source: Encyclopaedia Britannica, Book of the Year, 2014 (for Pakistan p. 691).
* Including mining and manufacturing.

Growth in the diverse sectors of the economy leads to a lowering of the share of agriculture in the GDP. In this case, Pakistan appears to be lagging behind, though it employs a relatively smaller labour force as compared to India and Bangladesh. Large farms and the use of machine have curtailed the labour requirement of Pakistan. A relatively high percentage of the contribution of agriculture to the national economy is no reflection on the efficiency in agriculture. It is more an indication of a slower growth of manufacturing and other sectors.

Progress during the last 25 years – During the last 27 years from 1981–1982 to 2008–2009, Pakistan has made unparalleled progress in agriculture. It increased the area under food crops marginally by 26 percent, but during the same period, the food crop production more than doubled, showing an increase of 116.3 percent. Similarly, during the same period, the area under wheat and rice grew by 26 and 49 percent, respectively, but the production of both these crops, the former increasing by more than 114 percent and the latter by 103 percent. This also leads to an inevitable conclusion that the yields of these crops also increased, notwithstanding the fact that the food crop yield in Pakistan is still low and needs improvement.

Much of the progress is achieved by harnessing the rivers, judicious water management, and standardised seeds and use of fertilisers. Of these, water management has been a primary concern. Double cropping is introduced in many parts of Punjab. Fresh areas have been brought under irrigation in the hilly areas of Balochistan's Quetta, Ziarat and Pishin hilly regions.

Despite the political instability in the state, agriculture has remained unaffected and buoyant, and an increase in yields of various crops may further strengthen the agricultural situation, providing food, fibre and other agricultural products including livestock products for domestic consumption as well as for exports.

****

Appendix: Area, Production and Yield of Some Important Crops of Pakistan (2019–2020)

Crops	*Area in Thousand Hectares*	*Production in Thousand Tonnes*	*Yield kg/hectare*
Wheat	8,825	24,946	2,827
Rice	3,034	7,410	2,442
Cotton	2,527	9.2 million bales	618
Maize	1,413	7,236	5.121
Sugarcane	1,040	66,880	64,308
Bajra	522	384	735
Jowar	199	120	603
Gram	940	545	580
Mung	187	133	711
Barley	53	55	1,038
Rapeseed and mustard	349	302	865
Tobacco	47	110	2,340

Source: Pakistan Economic Survey 2019–20, Finance Division, Government of Pakistan, p. 20–23.

Notes

1 Agricultural Statistics, Bureau of Statistics, Government of Pakistan, p. 193.
2 Most Government records in Pakistan give a figure of 796,100 km^2 as the area of Pakistan excluding the area of Azad Kashmir, Gilgit and Baltistan. In the above table, these areas are included hence a larger overall area.
3 Pakistan Economic Review (2011–2012) Finance Division, Government of Pakistan, p. 21.

12

ENERGY RESOURCES

Availability and Utilisation

Energy is the most significant component in the economic development of a country. In fact, it is the prime driver of a country's economy. Some countries are fortunate in having a natural endowment of huge energy resources like oil, gas, coal and uranium, while others are not so lucky. Pakistan, right since the days of its creation in 1947, has been struggling with the shortage of energy. The discovery of the Sui gas field, in Balochistan, in the fifties of the last century, was a redeeming feature in Pakistan's sustained effort to address the energy situation and generate a substantial amount of electricity. But, with an ever-increasing population, a growing economy and a rapid industrial development, the energy needs of the country could be seldom, if ever, met with the energy generation capacity of the country. It may be mentioned that till the recent discovery of the Thar lignite field in Tharparkar District (Sindh), the country had a very limited supply of coal, gas and oil and has been importing over two-thirds of its crude requirement, over 50 percent of petroleum products and more than half of the total coal that the country puts to different uses.

12.1 Energy Availability, Production and Utilisation

The most critical aspect of energy and fuel consumption is the inadequate generation of electrical energy, almost one-third short of the present requirement. The production of different sources of energy for Pakistan to give some idea at a glance is given below (Table 12.1):

Local Production and Import of Energy Fuels – While the entire amount of electricity is generated domestically, the fuel resources for electricity generation are partially imported. The total availability and consumption of

DOI: 10.4324/9781003510147-15

TABLE 12.1 Production of Energy Fuels in Pakistan 2020-2021

Fuel Category	*Domestic Production*		*Import*		*Total Availability*	
	2010–2011	*2020–2021*	*2010–2011*	*2020–2021*	*2010–2011*	*2020–2021*
Oil in thousand barrels	24,041	27,568	51,306	65,075	75,347	92,658
Gas in million ft^3	1,471,591	1,279,242	—	234,957	147,591	1,514,199
Petroleum products in thousand tons	8,911	10,070	12,371	10,117	21,282	20,187
Coal in thousand tons	3,450	9,230	4,267	18,850	7,817	28,080
Electricity (installed capacity MW)	22,471	36,485	—	—	22,471	36,488
Generation (GW/h)	94,653	135,073	—	—	94,653	135,073

Note. MW-Megawatt, GW/h – Giga watt hours.

Source: Pakistan Economic Survey 2021–2022, Table 14-1, p.138.

coal during the last decade (2010–2020) from 7.8 to 28 million tons, and the availability of about 20 million tons of coal in the last ten years is largely the result of a four-fold increase in the import of coal from 4 million tons in 2010 to over 18 million tons in 2020. Today over 65 percent of the total coal supply in Pakistan is imported. Similarly, of the 92 million barrels of crude that the country consumes, about 68 percent is imported. Pakistan is an important producer of natural gas. It produced 1,279,242 million cubic feet of gas in 2020–2020. Yet, it supplements its supply by importing gas from other sources. In 2020, the country imported 234,957 million cubic feet of gas, having dependence on the imported gas to the tune of 15 percent.

12.1.1 Sources of Electricity Generation

As anywhere else, Pakistan generates its electrical energy from different fuels and from the kinetic energy of flowing water and to a very limited extent from the nuclear power sources. The generation of electrical energy in the year 2020–2021 from different sources is given below (Table 12.2).

Thermal sources, principally, coal and gas remain the main fuel for the production of electrical energy.

12.1.2 Progress in the Supply of Energy

During the last 10 to 12 years, between 2010–2011 and 2020–2021, there has been a considerable increase in the supply of energy. Both, the installed capacity and actual generation of electricity, have increased. The installed capacity has increased from 22,471 MW in 2011 to 36,488MW in 2021, an increase of 64 percent, and the actual generation has also increased from 94,653 GW/h in 2011 to 135,073 GW/h in 2021, an increase of 43 percent.

Like several other countries of South Asia, Pakistan has not been able to generate an adequate amount of electricity to cope up with the increasing demand. The demand for electricity has increased not only with a rapid

TABLE 12.2 Contribution of Different Sources in the Generation of Electricity in 2020–2021

Source of Generation	*Installed Capacity Megawatts*	*Electricity Generated GW/h*	*Share of Each Source in Percent*
1. Thermal	24,461	87,847	65.27
2. Hydro-electric	8,673	33,556	24.93
3. Nuclear	1,430	9,346	6.90
4. Renewable (solar)	1,921	3,323	2.46
5. Imported	—	505	0.38
Total	36,485	134,557	99.94

increase in population but also equally with growing industrialisation and overall economic growth. The country has struggled to improve the energy situation and appears heading towards self-sufficiency, but the demand outpaces the supply, resulting in a regulated supply and even forced outages. The exploitation of the Thar lignite field and the generation of carbon-based energy is not climate-friendly and would not be a long-term solution. The completion of the Daimar-Bhasha Dam on Indus, about 350 km upstream of Tarbela, with a generating capacity of 5000 MW, would ameliorate the situation. Pakistan is very well placed in green energy resources, especially solar energy. In the coming years, the country may concentrate on solar and wind power to generate clean energy.

12.1.3 Consumption of Electricity

Against the generation of 134,577 KW/h electricity, the consumption of energy in Pakistan in the year 2020–2021 was 116,816 KW/h, which is over 50 percent increase in the ten-year period from 2010 to 2020. The level of consumption, it may be noted, is not because of lower demand. It is, on the contrary, circumscribed by the availability of power. A slight decrease in consumption vis-à-vis production, as is the norm, includes the loss in transmission and other leakages.

As can be seen (Table 12.3), more than 50 percent of the electricity energy is used for household purpose, followed by industries, which account for just around a fourth (25.6 percent) of the total consumption of electricity. Agricultural and commercial uses come next in the order. Besides, there are other minor uses like street lights or other government uses (Table 12.3).

TABLE 12.3 Consumption Pattern of Electricity by Usage (2020–2021)

Segment of Consumption	*Electricity Consumed (KW/h)*	*Percentage*
1. Household	58,722	50.26
2. Industries	29,954	25.64
3. Agriculture	10,238	8.76
4. Commercial	8,501	7.27
5. Street lights	413	0.35
6. Other govt. uses	8,621	7.38
7. General services	368	0.34
Total	116,817	100.00

Source: Ministry of Energy, Hydrocarbon Development Institute of Pakistan, Pakistan Economic Survey 2021–2022, p.138.

TABLE 12.4 Consumption Pattern of Different Fuels According to Their Use in Percent/Year

Consumption	*Total Consumption*	*Uses*	*In Percent*
Coal	28 million metric tons	Cement industry	36.260
		Power generation	32.820
		Brick kilns	20.900
		Household	0.005
Oil/petroleum	19,96 million tons	Transport	79.00
		Power generation	12.00
		Industry	0.15
		Household	0.06
		Agriculture	0.70
		Other uses	
Gas	1,463,002 million cubic feet	Power	30.00
		Fertiliser	21.50
		Household	21.40
		Industry	18.00
		Transport (CNG)	3.7
		Commercial	1.9
		Others	3.4

12.1.4 Consumption Pattern of Other Fuels

Different fuel resources have different patterns of consumption. Coal is consumed almost in equal proportion by cement industry, power generation and brick kilns, each consuming 9–10 million metric tons of coal annually, bringing the total annual consumption to 28 million metric tons of coal. Petroleum, on the other hand, is largely used by transport. Gas is used for a variety of purposes, like generation of power, as a raw material for the production of fertiliser, for household consumption, and in industry and transport. The utilisation of gas is more diversified than other energy fuels. The ease of handling and transport make gas the most preferred fuel (Table 12.4).

12.1.5 Per Capita Consumption of Electricity

For a comparative view, the per capita consumption of electricity is a convenient measure. The per capita consumption of electricity in Pakistan is still low despite a soaring demand. Limited availability resulting from inadequate generation of power is the main reason of low consumption. In terms of the per capita consumption of energy, the country is far below the developed countries of the world and even below some of the developing countries of Asia.

The total commercial energy consumption of electricity in Pakistan in the year 2020–2021 was 116,816 GW/h (Table 12.5).

TABLE 12.5 Per Capita Consumption of Electricity in Developed and Developing Countries

Country	*Yearly Per Capita Electricity Consumption KW/h (2019)*	*Country*	*Yearly Per Capita Electricity Consumption KW/h (2019)*
World average	3,081	Sri Lanka	616
Pakistan	557	Nepal	229
Japan	7,150	United Kingdom	4,496
China	5,885	Germany	6,306
Malaysia	4,608	France	6,702
India	1,208	United States	12,154
Indonesia	972	Bangladesh	433

Source: World Bank Data, 2019.

Pakistan's per capita electricity consumption per year is about 600 KW/h, one of the lowest in the world, and presents a serious bottleneck in its economic growth. The country has several projects underway, and it is hoped that in the coming decades, it will generate adequate electricity for its economic growth and social well-being.

12.2 Main Mineral Fuel Resources

The main fuel resources of Pakistan, as indeed of any other country, are coal, natural gas and crude oil. The total availability (domestic production + import) of these minerals for the year 2020–2021, for Pakistan, extracted from Table 12.1, is given below.

Mineral Fuel	*Total availability*
Coal	28.08 million metric tons
Crude Oil	92.65 million barrels
Gas	1514.billion cubic feet

12.2.1 Coal

In the overall scenario of energy production and consumption, coal has had only a marginal role accounting for only 5 percent of the total energy consumption in Pakistan.

The total production of coal in Pakistan was 18.85 million metric tons in 2020–2021. With an import of another 8 million tons of coal, the total supply comes to 28 million tons. The volume of coal production has gradually increased over the years. It was around 3 million tons in 1990, which increased to 3.45 million tons in 2010 and reached a level of 9.2 million tons in 2020. An import of over 18 million tons of coal (66 percent of the total supply) has increased the total availability to over 28 million tons of coal. It has to be noted that much of the coal produced in Pakistan is either bituminous or sub-bituminous with a relatively high content of volatile material and sulphur (Table 12.6).

The role of coal in the economy of Pakistan is very insignificant. One of the reasons is that coal is not used as a primary source of power either for generating electrical energy or heavy industries. This is not so much an index of Pakistan's relatively low level of industrialisation as of the non-availability of the massive amount of coal. The consumption of coal is largely used in brick kilns, cement industry and for generation of power. Traditionally, not much coal was used for the generation of power, but with the increasing demand and shortage of electricity, a number of power producing units have been started using imported coal. The situation of the consumption of coal in the year 2020–2021 was as follows (Table 12.7):

In an overall assessment, Pakistan had a limited amount of mineral fields, especially coal, till the recent find of the Thar lignite reserve. Though a low grade of fuel, the reserve appears to be enormous and may last the country a several centuries (Table 12.8).

TABLE 12.6 Supply of Coal in Pakistan (in Million Tons) During the Last Decades

Year	*Domestic Production*	*Imported Tonnage*	*Total*
2001–2002	3.321	1.081	4.402
2003–2004	3.275	2.789	6.064
2005–2006	4.871	2.843	7.714
2007–2008	4.124	5.987	10.111
2009–2010	3.481	4.658	8.139
2011–2012	3.613	4.057	7.670
2015–2016	4.142	4.885	9.027
2020–2021	9.230	18.85	28.150

Source: Bureau of Statistics, Government of Pakistan, Ministry of Petroleum and Natural Resources & HDIP (Hydrocarbon Development Institute of Pakistan). Pakistan Economic Survey 2020–2021.

TABLE 12.7 Coal Utilisation in Pakistan for Different Purposes in 2020–2021

Purpose for which Coal Is Used	*Amount in Million Metric Tons*	*Percentage of Coal Used*
1. Cement	10.184	34.42
2. Brick kilns	8.678	29.36
3. Power generation	9.215	31.15
4. Household	1.5	5.07

TABLE 12.8 Coal Reserve of Pakistan and Azad Kashmir (million tons)

Province	*Measured*	*Indicated*	*Inferred*	*Hypothetical*	*Total Reserve*
Sindh	3,339.00	11,835.00	56,646.00	113,637.00	185,457.00
Balochistan	79.45	150.45	183.50	45.30	458.70
Punjab	57.00	31.00	2.00	145.00	235.00
K. P.	3.00	5.75	109.24	5.00	122.99
Grand total	3,478.45	12,022.2	56,940.74	113,832.30	186,273.69

Source: Sadeq Malkani (2012) A Review of Coal and Water Resources of Pakistan, Sci, Tech. and Dev., 31 (3) 202–218.

According to the Economic Survey of Pakistan (2013–2014) "Pakistan has huge coal resources estimated to exceed 185 billion tons, including 175 billion tons, identified as Thar coal field (Sindh Province)".[1] Of this reserve 97 percent is lignite, and the remaining 3 percent are either bituminous or sub-bituminous.

12.2.1.1 *Coal Production and Distribution of Coal Mines in Pakistan*

Pakistan produces about ten million tons of coal annually and imports double that tonnage from overseas sources. There are, however, reserves that need to be exploited. The province-wise coal mines and coal reserves are given below (Table 12.9).

For long, Balochistan and Sindh have been the primary producers of coal. In Punjab, the Salt Range coal has not only limited reserve, totalling 213 million tons, but it has a high content of ash and sulphur, limiting its value. Similar is the case with Makarwal coalfield. Though one of the oldest coalfields where mining started as early as 1903, Makarwal coalfield has a very limited reserve of 22 million tons, and though low in ash, it has a high sulphur content.

The Khyber Pakhtunkhwa reserve of coal is also limited. The largest reserve in K.P. is in Hangu-Orakzai field. The thickness of coal seam ranges from 30 cm to 3 m. The measured reserve is only 1 million ton, though it is estimated that the total reserved may be as much as 80 million tons. This coal is sub-bituminous with high ash content (Figure 12.1).

12.2.1.2 *The Main Coalfields of Pakistan*

The most important coalfields are those of Balochistan and Sindh. In Balochistan, almost on the border of Khyber Pakhtunkhwa and Balochistan, and the area extending between Quetta and Musa Khel Bazar, are a number of interconnected coalfields. Of the five or six coalfields (1. Sor Range – Sinjidi Beghari, 2. Pir Ismail Ziarat – Margat Narwar, 3. Khost Shahrig Harnai, 4. Duki-Anambar, 5. Chamlang-Lunda-Losham and 6. Kingri Toi Nala), Chamlang has the largest reserve of over 100 million tons. Most of these coalfields are of early Eocene origin and occur in what is described in geological literature as *Toi* formation. Balochistan coalfields are by far the oldest. Coal in Balochistan was discovered for the first time in 1877 while the railway line to connect Quetta was laid.

The most important coalfield of Balochistan is the Duki-Anambar field in Loralai District, about 150 km from Quetta. Duki mines produce the largest quantity of coal in Pakistan. It is believed that Duki mines are the best privately run mines of Pakistan. Balochistan has a total reserve of around 460 million tons next only to Sindh.

TABLE 12.9 Province-wise Distribution of Coal Mines

Province	*Name of Coalfield*	*Location*	*Reserve (million tons)*
Balochistan	1. Mach-Abegum	70 km SE of Quetta	22.7
	2. Sor Range Sinjidi-Beghari	28 km E of Quetta	54.3
	3. Pir-Ismail Ziarat-Margat-Narwar	60 km E of Quetta	15.8
	4. Khost-Shahrig-Harnai	NE of Pir–Ismail field	86.4
	5. Duki-Anambar	In Lorlai District – Zhob dn	80.4
	6. Chamlang-Bahlol	On Punjab–Balochistan border	100.0
	7. Kingri	On Punjab–Balochistan border	82.0
Sindh	1. Thar	In Thar Parkar District on Rajasthan (India) border	175,506.0
	2. Badin	About 100 km SE of Hyderabad	850.0
	3. Meting-Jumpru	125 km East of Koraehi	181.0
	4. Lakhra	In Dadu District of Sindh	1,328.0
	5. Sonda-Thatta coalfield	Thatta District on both sides of Indus	3,700.0
	6. Jharuck coalfield	An extension of Sonda-Thatta field	1,823.0
	7. Ongar	In Indus delta region	312.0
	8. Indus East	In Indus delta region	1,777.0
Punjab	1. Salt Range coalfield	Parts of area in Chakwal Khushab, Jhelum and Mianwali Districts	213.0
	2. Makarwal coalfield	Trans-Indus region in Surghor Range	22.0
Khyber Pakhtunkhwa	1. Hangu-Orakzai	In Kohat and Hangu District	81.0
	2. Cherat-Dara Adam Khel	Nowshera District South of Peshawar Kohat Peshawar Road	11.5

The Sindh Province of Pakistan has a total coal reserve of over 185 billion tons. Much of it is lignite or sub-bituminous. Of this, the Thar coal reserve is over 175 billion tons. Before the discovery of the Thar coal in 1992 and the final assessment of the reserve, Sindh had a reserve of around 10 million tons of inferior variety of coal. The coal mines of Sindh can be grouped into three categories according to their location.

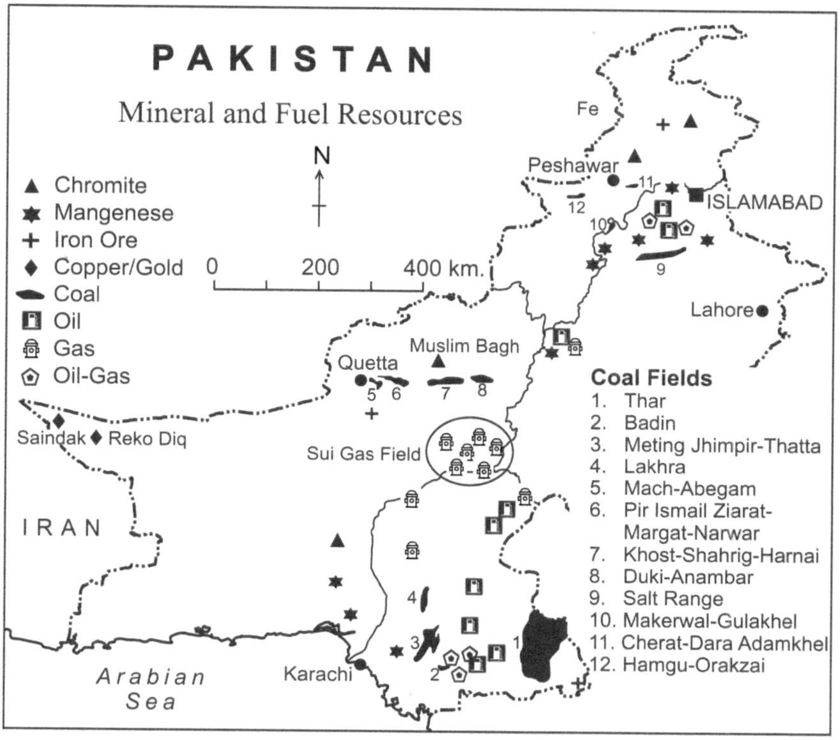

FIGURE 12.1 Pakistan: Minerals and Fuel Resources

1. Central Sindh	Lakhra coalfields in Dadu District
2. Delta region	in Thatta district with Sonda-Thatta Jherruck and Meting-Jhempur coalfields and Badin coalfield
3. Thar region	The south-eastern part of Sindh in Thar Parkar District with Thar coalfields

The Lakhra coal mines, located in Dadu District in West-Central Sindh – Hosted in Bara formation of Palaeocene (early Tertiary), the coal in Lakhra mines occurs at a shallow depth of 50–100 m. With the total reserve of 1,328 million tons, Lakhra mines are an important source of coal for brick kilns and cement in Sindh. The coal is either lignite or sub-bituminous, with a high content of ash and sulphur and a moisture content of around 25 percent.

The Indus Delta region mines in Thatta–Sonda – Thatta coalfields in the Thatta District occupy almost the entire delta of the Indus, and the coalfields in the Thatta District are grouped here as Indus delta region coalfields.[2] Sonda–Thatta coalfield—Lying in the Thatta District, it is the most important field. Coal occurs here in Bara formation of the early Eocene period. The coal seams vary in thickness from a few centimetres to over two metres. The Sonda–Thatta coalfield extends into the Jherruck coalfield and Indus East Coalfield. These are interconnected. The thickest seams occur in Jherruck coal mines. While the moisture content in these mines are highly variable from 10 to 40 percent, the ash and sulphur content is relatively low ranging from 5 to 10.00 BTU/lb (British Thermal Unit).

12.2.1.3 The Thar Coalfield

Discovered in 1992 by the Geological Survey of Pakistan, the Thar coalfields, with a reserve of over 175 billion tons lignite or sub-bituminous coal, have transformed the energy scenario of Pakistan. Spread over an area of 9,000 km^2, lying west of the international border with the Rajasthan Province of India, it is a major geological find. There are around 20 seams, each with a varying thickness ranging from 20 cm to 23 m. The minimum depth before a coal seam is encountered is 150 m. The overburden of 150 m is composed of 70 m of sand dunes, 70 m alluvium and roughly 10–15 m of bedrock that covers the coal seams. In some rare cases, the coal seams are overlain by alluvium. Bulk of the coal lies 50–120 m below the sea level, and an overburden of 100–200 m has to be removed.

The quality of the Thar coal can be judged by the following quality criteria, which include ash content, volatile material, moisture content, sulphur content and the heating value (Table 12.10).

Development of the Coalfield – The energy-deficient Pakistan State is all set to utilise the Thar coal for the generation of electricity. Though the lignite deposits were discovered by the Geological Survey of Pakistan in 1992, the paucity of fund and immediate non-availability of technical know-how

TABLE 12.10 Average Values of Different Properties of the Thar Coal

Constituent	*Value in Percent*
1. Ash content	5.14–5.56
2. Volatile matter	26.5–33.04
3. Moisture	43.24–49.01
4. Sulphur	0.92–1.34
5. Heating value in BTU/lb	
a. As received	5,780–6,398 BTU/lb
b. Dry lignite	10,723–11,353 BTU/lb

Source: Thar Coal Energy Board, Govt. of Sindh httpt://www//sindhcoal.gos.pk/thar-coal field/ accessed on 30th May 2015.

delayed the process of mining and generation of electrical energy. According to an estimate of the Geological Survey of Pakistan, Thar lignite, with a reserve of 175 billion tons, has an electricity generation potential of 100,000 MW consuming 536 million tons of coal a year, with a prospect of its lasting 200 years.

A serious effort to start development of the coalfield began in 2009 when the Sindh Government set up a joint venture group, a PPP (Public-Private Partnership) enterprise, involving Sindh Engro Mining Co. on a 40:60 ratio basis for the exploitation of block 2 in the Thar area. Similarly, Karachi Electric Supply Co. (KESC) signed an MOU with Oracle Coalfields in the United Kingdom for the exploitation and mining of coal. One of the major hurdles was the lack of initial infrastructure like transport, water and transmission lines. Of late, the project has progressed a great deal. Besides Sindh Engro Mining Co. and Karachi Electric Supply Co., Chinese groups have entered the fray. To cap the process, several MOUs were signed between officials of Pakistan and Chinese companies to develop the Thar project and generate electricity from the gas produced by the underground gasification plant.

Now Thar coalfield is divided into 12 blocks, each assigned to a government agency, a private company or a joint enterprise for the development and generation of power. Mention has to be made of the involvement of China, through several memoranda of understanding, finalising terms and conditions to facilitate the mining of 3–8 million tons of coal from block II annually, with Sindh Engro Coal Mining Co., in which finances are to be arranged by China Development Bank Corporation. The Chinese assistance would also extend to generate 660 MW (2x330 MW) in collaboration with Engro Powergen Thar Corporation, with finances arranged by China Development Bank Corporation.

The news is that the Government of Sindh has generated one MW energy from the Thar coal power plant, through an underground coal gasification process (The Express Tribute, May 30, 2015). The generation is done near Islamkot in the Thar Parkar District. It is expected that soon the coal-gas generators will be imported from China, as agreed with the Chinese Government, and the generation of electricity from lignite will multiply fast.

With the discovery and exploitation of the Thar coalfield, Pakistan is turned into a major producer of coal in the world. It is believed that the energy produced from Thar coal can meet the energy demand of Pakistan for centuries. The catch is that carbon-based fuels are no longer a preferred choice, in view of the global warming caused largely by Green House Gases, of which CO_2 is the principal gas, generated by burning fossil fuels. In view of the global outcry to cut down the generation and consumption of carbon-based energy, Pakistan may have to look for other renewable energy options in the long term.

12.2.2 Gas

Natural gas in Pakistan was discovered for the first time in 1952 at Sui in Balochistan, close to the tri-junction of Balochistan, Punjab and Sindh, and hardly 30 km north of the Sindh border and 15 km west of the Punjab border. Most of the remaining gas fields in this area occur in Sindh, in a narrow north-south belt that extends from Sui in Balochistan to Khairpur in Sindh, a distance of about 130 km.

Originally, the gas reserve of Pakistan was estimated to be 55.6 trillion ft^3. Till the year 2013, the cumulative production of gas was 33.9 trillion ft^3, which is roughly 56 percent of the total gas reserve. What is left now is 24.7 trillion ft^3 of gas, that is, 44 percent of the original reserve. The production and consumption of natural gas in 2020–2021 was 1.3 trillion ft^3. Though the consumption of natural gas has marginally declined, the present recoverable reserves are likely to be exhausted by 2035. In view of the limited reserve of natural gas, Liquefied Petroleum Gas (LPG) is also being used. At the present rate of 100 tons/per day production of LPG, the latter is contributing around 1 percent of the total energy supply of the country.

Much of the gas extraction in Pakistan is done by PPL (Pakistan Petroleum Ltd.), from ten gas fields, most of which, with the exception of Sui, are located in Sindh and are not equally productive. A summary of the gas fields in Pakistan is given below. The most important fields are Sui (Balochistan) and Qadirpur in northern Sindh, close to the city of Ghotki. The Qadirpur gas field is operated by Oil and Gas Development Co. (OGDCL) (Table 12.11).

The production by any individual gas field varies from year to year, but the total production as compiled by the Bureau of Statistics, Govt. of Pakistan, gives a reliable figure. Thus annual production of gas as given by the Bureau of Statistics, Government of Pakistan is as follows (Table 12.12):

In the absence of adequate energy resources, Pakistan has used its gas reserves rather speedily, if not extravagantly. The present reserve of gas, it is reported, is not going to last for more than 15 years, unless additional reserves are discovered. The discovery of 175 billion tons of coal reserve in Tharparkar is an important event in the energy history of Pakistan, and if judiciously utilised, the country will be rid of any concern about energy deficiency. The lurking fear of global warming and the climate change, more or less a reality, make the prospect of long-term exploitation of the Thar coalfield rather dim.

12.2.3 Oil

About a third of the total energy consumption of Pakistan is met by oil, of which only one-third is domestically produced, and the remaining two-thirds is imported. The country produces around 30 million barrels of crude oil annually. The quantity produced varies from year to year. In 2018–2019,

TABLE 12.11 Gas Fields of Pakistan

No.	*Name of Gas Field*	*Location*	*Year of Discovery*	*Approximate Daily Production (million ft³/day)*
1.	Sui gas field	Sui Balochistan	1952	660
2.	Qadirpur gas field	Ghotki Sindh	2012	432
3.	Kadanwari gas field	Khairpur Sindh	1989	60
4.	Sawan gas field	Khairpur Sindh	1998	135
5.	Kandhkot gas field	Kashmor dist. Sindh	1989	155
6.	Adhi gas field	30 km South of Islamabad	1978	41
7.	Mazarani gas field	Shahdkor Sindh	1959	5
8.	Chachar gas field	NE extension of Kandhkot gas field	1995	2
9.	Hala gas field	Saughar–Matihari dist. Sindh	—	11
10.	Kinza	Saughar dist. Sindh	2014	NA

Source: Pakistan Petroleum Co.

TABLE 12.12 Annual Production of Gas in Pakistan in Million ft^3. (2017–2018)

Province	*Natural Gas Production in Million ft^3 (2017–2018)*
1. Sindh	940,995
2. Balochistan	310,536
3. K. P.	151,177
4. Punjab	56,227
Pakistan	1,458,935 million ft^3 (i.e., 1,458 trillion ft^3)

Source: Bureau of Statistics, Pakistan Statistical Year Book, 2018, Govt. of Pakistan, p. 165.

Pakistan produced 32.5 million barrels of crude that declined to 27.6 million barrels in 2020. Besides, there is a large amount of import of crude oil that is more than double the amount of domestic production. Thus, the annual availability of crude is around 100 million barrels. The domestic production of oil is drawn, in variable quantity, from all the provinces of Pakistan, but, the two provinces of Khyber Pakhtunkhwa and Sindh have a dominant share. The province-wise crude production, as in 2018, is given below.

TABLE 12.13 Production of Crude Oil in the Provinces of Pakistan (2018)

Province	*Amount of Crude Produced (in Barrels)*	*Percentage of the Total Production of Pakistan*
Punjab	5,902,783	18.12
Sindh	10,280,844	31.57
Khyber Pakhtunkhwa	16,344,030	50.21
Balochistan	29,395	0.1
Total Production	32,557,052	100.00

Source: Pakistan Statistical Year Book 2018, Table 7.3 p. 165.

About 50 percent of the total crude production of Pakistan comes from the newly discovered oil fields of Khyber Pakhtunkhwa, located in the southern districts of Kohat, Karak and Hangu. With the discovery of oil fields in southern Khyber Pakhtunkhwa, the province has become an energy-rich province, producing 50 percent of crude, 15 percent of gas and 35 percent of the LPG produced in Pakistan. Another 30 percent of the domestic production of crude comes from Sindh, and the remaining 18 percent is contributed by the oil fields of Punjab.

12.2.3.1 Change in Oil Production Scenario

The discovery of oil fields in Khyber Pakhtunkhwa has completely transformed the energy scenario of Pakistan. With accelerated exploration in the Tertiary beds of Khyber Pakhtunkhwa, the situation may further improve. The originally known recoverable reserve of crude oil in Pakistan was 1,102.6 million barrels, of which 68 percent (731.5 million barrels) is already extracted. Pakistan has, thus, 371 million barrels of recoverable resource of crude oil. It was believed that the existing reserve would not last more than 15 years if the present rate of extraction of roughly 30 million barrels of oil annually continues. The situation, however, is no longer considered alarming as more oil fields may be discovered not only in Khyber Pakhtunkhwa but also in Balochistan and Sindh.

The Government of Pakistan has embarked on a very ambitious plan of oil exploration. It has allotted 50 blocks, with an area of over 100,000 km^2 for exploration of petroleum in all the provinces. The number of blocks, allotted in each province, is as follows:

1. Balochistan	21 blocks
2. Punjab	15 blocks
3. Sindh	06 blocks
4. Khyber Pakhtunkhwa	08 blocks

12.2.3.2 *The Existing Oil Wells*

Presently, oil production in Pakistan comes from three provinces, Sindh, Punjab and Khyber Pakhtunkhwa. Though the first oil field in Pakistan, after 1947, was discovered in 1952 at Sui in association with gas, not much oil could be extracted. Instead, Sui turned out to be a large gas field. The first successful oil exploration was undertaken in the 1960s when the Toot oil field was discovered in 1964 in the Potwar Plateau, about 130 km southwest of Islamabad. At its peak, the Toot oil field produced 2400 barrels of oil a day. In the 1990s, another oil field was discovered in Sindh, richer with a much larger reserve and greater production. The latest addition to the oil fields of Pakistan are the oil fields discovered in Khyber Pakhtunkhwa. With the discovery of these oil fields, the national production has jumped from 23 million barrels per year in 2009–2010 to over 32 million barrels in 2017–2018 (Table 12.14).

12.2.3.3 *A Word About the Recent Discovery of Oil in Khyber Pakhtunkhwa*

An intensive exploration initiated in the beginning of the current century has yielded result, and good oil reserves have been found in Kohat and Karak Districts of Khyber Pakhtunkhwa. It has to be noted that Kohat and Karak Districts form the southwest extension of the Potwar Plateau, which houses the famous Toot oil field. The discovery of oil and gas in Khyber Pakhtunkhwa is the extension of the same geological formation.

The oil fields of Khyber Pakhtunkhwa have also substantial volume of gas and form rich source of energy. The first oil field in K.P. was spudded in the year 1999, at Chanda in Kohat District by Oil and Gas Development Corporation Ltd. (OGDCL). With this beginning, a number of oil yielding wells have been discovered in the region, one after the other.

Despite the new discovery of oil fields, the country produces around 30 million barrels of oil a year, which is far below the country's needs, and it has

TABLE 12.14 Progress in the Production of Crude Oil in Pakistan (in Barrels in Thousands)

Year	*Sindh*	*Punjab*	*K. P.*	*Balochistan*	*Pakistan*
2001–2002	15,346	7,849	–	–	23,195
2003–2004	14,220	8,405	–	–	22,625
2005–2006	13,299	9,348	1,264	25.0	23,936
2007–2008	14,370	6,518	4,689	26.0	25,603
2009–2010	13,292	5,089	5,303	22.0	23,706
2017–2018	10,280	5,902	16,334	29.4	32,555

Source: The production figures of crude oil for 2001–2010 are based on Bureau of Statistics, Govt. of Pakistan, 2011, p. 164. For 2018–19, Pakistan Statistical Handbook 2018.

to import over 60 million barrels of oil every year, which costs the Pakistani exchequer a sum of US $4.3 billion. With the increase in the domestic production, the import of oil would definitely decline saving precious foreign exchange.

While the advantage of oil and gas over coal either as a transport driver, a source of power generation or for industrial purposes has been obvious, it is more so in the case of Pakistan where coal, like oil, is in short supply. The ease of transport and the handling of oil and gas give them an added advantage. The advantage of oil over gas and vice versa can be seen from a comparison of the two for different economic activities where energy is needed (Table 12.15).

While oil is used largely for power generation and transport, gas has a more diversified use. It is almost equally distributed between household use, power generation and industry.

Future Prospects – A very serious concern about energy has guided Pakistan to launch a massive hunt for oil and gas in the country. Exploration of these fuels is a priority. The country has succeeded in finding huge oil and gas deposits in Khyber Pakhtunkhwa. The tertiary beds of Pakistan are not fully explored, and the present drive to explore more oil and gas fields in a sustained manner would certainly yield result.

12.3 Nuclear Energy

The generation of nuclear power in Pakistan is planned and executed by Pakistan Atomic Energy Commission. Presently, Pakistan has a capacity of 2,530 MW of electrical energy from its six nuclear power plants. One of the advantages the country has in this respect is the indigenous production of uranium. Nuclear energy contributes 4.7 percent of the total electricity generated in Pakistan, and the nuclear reactors together produce 5,265 GW/h of electricity.

The first nuclear power reactor installed in 1971 was a Canadian Pressurised Heavy Water Reactor (PAWR) at Kanupp Paradise Point, about 25 km from Karachi, having a generating capacity of 325 MW. The second one was in Chashma in Punjab, with two pressurised water reactors, supplied by China. Now another unit, Chashma II, started in 2005 and completed in 2011 has a capacity of 300 MW. In addition, two more nuclear plants Chashma III and Chashma IV, each with an installed capacity of 340 MW have been commissioned. The latest addition to the nuclear power generation capacity of Pakistan has been the commissioning of Kanupp III and Kanupp Nuclear Plants near Karachi, each with a capacity of 1,100 MW. Thus, presently Pakistan has seven nuclear power generating units, three at Karachi and four at Chashma, having a total installed capacity of 3,635 MW. The details, outlining the generating capacity and the date of commissioning are given below (Table 12.16).

TABLE 12.15 Comparison of Oil and Gas as Sources of Energy for Different Activities (in Percent)

Consumed Source of Energy	*Purpose for which Energy Source Is Used in Percentage*								
	Power	*Transport*	*Industry*	*Agriculture*	*Household*	*Commercial*	*Fertiliser*	*Others*	*Total*
Oil	42.8	48.7	6.2	0.2	00.4	—	—	1.6	99.9
Gas	26.1	07.0 (CNG)	21.7	—	23.2	3.1	19.0	—	100.1

TABLE 12.16 Operating Nuclear Power Plants of Pakistan

Reactors	*Location*	*Type*	*Generating Capacity in MW*	*Year of Start of Construction*	*Beginning of Commercial Production*
1. Karachi-I (KANUPP)	Near Karachi (Sindh)	PHWR*	100	1966	1972
2. Chashma-I (CHASNUPP-I)	Near Chashma barrage (Punjab)	PWR**	325	—	2000
3. Chashma-II (CHASNUPP-II)	Near Chashma barrage (Punjab)	PWR	330	2005	2011
4. Chashma-III	Near Chashsma barrage		340	2011	2016
5. Chashma-IV	Near Chashsma barrage		340	2011	2017
6. Kanupp-II	Near Karachi		1,100	2015	2021
7. Kanupp-III	Near Karachi		1,100	2016	2022

Source: 1. World Nuclear Association – Nuclear Power in Pakistan updated April 2015 (http://www.world-nuclear.org/country profiles (accessed on 3rd May, 2015), 2. Pakistan Economic Survey, Statistical Appendix, Govt. of Pakistan, 2019–20, p. 166.

*PHWR**: Pressurised Heavy Water Reactor, PWR**: Pressurised Water Reactor, KANUPP: Karachi Nuclear Power Plant, CHASNUPP: Chashma Nuclear Power Plant.

As things stand today, there are seven nuclear reactors producing power (Kanupp-I, Kanupp II and Kanupp III near Karachi and Chashma-I and Chashma-II, Chashma III) having an installed capacity of 3,635 MW.

12.3.1 Future Prospects of Nuclear Energy

Nuclear energy is a clean, yet risky, source of power. There are countries in the world like France which produce nearly 75 percent of their requirement of power from nuclear sources. It may be added that besides the risk of accidental radiation that often force a country to abdicate the use of nuclear energy, an equally daunting feature is the procurement of nuclear fuel. With a set of international controls, it is difficult to procure nuclear fuel, especially uranium, in adequate quantity to feed the reactors.

Despite all these constraints, Pakistan has plans to generate 8,800 MW electricity from nuclear sources placed at ten sites. For this, Pakistan Nuclear Regulatory Authority has selected six new sites. These are:

1. Qadirabad-Bulloki Link Canal
2. Dera Ghazi Khan Canal, near Taunsa Barrage
3. Taunsa Panjnad Canal
4. Nara Canal near Sukkur
5. Pat Feeder Canal, near Guddu
6. Kabul River (site not fixed)

Many more nuclear power plants are being planned by Pakistan Atomic Energy Commission to supply at least one-fourth of the electrical power needed by 2030.

Enrichment of Fuel – The Government of Pakistan plans to produce 350 tons of uranium (U_3O_8) per year from 2015. Uranium-enrichment centrifuges are installed at Kahuta and one at Gadwal. Additional enriched fuel for the generation of power is imported from China. There are a number of enrichment plants in Pakistan for producing weapon-grade uranium, not discussed here.

As things stand today, there are seven nuclear reactors producing power (Kanupp-I, Kanupp II and Kanupp III near Karachi and Chashma-I and Chashma-II, Chashma III and Chashma IV, near Chashma barrage). A number of other reactors planned by the Pakistan Atomic Energy Commission (PAEC) may not come up, as the emphasis is shifted to producing clean and safe energy. Except the Karachi-I nuclear Pressurised Heavy Water Reactor supplied by Canada, the remaining reactors are supplied by China. China has also supplied enriched uranium, though Pakistan has its own uranium-enrichment plant.

12.4 Hydroelectric Power Generation

Generation of Electricity – The Pakistan Electric Power Co. (PEPCO) is an umbrella organisation, which supervises the generation of power by public and private enterprises. The generation of electricity in the public sector is largely confined to hydroelectricity generated from the hydroelectric projects like Tarbela on the Indus and Mangla on the Jhelum. The private sector companies generate power from thermal sources. These include Independent Power Plants (IPP), Small Power Plants (SPP), Captive Power Plants (CPP), Wind Power Generation (WPG), Hydel Power Generation (HPG), Hub Power Co. Ltd. (HUBCO), Karachi Electric Supply Co. (KESC), besides the NPPs (Nuclear Power Plants) that generate a small amount of electricity. Besides these generating entities, there are distribution companies like NTDC (National Thermal Distribution Company), which has a number of subsidiaries.

As of 2021–2022, the installed capacity of Pakistan's hydroelectricity is around 8,668 MW, generating over 33,556 GW/h of electricity. Besides, scores of hydroelectric projects are in the pipeline. Some, like Neelam-Jhelum project near Muzaffarabad, are under construction, some are ready to be constructed, while some others are either under feasibility study or at a stage of tendering construction or are simply proposed.

Among the projects that are operational and contribute to the national grid are the following (Table 12.17):

There are several other hydroelectric power stations with an installed capacity of less than 25 MW. Besides, there are some major projects nearing completion and some others at an elementary stage (Table 12.18).

TABLE 12.17 Operational Hydel Power Projects Contributing to the National Grid

Name of the Project	*Location*	*River Dammed*	*Installed Capacity MW*	*Date of Completion*
1. Tarbela	Haripur (K.P.)	Indus	3,478	1974
2. Ghazi–Barotha (Tarbela auxiliary)	Ghazi (K.P.)	Indus	1,450	1974
3. Mangla Dam	Mirpur (AK)	Jhelum	1,000	1967
4. Warsak Dam	Peshawar (K.P.)	Kabul	243	1960
5. Chashma Barrage	Mianwali (Punjab)	Indus	184	1971
6. New Bong Escape	Mirpur (privately owned)	Jhelum	84	2013
7. Khan Khwar	Besham Shangla	Khan Khwar	72	2012
8. Rasul Dam	Mandi Bahauddin (Dist.)	Jhelum	22	1968

Source: https://www.ppib.gov.pk/wp-content/uploads/2023/08/Hydropower_Resources-Book_2022.pdf

TABLE 12.18 Major Hydroelectric Power Stations at Different Level of Execution

Projects	*Location*	*River*	*Installed Capacity MW*	*Status*
1. Jhelum-Neelam Hydel Project	Muzaffarabad (A.K.)	Jhelum	969	To be operational in 2016, close to completion
2. Kalabagh Dam	Kalabagh	Indus	3,600	Ready to be constructed, entangled in interprovincial controversy
3. Diamer Bhasha Dam	Diamer-Bhasha (Gilgit)	Indus	4,800	Under construction
4. Bunji Hydel Power Project	Near Gilgit (600 km from Islamabad)	Indus	7,100	Under feasibility study
5. Dasu Hydel Power Project	7 km upstream of Dasu village	Indus (run of the river project)	4,320	Feasibility study
6. Thakot Hydel Power Project	At Thakot, 240 km N of Islamabad	Indus	4,000	Feasibility study
7. Pattan Hydel Power Project	At Pattan, 278 km N of Islamabad	Indus	2,000	Feasibility study

With the exception of Kalabagh Dam, which is mired in controversy, rest of the projects are either at a feasibility study stage or under construction.

But the more pertinent question is the limit to which Indus can be harnessed for electric power. With so many proposed dams, like Bunji, Dasu and Thakot and the existing dam at Tarbela and a series of barrages further south, will there be enough water left in Indus to reach the sea? Granted that these are "Run of the river hydroelectric projects", and the water in such projects flows back to the river, yet, the impounding a huge volume of water will certainly impact the discharge and the availability of water in the downstream regions. Sindh, in particular, is likely to be impacted, and the Sukkur Barrage and the Nara right and left canals may not receive enough water.

Second, impounding a large volume of water at close intervals may disturb the isostatic equilibrium of the region and may induce seismic tremors. The distance between Muzafferabad and Thakot, the former in Azad Kashmir and the latter in Khyber Pakhtunkhwa, where a huge dam is proposed, is not even 100 km from Muzafferabad, which had recently experienced a severe earthquake.

Care has to be taken that ambitious hydroelectric projects don't adversely affect the well-organised irrigation system, on which Pakistan's agriculture depends almost exclusively, and without which, the country would find it difficult to maintain an efficient agricultural system.

12.5 Non-conventional Energy (Alternative Energy Sources)

Among the non-conventional – referred to as alternative energy sources – wind and solar power find a place in the energy sources of Pakistan, though their contribution is very small. The generation of *wind power* is confined to Thatta, the deltaic district of Sindh. The two firms that generate wind energy are:

1. F.F.C. Energy Ltd. – 50 MW – Jhimpir, district Thatta
2. Zorlu Energy (Pvt.) Ltd. – 56 MW - Jhimpir, district Thatta

Besides these two firms, there are around 30 independent power producing companies, holding licenses to install wind power turbines to develop a generating capacity of 1,800 MW. A few that are under construction include:

1. Three Gorges First Wind farm, Pakistan (Pvt.) at Jhimpir, Sindh – 50 MW
2. Foundation Wind Power I Ltd. at Khuttikun, Gharo, Sindh – 50 MW
3. Foundation Wind Power II (Pvt). at Khuttikun, Gharo, Sindh

The deltaic region of Thatta has the advantage of receiving unimpeded westerly winds from offshore region and holds good prospects of wind energy.

Solar Energy – With a modest beginning of "Solar Village Electricity Programme" initiated for electrification of villages in the dry region of Tharparkar District in Sindh and Balochistan, Pakistan has moved to a more ambitious plan of installing 1,000 MW generating capacity of solar power at Quaid-e-Azam Solar Park in Bahawalpur District of Punjab. The first project of 100 MW has already initiated.

When completed, the addition of 900 MW installed capacity would go a long way in ameliorating the present energy situation of Pakistan.

12.6 Economy of Energy Production and Distribution

Electrical energy is the basic requirement of modern society and indispensable for the growth of the economy. In addition to the other fuels used for a variety of purposes, electrical energy is used not only for running industries and carrying out agricultural operations and transport but also for domestic uses. The Government of Pakistan, as several other countries in South Asia, subsidise the energy consumption. This is done not only to lighten the burden on the consumers but also to boost the economy, especially agriculture. By paying the compensation amount to electricity-generating companies, the Government incurs a huge burden of expenditure. Over a period of ten years, from 2003 to 2013, Pakistan had disbursed a sum of PKR 1.7 trillion of power-generating companies to subsidise the energy consumption. The situation is now changed, and a gradual withdrawal of energy subsidy is being introduced. But the most pressing undertaken of the Government is to improve the energy situation on priority basis.

The most significant step is the generation of coal-based thermal power, using lignite from Tharparkar as a fuel. There is an active participation of many companies in developing the coal mines and generation of electricity in the 12 blocks of the Thar coal mines allotted to different enterprises. The credit facility is extended by China, through the Exim Bank of China, the Industrial and Commercial Bank of China and the China Development Bank Corporation. The Chinese banks and the entrepreneurs are not involved only in the development of Thar coal, but they are also involved in executing several hydroelectric projects like the Karot Hydroelectric Power Project with an installed capacity of 870 MW.

The Government of Pakistan is striving hard to eliminate the power shortage, avoiding wastage of energy, rationalising tariff and generating additional energy. With many hydroelectric projects underway and the discovery of enormous reserves of Thar coal, the country is well on its way to produce enough electrical energy to the point of being self-sufficient.

The need for electricity is not static, and its consumption grows with the growth of the economy, involving greater industrialisation, especially heavy industry and manufacturing, a more efficient transport system, greater power demand from the agricultural sector and even greater domestic consumption with a rise in the standard of living with the rise in the per capita income. Since the developing countries of South Asia, including Pakistan, have still a very low level of per capita energy consumption, the demand will keep growing as the economy grows, as it must, and the countries have to keep enhancing their energy supplies, through different sources.

The recent shift in the generation of electricity from carbon-based thermal sources like fossil fuels (coal, petroleum and gas) to carbon-free generation of electricity has added additional responsibility and burden on all countries including Pakistan. Global warming because of the addition of CO_2 in the atmosphere as a result of burning fossil fuel is a reality, and all nations are involved in switching over to clean energy like solar and wind energy, besides the existing hydroelectricity. The latest entrant in clean energy is hydrogen energy. Despite the technology to generate electrical energy from hydrogen is known, its application in a scaled-up mode for different purposes is yet to be firmly established. Pakistan, like many other countries, is joining the international effort to make widespread use of hydrogen energy a reality.

Notes

1 Pakistan Economic Survey (2013–2014) Finance Division, Government of Pakistan, p. 228.
2 Pakistan Economic Survey (2013–2014) Finance Division, Government of Pakistan.

13

INDUSTRIAL LANDSCAPE OF PAKISTAN

Industries are the most important segment in a nation's economy. Unlike agriculture, which has a limited potential because of the limits to the availability of arable land, industries have unlimited potential that keeps growing with advances in science and technology. Every generation witnesses addition of new industries based on scientific research and the development of new technologies. The industrial landscape of the world has been changing over the centuries in complexity, typology and volume of production. And, no part of the world, and no sphere of human activity has remained untouched by successive waves of innovations in different fields. Universal use of machines and minimal use of manpower, mass production of goods and adoption of ever changing technology are the main features of modern industries. Second, each industry has its backward linkage, collecting inputs like raw material and power and a forward linkage to domestic and world market, navigating through a series of rules and laws regulating the global trade.

The economy of a nation is usually divided into three main sectors based on their evolution in history, operational complexities and the goods they produce or the services they render. These are 1. primary sector, 2. secondary sector and 3. tertiary sector. The primary sector consists of agriculture, fishing etc, the traditional economic activities that provided sustenance to mankind, early in history. Though agriculture and fishing have grown into a highly productive occupation, these and several other allied occupations are still included in primary sector because these occupations developed early, and are, even today, directly linked to land and water as the bases of their production. Secondary segment is composed of industries where processing

DOI: 10.4324/9781003510147-16

and manufacturing of goods are the main activity. Tertiary sector includes the services of all kinds like transport and trade, education, health and other administrative and managerial services.

The importance of these three sectors has varied in time and in their contribution to a nation's wealth. Agriculture was the most important segment in the non-industrialised underdeveloped countries. With the development of manufacturing industries, secondary sector, especially industries turned out to be an important contributor to a nation's wealth, commonly known as the Gross Domestic Product (GDP). As the human activities grew in volume and complexity, the service sector started appearing ever more prominent. The result is that in rich and developed countries, services have become as important as manufacturing, seen in terms of their contribution to national economy. Industries, however, remain the kingpin in a nation's economy. Today, the industrially advanced countries are also the rich countries of the world.

13.1 Manufacturing as an Industry

13.1.1 Importance of Industries in Pakistan's Economy

Industrial sector in Pakistan's economy contributes about 20 percent to its Gross Domestic Product (GDP) and employs roughly a fifth of its labour force. The Government of Pakistan has divided the industrial activities of the state into four major components, viz. (1) Quarrying and Mining, (2) Manufacturing, (3) Construction and (4) Electricity generation and distribution. Of these, manufacturing is most important accounting for roughly two-thirds of the total industrial production (in monetary value terms) and engages over 60 percent of the total labour force. Unlike agriculture and services, the other two important sectors of the economy, industry is better organised, yields greater revenue, offers gainful employment and the exports of its products earn valuable foreign exchange. In fact, the economic strength of a country is measured in its industrial development, which is the primary source of national wealth and foreign exchange reserve (Table 13.1).

13.1.2 The Value of Industrial Production and Its Contribution to the National Economy in Pakistan

The above table abstracted from the National Economic Survey includes the entire gamut of financial accounting, used by economists to calculate the Gross National Product, which includes the Net Income from Abroad. It may be noted that agriculture, for calculation the Gross National Product, includes crops, livestock, forests and fisheries. Industrial sector includes mining and quarrying, manufacturing including slaughtering, electricity

TABLE 13.1 Sectoral Contribution to Pakistan's Gross Domestic Product at Current Prices for 2019–2020

Sector of Economy	*Production Value (million PKR)*	*Contribution to Gross National Product in Percent*
1. Agriculture	9,469,395	24.36
2. Industrial sector	7,381,423	18.98
3. Service sector	22,027,662	56.66
Total of the three sectors	38,878,480	100.00
4. Indirect taxes	+3,168,063	
5. Subsidies	-319,860	
Total GDP	41,726,683	
Net income from Abroad	2,963,949	
Gross National Product	44,690,632	

Source: Pakistan Economic Survey 2019–20, Statistical Appendix, Table 1.5, Gross National Product at current Price, p.15.

generation and distribution and gas distribution and construction. The services sector includes wholesale and retail trade, transport, storage and communication, finance and insurance, housing services, general government services and other private services. Besides these, national product also includes indirect taxes and subsidies.

13.2 The Industrial Sector in Pakistan's Economy

The apparently low contribution (about 18.5) of industrial sector to Pakistan's economy should not be taken as the sole measure of importance of this sector. It has to be emphasised that a large part of service sector thrives on the working of industries. As the industries grow, the tertiary sector comprising services grows by a multiplier effect. In reality, all the professional as well non-professional services are directly or indirectly linked to industries, either as a part of production process or through services in marketing and publicity of the goods.

The industrial sector has a number of activities, of which manufacturing is most important. The industrial activities in Pakistan are divided into the following four sub-groups: (1) Mining and Quarrying, (2) Manufacturing, (3) Slaughtering and (4) Electricity generation and distribution and gas distribution (Table 13.2).

As seen above, manufacturing is the most important economic activity. It consists of large-scale and small-scale manufacturing. Almost two-thirds (65.28 percent) of the entire value of industrial production is claimed by manufacturing, which includes a variety of sectors. It contributes about 15 percent to the GDP of Pakistan and employs 15–20 percent of the labour force of the country. Manufacturing is grouped into Large-Scale Manufacturing (LSM)

TABLE 13.2 The Four Major Components of Pakistan's Industrial Activities

Industrial Group	*Value of Production in Million PKR*	*Percentage Value of the Industrial Sector Product*
1. Mining and Quarrying	1,081,703	14.65
2. Construction	784,739	10.63
3. Manufacturing	4,818,863	65.28
4. Electricity generation and distribution	686,118	9.29
Total	7,381,423	99.85

Source: Economic Survey of Pakistan 2019–20, Appendix p.15.

and Small-Scale Manufacturing (SSM). It is the Large-Scale Manufacturing that produces 80 percent of the goods, contributing 0.5 percent to the GDP. In value terms, Small-Scale manufacturing has around 15 percent of the total industrial production, though it has a higher employment potential and needs encouragement and promotion. Slaughtering, as an enterprise, claims 7percent of manufacturing. Construction as an industry is labour-intensive, and though its contribution to industrial production is 10 percent and to the national GDP is a meagre 2.4 percent, it employs over 7 percent of the labour force of the country. In contrast, mining is more capital-intensive and has a limited employment potential but contributes almost 3 percent to the GDP and claims 14.65 percent share of the industrial production of Pakistan. Electricity generation is another capital-intensive industry, with limited employment potential.

Employment in a developing economy like Pakistan is as important as its GDP, as the former, as an economic phenomenon, mediates in income distribution. If one looks at the ratio between the share of labour force employed in a specific industry and its contribution to GDP, mining turns out to be highly capital-intensive and employs minimum labour. Construction classed as an industry in Pakistan employs a large number of people.

13.2.1 Development of Industries in Pakistan

The industrial development of Pakistan during the last seven decades of its existence has been, to say the least, less than the potential it holds. A perpetual shortage of power, limited availability of basic raw materials like iron and the absence of large coalfields restrained it from producing steel and other basic products like chemicals on a large scale. The country, even today, mines just about 700,000 tons of iron ore and just over 9 million tons of coal, constrained as it is by limited domestic reserves of these minerals. Till lately, the country has been importing mineral fuels worth US $8 billion. The situation of coal has lately improved, as the country has discovered roughly 190 billion tons of lignite in Tharparkar, the desert district of Sindh. A massive plan to tackle the energy crisis is underway, and within a few years, the country may have enough installed capacity as well as generation of electricity.

The country has been producing roughly 30 million barrels of petroleum and imports 65 million barrels of crude annually. Pakistan has been fortunate in striking gas early in 1950, and the latter has played a vital role in meeting the energy requirement of the country. It has been also used as feed in fertiliser production. The possibility of finding additional fuel resources like oil and coal fields is not ruled out. Yet, the 21st century utilisation of coal and petroleum are not the optimal choices. Non-availability of large quantity of quality iron ore and even a larger tonnage of coal, one assumes, may have been the primary reason in the failure in the efficient functioning of the Pakistan Steel Mill, developed originally as a show piece in Pakistan's industrial landscape.

13.2.2 An Efficient Development of Agro-Based Industries

Pakistan has very aptly utilised the agro-based raw materials and developed a flourishing textile industry. As will be seen in the sequel, textiles form the core of Pakistan's industrial complex in terms of volume and the value of the product. Textiles have a high employment from cotton ginning, production of yarn in the spinning mills to the manufacture of textiles in integrated mills. The production of garments, an extension of textile industry, adds more value and employment potential to this industry. Sugar refineries are another agro-based industry, which has grown in volume over the decades.

Manufacturing as an Industry – Manufacturing is an omnibus term, which includes a variety of industries. These include iron and steel, textiles, cement, chemicals and fertilisers, automobiles, electronics, machinery, commodities, machinery and equipment and petroleum refineries, leather and leather products and several others. Only some of the important industries of Pakistan are discussed here. These include iron and steel, textiles, cement, automobiles, petroleum refining and chemicals.

Mineral-based Industries There are two major mineral based industries. These are (1) iron and steel and (2) cement. There are some additional industries like jewellery and chemicals and fertilisers.

13.3 Iron and Steel Industry

Though there are about 300 industrial units in Pakistan producing a variety of metal products, there are only eight mills producing iron and steel and ferro-alloys, and all are located in Sindh, taking advantage of the seaport at Karachi that facilitates import of iron ore and coal and necessary heavy equipment.

13.3.1 Constraints on the Development of Iron and Steel Industry in Pakistan

A major constraint in the development of iron and steel industry has been an inadequate reserve of iron. The total iron ore reserve of Pakistan is 430 million tonnes, not of consistently good quality and higher iron content. The total coal reserve of Pakistan is 186 billion tonnes including 185 billion tonnes of lignite deposits in Sindh. The measured coal reserves of Pakistan are roughly 3.5 billion tonnes, largely concentrated in Sindh and Balochistan. The country has been mining 4–5 million tonnes of coal and 5–600,000 tonnes of iron per year.

The Domestic Iron Deposits – While the location of coal mines and their reserves are given in the chapter on energy, the occurrence, distribution, and the reserves of iron ore in Pakistan is given here. There are four major iron ore reserves in Pakistan (Table 13.3).

The inadequacy of iron ore deposits has been a factor in an uneconomic production of steel in Pakistan. The following table gives the level of mining of iron ore and coal in Pakistan during the 2014–2015 to 2018–2019 years (Table 13.4).

The inadequacy of domestically produced iron ore as well as coal has led Pakistan to depend on imported iron ore and coal for its steel mills. This has led to a slow growth of the industry. The first major steel plant of Pakistan, the Pakistan Steel Mill, became operational in 1983. Subsequently, a number of small steel plants in the private sector have been added. The country, however, does not produce enough steel to meet its requirement. The story of the steel industry began with building a very ambitious steel plant in the public sector, the Pakistan Steel Plant, which never reached its designed capacity and has been since riddled with problems of all kinds. The below par performance of Pakistan Steel Mill led a number of entrepreneurs to enter into fray, and a number of steel plants, in private sector, have emerged, which add to the overall steel manufacturing capacity of the country.

TABLE 13.3 The Location of Iron Ore Deposits and their Reserves (Million Tonnes)

Name of Ore Deposit	*Location*	*Ore Reserve*
1. Pachim Koh-Chingendek	Near town of Nokundi, Dist. Chagai, Balochistan	100 million tonnes
2. Chiniot iron ore deposit	Rajuam. Chiniot District in Punjab	109 million tonnes
3. Dilband iron ore deposit	Near Quetta, near Kolpur railway station, Balochistan	200 million tonnes
4. Kalabagh iron ore deposit	Near Kalabagh on Indus	300 million tonnes

TABLE 13.4 Reserves and Extraction of Iron and Coal in Pakistan 2014–2015 to 2018-2019

Year	*Coal*		*Iron*	
	Reserves	*Production in Thousand Tonnes*	*Reserves*	*Production (Tonnes)*
2014–2015	185 billion	3,408	Over 430 million tonnes	328,708
2015–2016	"	3,740	"	432,156
2016–2017	"	3,954	"	501,664
2017–2018	"	4,478	"	667,206
2018–2019	"	5,372	"	626,646

Source: Bureau of Statistics, Govt. of Pakistan 2018, Pakistan Economic Survey 2019–20, Statistical Appendix pp. 43–44.

13.3.2 A Brief History of Pakistan Steel Mill

Occupying the pride of a place in the industrial landscape of Pakistan, the idea of Pakistan Steel Mill was conceived in the 1950s by the Pakistan Council of Scientific and Industrial Research. Though approved in the 1960s, the foundation stone for the mill was laid on December 30, 1973. The project was completed and finally inaugurated on January 15, 1985, at a total cost of PKR 24.7 billion. The Pakistan Steel Mill (PSM) was established as integrated steel mill, with technical and financial support of the USSR, in the public sector, the ownership resting with the Government of Pakistan. The plant consists of 20 component units, each a factory in its own right.

Located at a distance of 40 km, southeast of Karachi, near port Muhammad Bin Qasim, Pakistan Steel Mills is spread over an area of 75 km², including 42 km² for the main plant, 33 km² for the township and 0.8 km² for water storage. A coastal location and proximity to port Muhammad Bin Qasim has ensured better connectivity for the import of machinery and raw material and fuel – iron ore and coal – and access to overseas markets. The site is also linked to the road and railway network. The plant originally designed to produce 2.2 million tonnes of steel annually could not be completed to its full capacity and has presently a capacity to produce only 1.1 million tonnes of steel. A captive thermal power plant with three generators, each with a capacity of 65 MW, provides the required power. The main products of Pakistan Steel Mill are coke, pig-iron billets, cold rolled sheets, hot rolled sheets, galvanized sheets and coils.

Pakistan Steel Mill never attained full production capacity and ran into losses. It also needed huge infusion of capital, which the Government was not always willing to invest. In the first decade of the present century, in 2006, an attempt was made to privatise the plant, and in an open bidding process, 75 percent of stake in the mill was awarded to a consortium based in Saudi Arabia at a cost of PKR 21.6 billion (US $362 million). This decision of the Government was annulled by the Supreme Court of Pakistan, on a petition from the labour union, and the mill was reverted to the public ownership of the Government of Pakistan. In 2011, it was restructured with another bailout and infusion of capital.

Pakistan Steel Mill, still operating as a public centre enterprise, works to 30–50 percent of its capacity and employs around 10,000 people. The Mill is a major producer of steel in Pakistan.

13.3.3 Reasons for the Failure of Pakistan Steel Mill

Many may not agree that the most prestigious steel plant of Pakistan, once employing roughly 20,000 persons, is languishing and faces closure because of accumulated losses. The most important reason often advanced is the escalating costs involved in the import of iron and coal, inexperience in running a giant steel mill and inadequate availability of finance. The imported Russian technology also turned obsolete over the years. In fact, the Pakistan Steel Mill was a part of Pakistan's nationalisation programme, under the prime ministership of Zulfiqar Ali Bhutto, and was more a symbol of prestige than a hard-core business enterprise. The consequences are obvious. Yet, the enormous asset value of this steel mill is estimated at US $1.25 billion. With an expandable capacity to produce over 2 million tonnes of steel, it is just a matter of time before the Government of Pakistan should be able to energise this precious asset, invest capital and efficiency and turn it into a model steel plant in South Asia.

13.3.4 Other Steel Plants

Besides the Pakistan Steel Mill, located near Karachi, there are about 600 large and small steel producing units, producing 3.3 million tonnes of steel (Zia and Hina. 2021). Of the steel mills, besides the Pakistan Steel Mill, there are a dozen important units. These, as quoted by Zia and Hina (2021), are as follows:

1. Quality Steel Works,
2. Aisha Steel Mills,
3. Metropolitan Steel Corporation,
4. Aisha Steel Mills,
5. Crescent Steel and Allied Products,
6. Mughal Iron and Steel Industry,
7. Amreli Steel Ltd,
8. Dost Steel Ltd,
9. International Steel Ltd and
10. Agha Steel Ltd

Location and Distribution of Steel Mills in Pakistan – In situations where the raw material for an industry is imported, the choice invariably falls on a coastal location. In the case of steel, Karachi provides not only a coastal location but also a good domestic market by clusters of metal-using industries. Steel manufacturing in other parts of the world usually moves to coal/coke-producing areas or alternatively in areas with large iron reserves. Pakistan does not have large reserves of iron, and their coal mines are far inland. Coastal location gives the advantage of saving freight to an inland site; and in the case of Karachi, the industry has many other advantages, like transport, finance and market. The planners of Pakistan Steel Mill would have certainly weighed these factors in locating the mill close to Karachi. Other steel mills have chosen coastal location, as they are all based on imported iron and even the fuel resources.

Small steel factories, manufacturing all kinds of machines like tractors and automobiles and other metal goods, are distributed all over the country, as they benefit from the local market and ease of distribution.

13.4 Production of Steel in the Private Sector

Besides a large number of small units melting iron and producing steel, there are a couple of medium-size steel producing units in the private sector, which deserve mention.

1. **Tuwairqi Steel Mills Ltd.** – Located close to port Bin Qasim, in the same area as the Pakistan Steel Mills, not far from Karachi and spreading over an area of over 800 hectares is the Tuwairqi Steel Mills. It employs the latest technology, called DRI (Direct Reduction of Iron), used in Japanese steel plants, and claims to be very efficient and competitive. This steel mill is owned jointly by Al-Tuwairqi Holdings (Saudi Arabia), Posco (South Korea) and Arif Habib Steel Mill (Pakistan). The mill, started in 2013, has a capacity to produce 1.28 million tons of steel annually, and is expected to produce one million metric tons of steel per year.
2. **Aisha Steel Mills Ltd.** – It is a joint venture between Arif Habib group, Metal One Corporation, a subsidiary of Mitsubishi and Universal Metal. The plant is located at Bin Qasim, Karachi, producing 220,000 metric tons of steel annually. This steel mill expanded gradually and consists of a large number of small units, some of which are integrated.

Besides these three relatively large plants, there are a number of smaller units depending largely on scrap iron to manufacture a variety of steel goods.

Trend in Iron – Steel Production – The progress of industry can be gauged by comparing the annual production of basic steel production over a decade (2002–2012).

TABLE 13.5 Production of Coke, Pig Iron and Basic Steel Products (in Thousand Tonnes)

Year	*Coke*	*Pig Iron*	*Billets*
2001–2002	694.6	1,042.9	1,874.2
2006–2007	326.3	1,008.8	3,677.8
2010–2011	301.7	433.1	1,828.9
2014–2015	207.8	265.5	2,731.0
2018–2019	00.0	00.0	3,874.0

Source: Pakistan Economic Survey, 2018–19, Appendix table 3.5, p 50, and Pakistan Statistical Year Book 2018, p. 273.

As can be seen, neither the production of iron nor that of coke has increased, and the production of both these indispensable minerals seems to be stagnating. Second, the production of pig iron and billets is erratic and does not appear to be picking up year after year., The amount incurred on the import of steel has risen from PKR 62.4 billion in 2004–2005 to PKR 157 billion equivalent to US $1.57 billion in 20011–2012. It is estimated that Pakistan would need around 10 million tonnes of steel in the near future. At present, it produces one-third of its requirement. This calls for rapid expansion in the steel-producing capacity of the country; what is clear is the urgent need for the revival and expansion of the Pakistan Steel Mills, besides the establishment of more steel mills in the private sector.

13.5 Cement Industry

With huge limestone deposits in the tertiary formations, exposed in the north-western part of the country, Pakistan is well placed to have a flourishing cement industry. Yet, before 1947, as a part of British India, the region had only four small cement plants with a total production capacity of 0.5 million tonnes. During the 1960s, the country added some more capacity, but it was not sufficient even for domestic consumption. The country remained a net importer of cement till the early years of the 1990s. The privatisation of the industry in 1990 – it was earlier in the public sector – encouraged the setting up of a number of cement plants, leading to a rapid increase in the tonnage of cement production and the brightening of prospects of cement export to neighbouring countries. Today, the country has over 30 cement factories with a production capacity of 30 million tonnes of cement. These plants are owned by three armed force companies, four foreign and sixteen private companies. About 80 percent of the production capacity of cement is in the northern part of the country, near the limestone-rich mountainous regions.

Today, the cement industry in Pakistan is doing well and has shown rapid expansion, both, in the multiplication of the number of cement-producing

TABLE 13.6 Production of Limestone and Cement in Pakistan Between 1990 and 2012

Year	*Production of Limestone in Million Tonnes*	*Production of Cement in Million Tonnes*	*Export of Cement in Million Tonnes*	*Percentage of Export to the Total Cement Production*
1990–1991	9.009	7.76		
1995–1996	9.74	9.57		
2000–2001	10.87	9.67		
2005–2006	18.43	18.56	3.23	17.4
2009–2010	37.14	31.16	10.65	13.4
2011–2012	35.02	29.54	8.57	29
2012–2013	—	33.43	8.37	25
2013–2014	—	34.28	8.14	23
2017–2018	70.8	41.14		
2018–19	87.8	39.9		

Source: Basic Statistics, Govt. of Pakistan, Mining and Manufacturing, 2012–13, Table 3.1 & 3.4 7 for 2017 to 2019, Statistical appendix, Pakistan Economic Survey 2019–20, table 3.1 and 3.4.

units as well as the tonnage of cement production. Pakistan is among the top 20 cement-producing countries of the world and among the top five exporters. During the last quarter of a century (1990–2015), the production of cement has increased four-fold from a mere 7.7 million tonnes to 30 million tonnes in 2011–2012 (Table 13.6).

There is undoubtedly a clear correspondence between the volume of production of cement and limestone, as limestone is the basic and major raw material for cement. The dependence on limestone for the production of cement forces the cement industry to locate itself close to limestone sources, as the latter is a heavy rock material and can be mined only in limestone regions. The result is that most of the cement-producing industries, with a few exceptions, are close to the western and north-western mountainous regions, which have large outcrops of limestone deposits. The result is that the cement factories are concentrated either in the western fringe zone of Punjab or in Khyber Pakhtunkhwa. Sindh has only a fewer number of cement factories, one in the Khairpur District and the other in Thatta in the delta region of the Indus. The province-wise distribution of cement factories in Pakistan is as follows:

Province	No. Cement of Factories
Punjab	12
Khyber Pakhtunkhwa	9
Sindh	7
Balochistan	1
Total	29

Besides, there are a few smaller cement plants. Some of the large cement factories are quite well organised and can compete with the best in the world. The large industries are mentioned here.

1. *Lucky Cement Ltd.*, located in Darra Pezu, Lakki Marwat District of Khyber Pakhtunkhwa, with an operational capacity of three million tonnes of clinkers.
2. *D. G. Khan Cement Company Ltd.* with two plants one at Dera Ghazi Khan and another at Khairpur. Chakwal Dist. (Punjab), each with an operation capacity of two million tonnes of clinkers.
3. *Kohat Cement Co.*, located at Kohat (Khyber Pakhtunkhwa) with an operational capacity of over 2.5 million tonnes of clinker, and
4. *Fauji Cement Co. at Fatehjang* (close to Rawalpindi).

Lucky group has another factory on the Indus Highway, near Karachi. There are also small units like Thatta Cement at Thatta (Sindh), Fecto Cement Ltd. at Sangjani near Islamabad and Dandot Cement factory at Jhelum, each with the capacity of half a million tonnes of clinker.

13.5.1 Export Potential of Cement Industry

Pakistan has good prospects of exports of cement. Even today, about a fourth of the total production of cement is exported. It has huge market in Africa in countries like South Africa, Tanzania, Djibouti, Mozambique, Sudan and Kenya. The neighbouring countries of South Asia viz. Afghanistan, India and Sri Lanka also import cement from Pakistan. In 2011–2012, over six lakh tonnes (605,453) of cement were exported to India.

There is scope for expansion as the economy of other South Asian and African countries pick up. India and Bangladesh could be good export market, as the latter being an alluvial lowland does not have the prospect of developing its own cement industry, and the other because of proximity and easy sea route that could enable Pakistan to ship large volume of cement to

TABLE 13.7 Pakistan: Capacity, Production and Export of Cement (Million Tonnes)

Year	*Production Capacity*	*Production*	*Local Use*	*Export*	*Capacity Utilisation in %*
2009–2010	45.47	34.22	23.57	10.64	75.25
2010–2011	42.50	31.33	22.00	9.43	73.95
2011–2012	44.77	32.31	23.95	8.57	72.62
2017–2018	44.7	41.14	—	—	92.04
2020–2021	69 .0	—	—	—	—

Source: Economic Survey of Pakistan 2012–13, p. 43. Note: The export data available with the authors is in value term of Pakistan rupees. For example, the export value of lime, cement and fabricated building material in 2017–2018 was PKRs 25,494 million.

the western central states of India like Gujarat and Maharashtra. It seems, however, the geo-political discord between the two major South Asian Countries has temporarily come in the way of trade. Second, there is need to diversify the production of cement into cement-based products like different types of tiles and cement blocks and other flooring and wall material. A further diversification into cement-based products in the coming years is certain to create a larger market to provide impetus to the industry.

13.6 Textile Industry

The textile industry stands out as a shining star in the industrial landscape of Pakistan. Based largely on locally produced cotton, promoted by local entrepreneurs, run by a domestic work force with highly skilled artisans and craftsmen and financed by Pakistan banks, the industry forms a major support base for Pakistan's economy. In terms of its contribution to GDP as a valuable foreign exchange earner and employer of a large labour force, it stands out as a dependable enterprise of which the nation could be proud of. The textile sector contributes nearly one-fourth of the industrial value added, provides employment to 40 percent of the industrial labour force, consumes 40 percent of banking credit to manufacturing sector and accounts for 8 percent of the GDP[1] The textile exports form roughly 55 to 60 percent of the value of national exports. It ranks twelfth among the textile exporting nations of the world.

The clearest advantage the country has is the irrigation facility and sunshine that favour the production of cotton. Pakistan has been a traditional producer of cotton, but in undivided India, the cotton produced in Pakistan was used to cater to the cotton textile mills of Ahmedabad and Bombay (now Mumbai). It was only after 1950 that the cotton textile industry in Pakistan picked up.

The history of textile industry in Pakistan is one of a remarkable progress and expansion despite several periods of ups and downs. In 1947, the year Pakistan came into existence, the country had three mills with 177,000 spindles. The number of mills as well as the area under cotton steadily increased. The production of cotton increased from just over a million bales in 1947 to 9.6 million bales in 1990–1991, 10.7 million bales in 2000–2001 and 13.5 million bales in 2011–2012. Spurred by the demand emanating from the textile industry, Pakistan has become the fourth-largest producer and the third-largest consumer of cotton in the world (Table 13.8).

During the last 19 years (2001–2019), there has been a quantum jump in the growth of textile industry, the number of mills, installed capacity as expressed by the number of spindles and looms as well as the quantity of production. Today, according to Textile Commissioner Organization (TCO), the country has the following number of textile-producing units (Figure 13.1) (Table 13.9).

TABLE 13.8 Number of Textile Mills, Their installed Capacity, Consumption of Cotton and Production of Yarn (in Million kg) and Cloth (Million m²) Between 1947–1948 and 2014 and During the Latest Available Year 2019–2020

Year	*No. of Mills*	*Installed Capacity*		*Consumption of Cotton (in Million kg)*	*Total Yarn Produced (in Million kg)*	*Total Production of Cloth (Million Sq. Metre)*
		No. of Spindles (000)	*No. of Looms (000)*			
1947–1948	30	177	NA	NA	NA	NA
1990–1991	247	5,493	15	1,197.5	1,041.2	292.9
1995–1996	349	8,493	13	1,661.9	1,495.1	327.0
2000–2001	353	8,594	10	2,070.1	1,721.0	490.2
2005–2006	437	11,834	8	2,932.6	2,556.3	915.3
2010–2011	439	11,392	7	3,405.7	2,938.6	1,029.8
2014–2015	411	13,184	8	2,732.7	3,369.7	1,036.9
2018–2019	408	11,338	9	2,735.2	3,431.3	1,044.1
2019–2020	408	13,409	9	1,929.4	2,414.8	737.9

Source: 1. Bureau of Statistics, Govt. of Pakistan, 2012–13. Table 13.2, pp. 279–81.

2. Economic Survey of Pakistan 2012–13, Manufacturing and Mining Report, Ch. 3, p.37.

3. Economic Survey of Pakistan, 2019–20, Statistical Appendix, p.46.

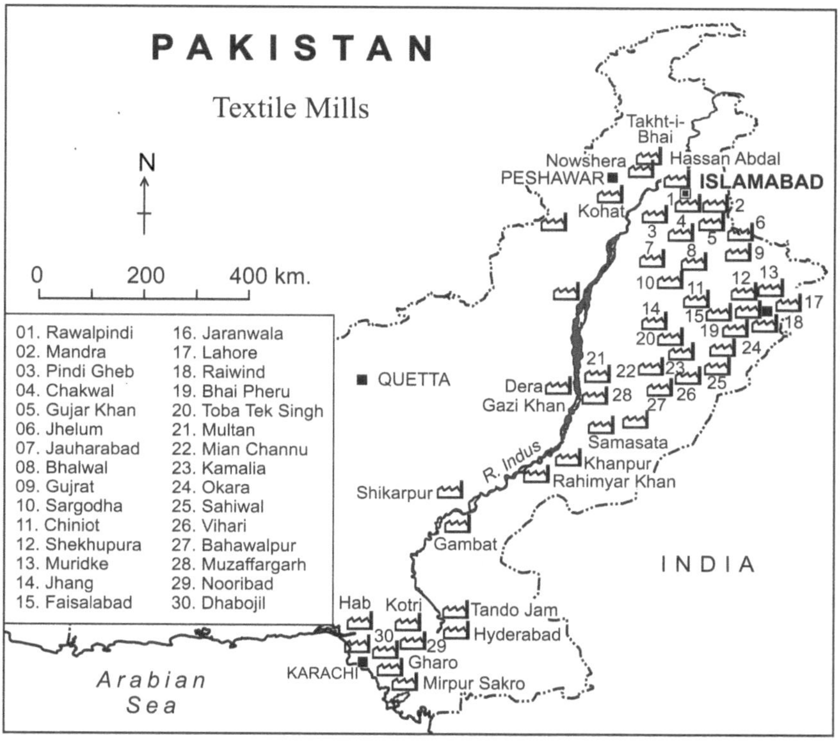

FIGURE 13.1 Distribution of Textile Mills in Pakistan

TABLE 13.9 Total Number of Different Textile-producing Units

1. Mills	Total no.	517
	a. composite units (spinning and weaving)	40
	b. spinning units	477
2. Spindles	a. installed	13.414 million
	b. in operation	11.338 million
3. Rotors	a. installed	198,801
	b. in operation	126,583

Source: Economic Survey of Pakistan 2019–20, p. 51.

13.6.1 Geographical Distribution of the Industry

Textile industry, especially the cotton textiles, is largely concentrated in the cotton-producing areas of Punjab and Sindh. Lahore, Multan and Faisalabad show some of the largest concentration of industries. About two-thirds of the textile mills are in Punjab, and about a third in Sindh, a couple of dozen mills are located elsewhere. Cotton is a bulky commodity, though it is not so heavy. The reason why textiles are largely concentrated in the plains of Punjab and Sindh is its labour-intensive character and the cost of transporting un-ginned cotton. From the picking, spinning and weaving, the textiles involve a production chain that requires cheap labour for picking cotton and even in ginning and highly skilled labour in the manufacture of textiles. Adequate labour is provided by these two high-density provinces. Second, for export, a good transport system and proximity to Karachi are other considerations that weigh with the industrialists in locating their mills in the plains, by the side of major transport arteries and close to Karachi, the port city.

13.6.2 The Nature of Textile Mills and the Variety of Textile Goods

There are three kinds of mills: 1. Integrated (spinning and weaving), 2. weaving units, and 3. power loom units. It may be mentioned that the non-mill sector of the industry consisting largely of power looms produce six to seven times more cloth, as measured in square metres, than the mills. They produce large quantities of coarser cloth and other items of textile "made-up sector", which produces textile goods like towels, canvas, bed wear, hosiery and knit-wear. Textile industry has the largest production chain with potential for value addition and employment at every stage, from cotton picking to ginning, spinning, fabric making, dyeing and finishing, besides "made-ups" and garments. As mentioned earlier, spinning is the core and the backbone of the textile industry and employs the largest number of people. The total production of cloth in Pakistan was 1,044 million cubic metres in 2019–2020.

13.6.3 Export of Textiles

The textiles form a major item of export. The export of textiles has grown over the years in value terms, from US $10.7 billion in 2007–2008 to US $23 billion in 2013–2014, that makes 55 percent of the total Pakistan's export of the value of US $25 billion in 2019–2020, which is 60 percent of the total export. The value of exported textiles has remained more or less constant in recent years, as during the year 2019–2020, Pakistan exported textiles worth US $13.4 billion, which is 60 percent (59.94 percent) of the total export of Pakistan. This is similar to the export value of textiles seven years earlier in the year 2011–2012. Of the total export of textiles, 95 percent is accounted for by cotton textile, around 4 percent by synthetic textiles and the remaining 1 percent by wool and woollen textiles (Table 13.10).

Being labour-intensive, most of the developing countries with relatively cheap labour have entered the field of garment manufacturing and export,

TABLE 13.10 Value of Export of Textiles from Pakistan (in Million US $) (2007–2008 to 2018–2019)

Item	*Years*			
	2007–2008	*2009–2010*	*2011–2012*	*2018–2019*
1. Cotton and cotton textiles	10,071	9,754	11,778	13,031.0
2. Synthetics	490	446	546	297.809
3. Wool and woollen textiles	216	137	121	67.265
4. Total textiles export	10,777	10,337	12,445	13,396.074
5. Total export	19,224	19,290	23,624	22979.325
6. Textiles as percent of export	56%	54%	53%	59.94%

Source: Pakistan Economic Survey 2014–15, Govt. of Pakistan, Ch. 3, Manufacturing and Mining, p. 51 7 Pakistan Economic Survey 2019–20, p. 51.

TABLE 13.11 Export of Textile and Clothing (Billion US $)

		Years		
		2007	*2010*	*2013*
World	Textile	240.4	250.7	306.0
	Clothing	345.8	351.5	460.0
Total world export		586.2	602.2	766.0
Pakistan	Textile	7.4	7.8	9.3
	Clothing	3.9	3.9	4.5
Total Pakistan export		11.2	11.8	13.8
Pakistan's share in world trade		1.9	2.0	1.8

Source: Pakistan Economic Survey 2014–15, Govt. of Pakistan, p. 50.

as the developed countries find it economical to import than to process and manufacture piece-meal garments. In Asia, besides Pakistan, India, Bangladesh, Taiwan, Vietnam, Indonesia, Thailand and Sri Lanka have entered the textile sector in a big way as their important foreign exchange earner. Though textiles are a leading industry of Pakistan, especially in terms of its contribution to country's export, on the global scale, the volume of textile from Pakistan appears insubstantial. This can be seen from the comparative figures given in the following Table 13.11:

Pakistan's export of textiles does not amount to even 2 percent of the world export trade. A characteristic feature of Pakistan's export of textiles is the dominance of textile fabrics, which form over two-thirds of the exported textile goods. In contrast, the global trade in textiles consists largely of clothing, which accounts for 60 percent of the total export trade. What is obvious is that Pakistan deprives itself of the advantages of processing the fabric into clothing and commanding higher price. The principal importers for Pakistani textiles are Europe, the United States, Middle East, Hong Kong, Japan and African countries.

Though textile industry is the best performer in the industrial landscape of Pakistan, several measures have been recommended by specialists to improve the competiveness of the industry. An economic size with not less than 25,000 spindles and 500 looms, as a norm, was recommended by the Textile Commission of Pakistan in 1960. But such recommendations are often circumvented as to accept and implement them would mean diverting resources from other sectors of economy.

13.6.4 Stagnation in Cotton Textile Industry

As seen from the value of production over several years, the textile industry in Pakistan, despite its overwhelming importance, appears to be stagnating. This is attributed to several factors, both external and domestic. Distortion in the cost of production by huge subsidy to farmers, by lowering the price of the finished goods, thereby increasing their competitiveness by several exporting countries, and stringiest conditions and demands on quality by the importing countries are factors to reckon with in the export trade of textiles. Besides, occasional recession in the international market affects the export trade. Domestically, not sticking to international quality standards of grading and standardisation of cotton bales by ginners and spinners alike has discounted the value of Pakistani cotton in the international market. The cotton production in the country has remained stagnant at around 13 million bales annually. The most sluggish sector is that of value-added garments. This sector has not grown and its demand in the export market, as opposed to the normal trend, has remained constant or even declined.

Every sub-sector of the textile industry in Pakistan is reeling under imminent decline and each has its grievances either against the government's textile policy or the unhelpful attitude of the government. The fact, however, is that the industry suffers from obsolescence in technology and a lack of enterprise in adopting the latest technology. The All Pakistan Textile Mills Association argues that "factors such as high interest rates and cost of inputs, non-conducive government policies and non-guaranteed energy supplies hinder their competiveness". There may be some element of truth in this. But stagnation in the industry has resulted from a variety of factors including low profitability. One reason on which there is a consensus is that the inconsistent electric supply adversely affected the industrial growth of Pakistan including the textile industry. An all-out effort at modernisation of the plants and machinery and removing other constraints like irregular supply of energy would certainly make Pakistan textiles more competitive leading to further growth.

The Textile Policy of the Government of Pakistan – Right from the late fifties of the last century, the Government of Pakistan, realising the potential of the textile industry, took the necessary initiative to promote innovation and research in textiles. Towards this end, they established an Institute of Textile Technology at Hyderabad. This institute was later upgraded into National Textile University in 2002. This university's primary task is to train the necessary skilled manpower and keep abreast of the latest technology in textile-related research and innovation and help the country adopt the latest technology. The university's research wing is engaged in improvement in all segments of the industry, from the quality production of cotton to ginning, spinning and weaving and even in market research at the international level.

The textile policy of the government intended to be implemented during 2014–2019 period was, to say the least, very ambitious. It aimed at doubling the production and export of textile in a period of five years, a task not easily achievable. Towards this end, it intended to achieve US $5 billion during this five-year period. The principal feature of this current policy is enhance the competitiveness of the industry by adopting technologies of global standard and production of high-quality textile goods.

13.7 Petroleum Refineries

Pakistan produces roughly 32.5 million barrels of crude every year. Till about ten years ago, the crude production in the country was about 25 million barrels, but the exploration and addition of a number of oil wells have added another six million barrels to the annual production of crude in the country. This, however, is not enough, and Pakistan imports roughly 51 barrels of crude annually. The import is largely from the Middle East countries. Much of the crude, except a very small fraction, is refined before being used for transport or aviation fuel or any other purpose.

The country has a number of oil refineries to process the local and imported crude oil. There are, in all, seven oil refineries in Pakistan, some of which are not yet operational. Engaged in refining largely imported crude, most of these refineries are located on the coast, close to Karachi to save the cost of handling and distribution. The following is the list of petroleum refineries in Pakistan with their refining capacity (Table 13.12).

Attock Refinery – Located close to Rawalpindi, this is the oldest refinery of Pakistan. The refinery was started in the beginning of the 20th century to process the locally produced crude. The refinery suffered from some degree of obsolescence, but a modernisation programme undertaken with the help of subsidy revived the company. The refinery has also enhanced its capacity, and, in addition, acquired the National Refinery located at Karachi. This has further increased the overall refining capacity of the plant. Of the total crude refined in the refinery, around 70 percent consists of local crude

TABLE 13.12 Petroleum Refineries in Pakistan, Location and Their Refining Capacity

Nr.	*Name of the Refineries*	*Refining Capacity*		*Location*
		bbl/d	*m^3/d*	
1	Attock Refinery Ltd.	46,000	7,300	Rawalpindi
2	Pakistan Refinery Ltd.	50,000	7,900	Korangi, Karachi
3	National Refinery Ltd.	64,000	10,200	Korangi, Karachi
4	Byco Petroleum Pakistan Ltd.	150,000	24,000	Karachi
5	Pakistan Arab Refinery Ltd.	100,000	16,000	Multan
6	Enar Petroleum Refinery Facility – Karachi	3,000	480	Near Karachi
7	Indus Oil Refinery (not yet operational)	100,000	16,000	Port Bin Qasim
8	Grace Refinery Ltd. (under construction)	120,000		Hub Balochistan

Note: bbl/d = barrels per day, m^3/d = cubic metre per day.

and the remaining of imported crude oil. Attock Refinery is by far the only important inland refinery.

Pakistan Refinery – Located near Karachi, Pakistan Refinery was started in Public Sector in 1962. Situated at Korangi, an area in the vicinity of Karachi, the refinery has a huge petrochemical complex. The refinery refined largely the imported crude to the extent of two million tons per annum. It produces high speed diesel, furnace oil, naphtha, aeration fuel and liquefied petroleum gas. Almost all the products are marketed domestically, except naphtha, which is also exported.

National Refinery – This is the second-largest refinery of Pakistan and, like Pakistan Refinery, is located at Korangi, near Karachi. The plant produces a variety of fuels like furnace oil, high speed diesel (HSD), jet fuels, liquefied petroleum gas, gasoline and naphtha. The company running the refinery has a 50-year long history. Incorporated in 1963, the company constructed its first refinery, a lube refinery, in 1966, the fuel refinery in 1977 and a petrochemical plant was added in 1979. This was the first petrochemical plant of the country. While the capacity was increased in 1985, a second lube refinery was commissioned in 1992. A significant turn in the management of the refinery took place when it was privatised with the Attock group of refineries that acquired over half of the shareholding of the company.

The transformation of the refinery, from one in the public sector into a private enterprise, has infused efficiency in the refining process. With 64,000 gallons a day, the National Refinery is the third-largest among the existing refineries, though the addition of larger capacity refineries like the Grace and the Indus refining will leave it far behind.

Byco Petroleum Pakistan – It was previously known as Bosicor Petroleum Pakistan. Besides the previously existing capacity of refining 35,000 bbl/d, Byco International has incorporated another 120,000 bbl/d facility in 2012, making the group the largest refinery of Pakistan. Owned by the two governments of Pakistan and Abu Dhabi, it is considered a public sector enterprise.

The refineries of the company are located at Hub in Lasbela District of Balochistan. Though physically lying in the province of Balochistan, the complex of refineries is located just around 25 km northwest of Greater Karachi and not more than 35 km inland, connected to the sea by the river Hub, thus garnering all the advantages of coastal location.

Pak-Arab Refinery (PARCO) – This large capacity refinery (100,000 bbl/d) is a joint venture between the Government of Pakistan and the Emirate of Abu Dhabi, with Pakistan owning 60 percent of the share capital. The Pak-Arab Refinery Co. is an integrated holding company with oil refining, transportation, storage and marketing of petroleum products, with an annual turnover of 250 billion PKR (Pakistani rupees), it is the second largest company in the country's corporate sector. PARCO is the strategic fuel supplier for Pakistan with a large operational structure. The refinery produces oil, high-speed diesel (HSD), aviation fuel and other petroleum products, and it generated a revenue of PKR 400 billion (2014). With continued modernisation, the refinery is able to meet international standards. For distribution, the refinery has a network of 2,000 km of pipeline across Pakistan. Lately (in 2012), the company has acquired 100 percent assets of SHV Energy Pakistan that was the largest LPG marketing company of Pakistan and that was a strategic partner of PARCO till 2001.

Enar Petroleum Refinery facility – It is a small refinery near Karachi with a refining capacity of 3,000 bbl/d.

Indus Oil Refinery – Located very close to port bin Qasim on the national highway. Though with a planned huge refinery capacity of 100,000 bbl/d, the refinery is not yet operational, but it has all the advantages of a location of a port.

Grace Refinery – Following the example of Byco Petroleum, Grace Refinery incorporated in 2015 is planning to establish its refinery in Hub region of Balochistan, with a refining capacity of 120,000 bbl/d. Once completed, it will meet roughly one-fifth of the need of the refined oil of Pakistan.

Some General Observations – Pakistan's total production of crude is around 32 million barrels. Another 50–60 million barrels are imported, thus making available roughly 90 million barrels of crude annually. The present refinery capacity of the six oil refineries,[2] operational today, is 400,000 barrels a day and with the addition of two more high capacity refineries (Indus Oil refinery 100,000 bbl/d and Grace Refinery Ltd. 120,000 bbl/d) Pakistan will have enough refining capacity to process more than 100 million barrels of crude, presently available.

Most of the refineries have a coastal location to refine the imported crude, which is twice the local production. The fluctuations in the crude price internationally is certainly not to the advantage of crude importing countries including Pakistan, but it is extremely difficult to switch over to alternative fuels, as petroleum scores over other fuels like coal because of the ease in handling and transport. In coming years, the share of oil in the country's power supply may increase from 32 percent to a much higher value, as part of the crude may be used unrefined for generating electricity.

The future energy scenario is, however, uncertain and fast-changing across the world because of global warming and the resultant climate change. The need to do away with carbon-based energy sources like coal, petroleum and gas is urgent. Though petroleum is far less polluting than coal, and gas is still less polluting than petroleum, all these fuels release CO_2 in the atmosphere in a greater or lesser degree. The urgency to switch over to carbon-free energy is universally acceptable. Yet, the switch over is very difficult. It involves not only research and innovation to find alternative sources of energy but also a huge investment, which developing countries can scarcely afford. Since the developed countries have pumped in disproportionately large amount of CO_2 in the atmosphere over the centuries and created this malaise of climate change, they should legitimately bear the burden of switching over to non-polluting energy resources by contributing to clean energy programme, especially in developing countries. The answer lies in the future.

13.8 Automobile Industry in Pakistan

One of the earliest industries of Pakistan is the automobile industry. Transformed over the decades, it produces cars, light commercial vehicles (LCVs), jeeps, buses, trucks, tractors and a variety of two- and three-wheeler vehicles.

The industry had a very early start, producing its first vehicles in 1953 when National Motors Co. assembled the Bedford trucks at Karachi. The plant also assembled cars and trucks. The assembled vehicles of National Motors had a long lease of life till the beginning of 1990s when the deregulation of the industry attracted several car and truck manufacturers, largely from Japan, and a number of collaborative projects were started. These joint-partnership companies with the year of their incorporation are mentioned below.

1. Pak-Suzuki (1982)	A joint venture of Pakistan Automobile Corporation (PACO), a government owned subsidiary, and Suzuki Motors of Pakistan.
2. Honda Atlas Cars	Produced jointly by Honda Motor Co. of Japan and Atlas group of companies of Pakistan.
3. Indus Motors	A joint venture between the House of Habib of Pakistan and Toyota Motor Corporation of Japan.
4. Gandhara Nissan Ltd.	Originally a sales company distributing Nissan vehicles, entered into technical collaboration with Nissan Co. of Japan (1992) to assemble cars, LCVs, and trucks. The plant is located near Karachi in the vicinity of port Qasim.
5. Hinopak Motors	It is a collaborative project between Hino Motors and Toyota Tsusho Corporation. Hino Pak produces over half of the total number of trucks and buses produced in Pakistan.
6. Ali Ghazi Tractors	It is a Pakistani company manufacturing tractors and generators in collaboration with Fiat New Holland.
7. Master Motors.	A part of the Master group of industries, the company produces buses, trucks in collaboration with Mitsubishi Motors of Japan.

Manufacture of Motorcycles – There are three leading motorcycle manufacturers in Pakistan. These are Ghani Automobile Industries, Suzuki Motorcycles and Yamaha Motorcycles. The Ghani group has its factory in the vicinity of Lahore. Yamaha has a joint venture with the Dawood Group of companies, called Dawood Yamaha Ltd. (DYL). The plant is located in Hub Chowki producing many parts, and assembly is done at Uthal, both in Balochistan.

The Present Status – The automotive industry contributes 28 percent to the GDP of Pakistan, and its share in manufacturing in monetary terms is 16 percent. That may increase in the future. The industry employs around 200,000 people directly in manufacture, transport and marketing and about a million people indirectly and contributes PKR 63 billion in indirect taxes to the national exchequer.

The country has, like many of its South Asian neighbours, remained largely an assembler of cars from imported cars or locally produced parts in collaboration with Japanese automobile manufacturers. The quality standards are not the latest. With the lowest density of cars (10 cars per 1,000 persons), Pakistan has the prospect of expanding its automobile industry and with increasing population and demand, will be able to develop indigenous models to compete with foreign brands and may emerge as the auto-hub of South Asia.

13.8.1 Production and Prospects of Automobile Industry

Though the economy of Pakistan is growing at a moderate 4 percent as seen by growth in its GDP, the automobile industry is not a fast-growing industry. It demonstrates a sluggish growth, and in the case of buses, it is marked by decline. One can have a good idea of the growth of the industry by looking at installed capacity and production of automobiles in different categories over a decade (Table 13.13).

It appears there was a rapid increase in automobile industry between 2001 and 2005, the years, during which the economy of Pakistan grew about 6 percent. After 2006, the production in most categories of automobiles has declined. The only exception is motorcycles. There has been a 12-fold increase in the production of motorcycles in Pakistan between 2001 and 2010. While the production of motor cycles is expected to reach a three million mark, and that of cars has also increased substantially, there is a virtual stagnation in the production of tractors. Being linked to agriculture where the cultivated land is almost fixed, the need for tractors is virtually constant, and additional production is meant largely for replacement. In case of cars and motorcycles, a growing demand in their number reflects the increase in individual's income. The demand and production of cars and even motorcycles may further increase as the national and individual's income level grows higher.

The story of the production of motorcycles started with the entering of Yamaha of Japan into collaboration with the Dawood group of companies in Pakistan, incorporated as Dawood-Yamaha Ltd. Together they introduced a number of 100 cc or sub-100 cc models of motorcycles, popular with the youth or young college going students. Some of the models like "YD-70 Dhoom" or "Y-D 100 Jannon" have a good appeal. The manufacturing plant of Dawood-Yamaha motorcycles is located at Hub-Chowki, 30 km northwest of Karachi. The assembly plant to turn out the motorcycles is located at Uthal, 100 km further northwest of Hub-Chowki. Both these plants, though close to Karachi, are administratively in Balochistan.

Lately, Yamaha Motor Pakistan (YMPK) has established a Pakistan Motorcycle Manufacturing Base (PMMB) in Bin Qasim Industrial Park, Karachi. With total ownership, the company plans to produce three (300,000) to four hundred thousand 400,000 motorcycles annually. Importing 75 percent of the parts from Japan and manufacturing locally the rest 25 percent, with a gradual shift towards complete indigenisation.

Suzuki Motors also produced a few brands of motorcycles, notably Suzuki Sprinter, Suzuki's Sprinter Standard and Suzuki Raider. The latest entry in the motorcycle industry is the Ghani Automobile Industries. In collaboration with the Chongqing Yingang Science and Technology Group of

TABLE 13.13 Installed Capacity and Production of Different Categories of Automobiles in Pakistan (2001–2011)

Year	*Cars*	*Jeeps*	*LCVs*	*Trucks*	*Buses*	*Motorcycles*	*Tractors*
2001–2002	40,601	570	8,491	1,141	1,089	133,334	24,331
2005–2006	60,642	2,472	29,581	4,518	825	751,667	49,439
2010–2011	1,33,972	883	19,142	2,810	490	1,637,571	70,855
2017–2018	2,17,774	13,364	29,055	9,187	784	2,825,071	71,894

Source: Bureau of Statistics, Govt. of Pakistan, Pakistan Statistical Year Book, 2018, p 274.

China, the company started production of motorcycles in 2005. The plant is located close to Lahore, about 50 km away on Multan Road. They produce motorcycles ranging from 70 to 150 cc engine capacity.

The automobile industry in Pakistan, as much as in other South Asian countries, is dominated by automobile manufactures from Japan and other countries of Western Europe and the United States. Except a very few, most of the plants in these countries manufacture parts in technical and financial collaboration with the overseas manufactures. Indigenous designed and produced parts are gradually replacing the imported ones. A fully indigenously designed and produced car would soon be reality with domestic automobile manufacturers upgrading their production.

13.9 Fertiliser Industry

In a country where agriculture still accounts for 22 percent of the GDP and provides employment to about half the working population, fertilisers occupy an important place in the national economy. In fact, fertiliser industry is an allied supporting industry of agriculture like irrigation. Application of fertilisers increases the yield of crops by 30–50 percent, and is, thus, a key factor in agricultural development.

Today, there are 23 fertiliser-producing plants in the country, having an installed capacity to produce 10 million tons of fertilisers annually. The actual production in 2017–2018 was roughly 8 million tons. Today, the plants operated by National Fertilizer Corporation have been privatised, and a number of private players have emerged. Today, the fertiliser industry is largely in the private sector with accelerated growth, competiveness and innovation.

The 1991 Fertiliser Policy of the Government incentivised the setting up of fertiliser factories and production of fertiliser. The concessions given included tax holidays for 8 years and duty-free import of plants and machinery, guaranteed supply of gas for at least nine months of the year and freezing the prices of feedstock gas for ten years. These concessions led to the increase in fertiliser production by 25–50 percent.

The revised policy of the Government witnessed an increase. The number of different kinds of fertiliser-producing plants and their annual production during the year 2017–2018 was as follows (Table 13.14):

As can be seen, urea, the most commonly utilised fertiliser, accounts for 75 percent of the total production of fertilisers in Pakistan, followed by Di-ammonium phosphate, Ammonium Nitrate and Nitro-Phosphate.

Natural gas is the main feed for anhydrous ammonia. In fact, of the total gas produced in Pakistan, around a quarter (26 percent) is utilised for the production of fertiliser. Phosphate is the raw material for the production of Di-ammonium phosphate (DAP). Pakistan does not produce much of phosphate, and phosphate rocks have to be imported entailing an increase in the

TABLE 13.14 Fertiliser Plants and Their Annual Production During 2017-2018

Fertiliser Category	*Number of Factories*	*Production in Tonnes*
1. Urea	10	5,405,194
2. Di-ammonium Phosphate	1	758,379
3. Nitro Phosphate	3	471,352
4. Super Phosphate	4	65,194
5. Ammonium Nitrate	2	518,947
6. Blended NKP	2	
Total Production during 2017–2018		7,219,066

Source: Pakistan Statistical Year Book 2018, table 13.2 p. 270.

cost of production. Phosphate as a raw material for DAP accounts for only 28 percent of the cost of production; and hence, its import is not so burdensome. Natural gas is the feedstock for both urea and anhydrous ammonium.

13.9.1 Gas Fields That Supply Gas for Fertiliser Industry

The two principal gas fields of Pakistan are 1. Sui gas field and 2. Mari gas field.

*The **Sui Gas Field*** – Located close to the tri-junction of the provinces of Pakistan – the Punjab, Sindh and Balochistan, the Sui gas field lies in Balochistan. The three fertiliser plants viz. 1. Dawood Hercules, 2. Pak-Arab Fertilizers and 3. Pak-China Fertiliser, depend for their feedstock on the Sui gas field. The gas produced in the Sui gas field has a very high count of methane, roughly 95 percent, and is very suitable for the production of urea. Though supply of gas from the Sui field is not very regular, and shortages are experienced occasionally, this drawback is compensated for by the high methane content of gas.

*The **Mari Gas Field*** – Located at Daharki (District Ghotki, Sindh), this gas field ensures a sustained supply of gas, though the gas in Mari field has low content of methane, which is around 75 percent by volume. Thus, more gas is needed to produce a ton of urea as compared to the gas of Sui field. The Mari gas field feeds the following three chemical fertiliser units.

1. Fauzi Fertilisers
2. Engro Chemicals
3. Pak-Saudi Fertilisers

The supply and transport of gas is not as difficult as that of coal or other fuel resources. The initial cost of laying the gas pipe lines may appear exorbitant, but in the long run, transport of gas by pipelines is far more economic. This explains the reason for locating the fertiliser plants at places far away from the source of gas (Table 13.15).

13.9.2 Location of Fertiliser Plants

TABLE 13.15 Location of Fertiliser Plants in Pakistan

Name of the Factory	*Source of Gas*	*Location*
A. Dawood Hercules Fertilisers Ltd.	Sui gas field	Sheikhupura (Punjab) 28 km from Lahore
B. Plants of National Fertiliser Corporation		
1. Pak-Arab Fertilizers (largest fertiliser Co. in Pakistan)	Sui gas field	Khanewal (Punjab) near Multan
2. Pak-China Fertilisers	Sui gas field	Haripur (K.P.)
3. Pak-Saudi Fertilisers	Mari gas field	Mirpur Mathelo (Sindh)
4. Pak-American Ltd.	–	Daud Khel (Punjab)
5. Hazara Phosphate Fertilisers	–	Haripur (K.P.)
6. Lyalpur Chemical & Fertiliser Co.	–	Faisalabad (Punjab)
All the six fertiliser plants of National Fertiliser Corporation have been privatised		
C. Fauzi Fertiliser Co. Ltd. (FFC), Established in 1978, the Fauzi Foundation owns Fauzi Fertiliser Co.		
1. First plant commissioned 1982	695,000 metric tons per annum	Goth Machhi, dist. Rahim Yar Khan (Punjab)
2. Second plant commissioned 1992	639,000 metric tons per annum	Goth Machhi
3. Third plant by acquisition of Pak-Saudi plant	5,74,000 metric tons per annum Upgraded to 7,18,000 metric ton per annum	Mirpur-Mathelo (dist. Ghotki)
The total production of FFC Fertiliser Co. is 2 million metric tons of fertilisers per annum		
D. Engro Chemical and Fertilisers (earlier known as Exon) is the largest producer of urea fertiliser		
1. First plant		Dharki (Sindh)
2. Second plant		Bin Qasim (Karachi)

Source: Pakistan Economic Survey, Govt. of Pakistan 2014–2015, p. 17.

13.9.3 Production and Consumption of Fertilisers

According to *Economic Survey of Pakistan 2019–2020*, the total installed capacity of all fertiliser factories in Pakistan is 9.2 million tons per annum. But this installed capacity is seldom reached. The principal cause for this below par production is the interrupted supply of gas, the main feedstock of fertiliser industry. Fertiliser is the second-largest consumer of gas in the country, next only to power, and any interruption in its supply affects the production target of the fertiliser plants.

Despite all the odds, the fertiliser industry has made considerable progress during the last two decades. To cater to the growing demand, its production has increased from a mere 3.8 million tonnes in 2000–2001 to 7.2 million tonnes in 2018–2019, a two-fold increase. There is certainly a need for increased production of fertiliser.

The consumption of fertiliser varies from crop to crop, and the cropping pattern and the acreage under different crops cause a variation in the consumption of fertilisers in different provinces (Table 13.16).

As can be seen, Punjab is the greatest consumer of fertilisers, accounting for 70 percent of the total consumption of fertilisers followed by Sindh, which utilises less than one-fourth of the total consumption of fertiliser of the country. Khyber Pakhtunkhwa and Balochistan, each consumes just over 3 percent of the total fertiliser consumption in Pakistan. A closer picture

TABLE 13.16 The Consumption of Fertiliser in Different Provinces of Pakistan for 2009–2010

Province	*Consumption in Thousand Tonnes*	*The Three Nutrients*			
		Nitrogen	*Phosphate*	*Potash*	*Total*
1. Punjab	a. consumption	2,353.5	924.8	37.1	3,315.4
	b. percentage	68.52	72.34	74.5	69.61
2. Sindh	a. consumption	852.0	265.9	8.3	1,126.2
	b. percentage	24.8	20.79	16.6	23.64
3. Khyber Pakhtunkhwa	a. consumption	124.9	46.6	2.7	174.2
	b. percentage	3.63	3.64	5.42	3.65
4. Balochistan	a. consumption	103.9	41.1	1.7	146.7
	b. percentage	3.02	3.21	3.41	3.08
Total consumption in Pakistan	a. consumption	3,434.3	1,278.4	49.8	4,762.5
	b. percentage	99.97	99.98	99.93	99.98

Source: Pakistan Statistical Year Book -2018, table 1.11 p.62.

Note: a. Is the consumption of fertilizers in thousand tonnes: b. Is the consumption in percentage terms of each province of the total consumption of fertiliser in Pakistan.

TABLE 13.17 Province-wise Cropped Area of Pakistan Compared to the Percentage of Nutrients Used

State/province	*Cropped Area in Million Hectare*	*Expressed as Percentage of the Total Cropped Area of Pakistan*	*Total Use of Fertilisers (nutrient thousand tonnes)*	*Percentage of Total Consumption of Fertilisers*
Pakistan	22.10	100	4,762.5	100
Punjab	16.10	72.8	3,315.4	72.0
Sindh	3.16	14.29	1,126.2	19.0
Khyber Pakhtunkhwa	2.01	9.5	174.2	5.9
Balochistan	0.83	3.75	146.7	2.8

would emerge if cropped area in each province expressed in percentage of the total cropped area of Pakistan is compared with the percentage of fertiliser used (Table 13.17).

As warranted by the percentage of cropped land, Punjab and Sindh, both seem to be utilising their share fully, but Khyber Pakhtunkhwa utilises much less fertiliser. Khyber Pakhtunkhwa and Balochistan have 9.5 and 3.75 percent of the total cultivated area of Pakistan, respectively, but utilise only 6 and 2.8 percent of the total consumption of fertiliser in the country. It is the mountainous terrain on the one hand, and a noticeable lack of irrigation facilities, on the other, that account for such a moderate use of fertiliser in these provinces. In the absence of adequate rainfall as well as irrigation, their capacity to use larger volume of fertiliser remains restrained.

It has to be noted that Pakistan uses 98 kg of fertiliser per hectare in contrast to 600 kg/ha in the Netherlands, 321 kg/ha in the United States and 405 kg/ha in Egypt. The neighbouring country India, however, uses only 77 kg of fertiliser per hectare of cropped area. A wet climatic region in temperate latitudes, as west European countries, or an agricultural tract in the sub-tropics with very good facility of irrigation, like Egypt, can use a high dose of fertiliser. A country like Pakistan or India with very high summer temperatures suffers from a very high level of evaporation, causing accumulation of residual salts in the soil. In heavy rainfall areas, these salts are washed away. Pakistan has the problem salt accumulation in the soil in the very dry irrigated tracts of Punjab and Sindh. The country, however, is able to solve this problem to a great extent by draining channels, especially in Punjab, and by crop rotation.

13.9.4 Agricultural Subsidy and Fertiliser Industry

In most countries of the world, agriculture is subsidised to make it more competitive in the world market and soften the impact of the floods, unseasonal rains, droughts or any other natural calamity that frequents farming areas.

The subsidy is both direct and indirect. Subsidising the cost of fertilisers to ensure a minimum return to the farmers is a direct subsidy, but indirect subsidy is given by providing power at subsidised rates. Of all these subsidies, subsidising the production of fertiliser by supplying cheaper natural gas, the principal feedstock, is most important. The subsidy on natural gas for the manufacture of urea in 2013–2014 was PKR 41.3 billion (PKR 4130 Crore). This is a huge amount yet in the interest of agricultural production, especially food production, the subsidy may continue for some more time.

The installed capacity of fertiliser production of the 19 fertiliser units can adequately meet the domestic demand if the installed capacity is fully utilised and the factories are able to produce the maximum tonnage of fertiliser as required by their installed capacity.

13.10 Sugar Industry

Pakistan today has 85 sugar mills producing roughly seven million tonnes of sugar annually. The sugar factories are largely concentrated in Punjab and Sindh following the cultivation of sugarcane.

Province-wise distribution of sugar mills

Punjab	45
Sindh	32
Khyber Pakhtunkhwa	8
Total	85

In the later forties of the last century, following the establishment of Pakistan, the country had only a couple of sugar mills, with limited crushing capacity. Together, these two mills (one at Rahiwal and the other at Takht-i-Bahi, near Mardan) crushed around 1,800 tonnes of cane per day, producing 10,000 tonnes of sugar annually. By 1956, six more sugar mills were established, increasing the crushing capacity to 132,000 TCD (tons per day). During the 1960s, three more mills were added, but the greatest expansion of industry took place during 1970–1980 when 15 new sugar mills were added. The growth of mills has been steady, and more mills are still being added (Table 13.18).

As the number of mills has increased, so has the production of sugar (Table 13.19).

13.10.1 Location and Areas of Sugar Mills

Sugar mills are tethered to the areas of sugarcane production, which is very wide spread. It may be mentioned that cane is grown not only for the production of sugar, but also for the production of *Jaggery*, a common sweetening agent in rural Pakistan. In the latter case, sugarcane juice is solidified into

TABLE 13.18 Growth in the Number of Sugar Mills From 1948 to 2017–2018

Year	*No. of Mills*	*Year*	*No. of Mills*
1948	2	1990	51
1956	8	1997	75
1970	11	2015	84
1980	26	2018	84

TABLE 13.19 Increase in the Production of Sugar between 1986–1987 and 2017–2018 (in Million Tons)

Year	*Production of Sugar in Million Tons*	*Year*	*Production of Sugar in Million Tons*
1986–1987	1.26	2004–2005	3.0
1990–1991	1.98	2007–2008	4.7
1995–1996	2.5	2010–2011	4.2
1997–1998	3.6	2011–2012	4.6
		2017–2018	6.6

Source: Bureau of Statistics, Govt. of Pakistan, Pakistan Statistical Year Book. 2018, table 13.2, p.264.

TABLE 13.20 Distribution of Sugarcane Cultivation and the Production of Sugarcane, Province-wise (2017–2018)

Province	*Area of Sugarcane in Thousand Hectares*	*Production of Sugarcane in Thousand Tonnes*
Punjab	859	55,067
Sindh	333	20,612
Khyber Pakhtunkhwa	149	7,611
Balochistan	01	43
Total	1,342	83,333

jaggery, which has been a traditional item of food in the country. *Jaggery* production is a small-scale enterprise of individual farmers or a group of farmers, whereas sugar industry is a large-scale enterprise producing sugar in sugar mills (Table 13.20).

In 2017–2018, Pakistan grew roughly over 1.3 million hectares of sugarcane and produced over 83 million tonnes of cane, giving a yield of 62 tonnes per hectare. Sugarcane, being a tropical crop with large vegetative growth, needs high temperature and plenty of water. It is, thus, confined to the irrigated tracts of land in Punjab and Sindh.

The sugarcane cultivation in ***Punjab*** is largely concentrated in the north-central districts of Faisalabad, Jhang, Sargodha, Rahim Yar Khan in the central and south Punjab and some other districts like Toba Tek Singh and Kasur. Following this spatial pattern of sugarcane cultivation is the location of the sugar mills. Faisalabad and Jhang each has six sugar mills, Rahim Yar Khan has five, Sargodha has four and Muzaffargarh and Kasur, each have three sugar mills. Thus, these six districts of Punjab have more than 80 percent of the total number of sugar mills in Punjab.

Sugar Mills in Sindh – In Sindh, sugarcane production and the sugar mills are in located in two clusters. These are 1. The deltaic region of Badin-Thatta, the delta hinterland of Tando Muhammad Khan and Tando Allahyar Khan and Hyderabad and 2. the northern cluster of Khairpur and Naushehro Feroze. Both the regions, the deltaic region of south Sindh and the dry and hot interior of northern Sindh in Khairpur District, have benefitted from canal irrigation and are good areas for sugarcane growth. The humidity caused by the proximity to the sea provides a moderating effect against excessive evaporation in the delta region, whereas the conditions in the northern part of Sindh are different and demand more intensive irrigation.

In the deltaic region, Thatta and Badin have four sugar mills, but the largest concentration is in Hyderabad District, possibly because of its centrality and ease of distribution to the consuming centres in the countryside and all over the province (Figure 13.2).

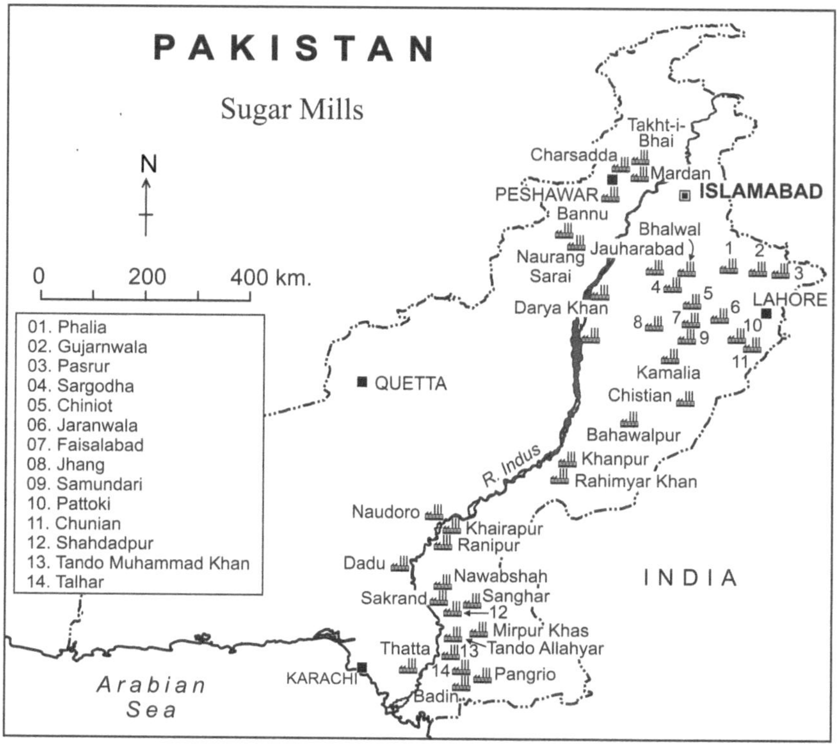

FIGURE 13.2 Distribution of Sugar Mills in Pakistan

Sugar Mills in Khyber Pakhtunkhwa – In this frontier region, there are eight sugar mills, four in Dera Ismail Khan, two in Mardan, one in Peshawar and one in Bannu.

The largest concentration of sugarcane cultivation in Khyber Pakhtunkhwa is in Charsadda and Mardan located, respectively, in the narrow alluvial plains of river Kabul and its tributary the Swat, joining it from north. While Charsadda has benefitted from the canals emanating from Warsak reservoir in Kabul, Mardan enjoys irrigation from Swat canal. Assured of water, both these districts, Charsadda and Mardan, have together 60,000 ha of land under sugarcane, comparable to some districts of Punjab. Dera Ismail Khan is the only district of Khyber Pakhtunkhwa, which lies in the Indus Plain. The total area in the district under sugarcane is just around 10,000 hectares that supports four sugar mills.

It must be emphasised that one of the earliest sugar mill of Pakistan was the Frontier Sugar Mills Ltd., set up in 1949. It is an industrial unit of the Premier Group of industries. The Frontier Sugar Mills Ltd. is currently not operating, though its sister mill, the Premier Sugar Mills Ltd., located at Mardan is doing well.

13.10.2 Sugar Production

The area under sugarcane, the number of mills and the production of sugar have steadily increased. The production of sugar has increased from two million tons in 1990 to over seven million tons in 2017–2018. Presently, Pakistan has over 1.3 million hectares of land under sugarcane, producing 83.3 million tonnes of sugarcane. And, with 84 sugar mills, the country produces around seven million tonnes of sugar annually.

The importance of sugar industry lies not only in meeting the domestic needs of sugar, but also in creating forward and the backward linkages that have a multiplier effect, giving rise to ancillary industries and generating employment. Being an agro-based industry, it provides employment to a large mass of agricultural labour. Cultivation, harvesting and transportation of sugarcane to sugar mills require manpower, especially for harvesting the cane. The industry provides employment to approximately four million people. Besides, it has transformed the rural landscape. Wherever a sugar mill comes up, it forms the nucleus, around which markets, transport, and a number of subsidiary facilities develop. As a result of the income generated from sugarcane as a cash crop and sugar as an important marketable commodity, the rural landscape with sugar mills appears prosperous with infrastructure facilities like roads, schools and dispensaries. Sometimes, a dairy unit also appears as an adjunct industry.

Sugar Industry operates seasonally for six months depending on the supply of cane, normally estimated to be around 180 days. In reality, the mills operate much below their capacity, and a working period of 150 days is more common. The industry, though working at a moderate level of efficiency, produces some surplus sugar and exports its main by-product, the molasses, worth thousands millions of rupees.

Problems of Sugar Industry – Despite being the second-largest industry, next to cotton textiles, the industry faces several problems some related to cane and others related to sugar mills and the policy of the government. Low yield, with low sucrose content, except in the case of Sindh, under-utilisation of crushing capacity largely because of inadequate supply of cane, competition from other sugar producing countries and delayed payment of cane price to the farmers are some of the hurdles that need to be overcome.

Located in the countryside and employing a large manpower, a symbol of rural industrialisation, sugarcane and sugar mills provide relief and the much needed cash to the farmers while promoting regional development.

Other Industries – Besides a few industries discussed above, there are a large number of medium- and small-size industries. These include electrical goods, machinery, pharmaceuticals, paper and paper products, sports goods, manufacture of leather goods and several others. Some of the medium- and small-scale industries like sports goods, processed fish and meat products earn valuable foreign exchange for the country.

13.11 Tourism

An important aspect of Pakistan's economy is revenue from tourism. The tourism industry in the country is largely based on visits to significant pre-historic and historic sites on the one side and the adventure and eco-tourism oriented towards the high mountainous areas of the country on the other. Pakistan has some of the most important ruins of pre-historic sites like the ruins of Mohen jo Daro in Sindh and those of Harappa in Punjab, both tracing their existence to pre-historic Chalcolithic period. Besides, the country is replete with very early historic monuments, especially in Khyber Pakhtunkhwa where remnants of Gandhara culture and the ruins of Buddhist monuments attract tourists, especially those with an interest in the early history of Pakistan. The Buddhist stupas and the Ashokan inscriptions dating back to several centuries before Christ, rather well preserved, are a great draw. The country has also some of the architectural marvels of the medieval period, preserved in mosques, forts and mausoleums, spread all over the country, but mainly in large cities like Lahore, Multan and Peshawar. The newly built federal capital, Islamabad, has its own charm.

It may be mentioned that Pakistan has six World Heritage sites (Mohen-jo-Daro, Takht-i-Bahi, Rohtas fort, Lahore fort, Taxila, and Makli hill), and these attract tourists from all over the world.

Eco and Adventure Tourism – Away from the monuments located in the plains, Pakistan is a rich depository of scenic landscapes embedded in the folds of the Himalayas. These include glaciers, beautiful lakes, high altitude pastures camping sites, trekking paths and fishing river sites. Besides, there are forests and fruit orchards.

Most of the sites for eco and adventure tourism lie in the northern part of the Khyber Pakhtunkhwa, one of the provinces of Pakistan. Thus, much of the adventure and mountain tourism is based in Khyber Pakhtunkhwa, which is a major beneficiary of this aspect of economic activity.

Some Important Eco-tourism Sites – Most of the sources quote ten principal eco-tourism sites from the hilly areas of Khyber Pakhtunkhwa. These include 1. Kumrat Valley in upper Dir District, 2.Takht-i-Bahi, a Buddhist monument not too far from Peshawar, 3. Tarbela Dam, a dam on the Indus, 4. Kaghan Valley, a left bank tributary of the Indus in Mansehra District, 5.Tirich Mir, a glaciated peak (7,708 m ASL) in Chitral District, 6. Mahodand Lake, 2,865 m, at the foot hills of Hindukush mountain, a good camping site, 7. Shandur Pass (3,720 m) in Ghizer District of Gilgit-Baltistan, often called the roof of the world, known for trout fishing and polo matches during summer, 8. Khyber Pass, 9. Lulusar Pass and 10. Ansoo Lake. Besides, there are other high altitude sites that attract mountain climbers, hill trekkers and high altitude tourists.

Some recent additions to places visited by tourists are the Hingol National Park and not long ago excavated pre-historic site of Mehrgarh, both in Balochistan. The former is a model of naturally sculpted landscape in a clayey terrain, and the latter 5,000 BC old Neolithic site. Mehrgarh archaeological site, not too far from Quetta, is known for abundance of artefacts and its stunning beautiful pottery and is a cultural site that speaks of the glorious past of the region.

Tourism is both an economic as well as a cultural activity. For the tourists, it is a cultural activity including entertainment and even education, but for governments, in general, it is an economic activity, generating revenue, specifically foreign exchange. Most governments, across the board, try to promote tourism, and most have tourism promoting agencies. In Pakistan are, deeply involved in promoting tourism, the Pakistan Tourism Development Corporation, a Government agency and Pakistan Tourism Development Board.

Numerical Assessment of the Growth and Decline of Tourism – A basic measure of the status of tourism as an economic activity in any country is the number of tourists, both domestic and foreign. Data available of the number of tourists visiting the country are not very bright and show a declining trend (Table 13.21).

As can be seen from the table above, there has been a steep decline in tourism in general and in tourists arrival in particular in Pakistan from around

TABLE 13.21 Number of Tourists visiting Pakistan Between 2005 and 2018

Year	*Number of Tourists*	*Year*	*Number of Tourists*
2005	798,300	2006	897,600
2007	839,500	2008	822,800
2009	485,326	2010	440,064
2011	285,576	2012	149,209
2013	32,905	2014	6,018
2015	6,205	2016	8,352
2017	7,472	2018	20,689

Source: Pakistan Statistical Handbook 2019, table 2030 p, 495.

800,000 in 2005 to almost a trickle, around 1 percent of the previous high. The reasons for this decline are many. A lurking general depression in the beginning of the century may have affected the overall climate for tourism. But domestic conditions in Pakistan could be equally responsible for this extreme decline of this industry. Political instability on the one hand and a series of violent incidents, including war in the neighbouring Afghanistan, and indirect involvement of Pakistan may have contributed to this sudden drop in the arrival of tourists from all over the world.

Foreign and Local Tourists – Most governments pay considerable attention to foreign tourists, as they are the principal source of foreign exchange earnings. The largest number of foreign tourists was recorded in 2006 when 897,600 foreign tourists visited Pakistan. It may appear paradoxical, as during this period, the country was under a military regime, and war in Afghanistan was still on. The trend continued till 2012 when there was a sudden drop in the number of foreign tourists, their number going down to 150,000, around 20 percent of the number in 2008.

It seems, the most important factor was security and the military regime guaranteed security to the tourists. Around 40 percent of the tourists came from Europe, principally from the United Kingdom, another 15–20 percent arrived from the United States, Canada, Australia and China were other countries that contributed to the stream of tourists.

Revenue from Tourism – Regular yearly data of the revenue collected from foreign tourists are not available. At the height of the flourishing tourism industry, the foreign exchange receipt from the tourists was US $241 million in 2009. After that, it is a gradual decline till today when the country badly needs foreign exchange reserve,

Revival of Tourism in Pakistan -- Once a flourishing industry, the decline of this industry could be attributed to the security climate of the country. The first thing a tourist needs is security. Presently, Pakistan is embroiled in all kinds of controversy. Political instability, economic meltdown and societal conflicts have all congregated to worsen the situation. Only when the country emerges from the present economic crisis and restore social harmony, will it be able to inspire confidence among the intending tourists, the foreign tourists will return to the shores and heights of Pakistan.

13.12 Overview of Pakistan's Industrial Landscape

Despite the best efforts of the Government of the entrepreneurs, Pakistan has not been able to transform itself into a buzzing industrial country. The reason is both natural and man-made. To start with, the country does not have the raw material for basic industries like iron and steel. It works its steel mills, the Pakistan Steel Mill, the largest of the steel mills in Pakistan, and a number of small plants with imported iron and coal. The discovery of Sui gas in the early fifties was a God-send, and the discovery of the gas fields of northern Sindh added to the supply. But there were, and still are, many competing uses. Despite the increase in supply, the demand from multiple users, like fertilisers, power stations, transport and other users, always outpaced the supply and could not be met. The recent discovery of Tharparkar lignite, estimated to be 190 billion tons, has come at a wrong time in the sense that a highly polluting carbon source of energy like lignite does not fit in the international effort to cut down the burning of fossil fuels to stop the global warming and climate change. Non-availability of adequate energy, electrical or fuel for the industries, has been a recurring problem in Pakistan's industrial development.

Cotton textile industry in Pakistan is one, of which the country can legitimate proud, but the irregular supply of power often harms this very vital sector of Pakistan's economy. Even textile industry is facing a stiff competition from countries like Bangladesh and Vietnam. The harnessing of solar power may improve the energy position in Pakistan, attracting both light manufacturing like electronics and other services to the country.

Acceleration in investment, public and private, could boost manufacturing and service growth. An important component could come from Foreign Direct Investment (FDI). But an investment friendly climate requires an enabling public policy, but also an environment of social harmony. Unstable security conditions are detrimental to attracting foreign investment in any part of the world.

Pakistan has managed its water resources and agriculture rather well. One hopes, it turns into an industrially advanced country as well. It is beyond the scope of this text to examine and evaluate the reasons for Pakistan's sluggish industrial development. But one thing is certain, a stable political situation, a conflict-free society, and a secure economic environment will unfailingly attract foreign investors and investment. The recent agreement with China, the China Pakistan-Economic-Corridor (CPEC), involving a massive investment of US $50 billion, will certainly improve the country's infrastructure.

This is not enough and a much larger investment from domestic and foreign sources is needed. The Government of Pakistan and the people of Pakistan, being aware of the problem, should be able to find proper solution.

Notes

1 Source: Economic Survey of Pakistan, Government of Pakistan (2012–13) Finance Division, Government of Pakistan p. 37.

2 These are 1. Pak-Arab Refinery (100,000 bbl/d), 2. National Refinery (86,000 bbl/d), 3. Attock Refinery (46,000 bbl/d), 4. Byco Petroleum Pakistan (150,000 bbl/d), 5. Pakistan Refinery (50,000 bbl/d), 6. Enar Petroleum Refinery (3000 bbl/d).

14

TRANSPORT AND TRADE

Transport and communication are an essential element in the infrastructure of a nation. An efficient transport network is an indispensable precondition to growth in a country's economy. Pakistan has an elaborate network of railroads and roads that are able to promote a smooth flow of goods and services. Much of the transport system in Pakistan rests on the rail and roadways. Water transport has a limited scope, and air transport, though critical in emergency situations, can handle only a moderate volume of traffic.

14.1 History of Transport in Pakistan

Prior to the coming of the railways, which revolutionised the overland transport system in the world, transport in Pakistan, as in other parts of the world, rested on slow-moving wheeled transport, like bullock carts and horse-drawn carriages, or on some beasts of burden like donkeys, mules, camels or even horses. It is worth mentioning that wheeled transport was known to the region very early in history, as witnessed by "the toy carts with solid wheels" found in Indus Valley excavations. Slow-rumbling bullock carts have been the traditional mode of transport in the region, as in other countries of South Asia, supported by water transport through perennially navigable rivers. Early in history (326 BC), Alexander, the Macedonian king, had sailed down the Indus in a flotilla of small vessels to the Arabian Sea.

The coming of the railways in the latter half of the 20th century revolutionised the transport system of the region. Introduced by the British colonialists, to service their own interests, the railways provided a speedy and safe transport to export food grains and precious raw materials, like cotton, for British industries, and to import finished goods from the United Kingdom

DOI: 10.4324/9781003510147-17

for distribution in the huge Pakistani market, the railways became the principal mode of transport of India. Besides, the British required a rapid mode of transport like the railways for security reasons to rush troops to quell any resistance anywhere in the country. One must, however, appreciate that regardless of the motives behind introducing railway transport in the country, the British, while leaving, left behind an elaborate functioning network of railways, which has since been expanded, modernised and made more efficient.

14.2 The Pakistan Railways

Railway transport began in Pakistan in 1855 with the establishment of Scinde Railway Company and the laying of the first railway track between Karachi and Kotri, a distance of 174 km, which became operational in May 1861. This was connected upstream, between Kotri and Multan, by a steamer service. Soon after, there was a large expansion of railways and many independent companies like the Scinde, Punjab and Delhi Railway companies merged into what came to be known as North-Western State Railways.[1]

Sub sequently, the line was extended to Peshawar before the end of 1883. In the early stages of colonisation, the British had apprehension of the Russians approaching India via Afghanistan. Hence, they laid a railway in a very hilly terrain of Balochistan, negotiating the very difficult Bolan Pass up to Quetta and beyond, through Khojak Pass (2290 m ASL), up to Chaman. The line is known today as the Rohri–Chaman railway. Known for an engineering feat, the laying of the railway through Khojak Pass involved making a 3.9 km long tunnel through Achakzai range, connecting Quetta to Chaman on the Pakistan–Afghanistan border. Another line known as Trans–Balochistan line was laid from Quetta via Nuski and Dalbandin up to Kuh-i-Taftan on Balochistan–Iran border, extended right up to Zahedan in Iran. These two railway lines Rohri–Quetta and Quetta–Kuh-i-Taftan were laid largely, if not exclusively, as a strategic consideration to ward off any attack from the west.

With the laying of many more railway lines in the late 19th and early 20th century, like the Delhi–Lahore–Rawalpindi–Peshawar Railway, the Indus Valley railway and opening of a number of additional routes, like the Karachi–Lahore route, or the Karachi–Quetta route, and a number of small branch lines like, the Jammu-Sialkot and the Kasur–Lodhram branch, a dense network of railways emerged. After 1947, the year the country came into existence, Pakistan added a few more railway lines, changed the gauge of the lines from a metre to a broad gauge, facilitating connectivity, and further expanding the railway network. After 1947, the entire network of railways in Pakistan began to be known as Pakistan Railways (PR); divided into two divisions, viz. Pakistan Western Railways and Pakistan Eastern Railways covering the areas of West and East Pakistan respectively. With the secession of East Pakistan (Bangladesh), the West Pakistan Railways was christened as Pakistan Railways, a name that continues till today.

14.2.1 The Network of Pakistan Railways

The rail network of Pakistan consists of a number of main lines and many branch lines. There are half a dozen main lines connecting most parts of the country. The most important of these are (1) Karachi–Peshawar Line, (2) Kotri–Attock Line, (3) Rohri–Chaman Line and (4) Quetta–Taftan Line.

1. *Karachi–Peshawar Line* – The railway line handling the largest volume of passenger and goods traffic is the Karachi–Peshawar line, starting from Karachi port and terminating at the frontier city of Peshawar, the capital of Khyber Pakhtunkhwa. This line traverses the entire province of Sindh, Punjab and Khyber Pakhtunkhwa, passing through some important cities like Hyderabad–Kotri, Dadu, Larkana, Sukkur and Ghotki in Sindh, and cuts cross Punjab via Rahim Yar Khan, Multan, Sahiwal, Lahore and then turning northwestward, touches Gujranwala, Rawalpindi and crossing the river Indus at Attock, finally reaches Peshawar via Nowshera Junction. In railway parlance it is known as Mainline 1 (ML-1).
2. *Kotri–Attock Line* – This also starts from the twin city of Kotri–Hyderabad, located at a distance of 180 km northeast of Hyderabad and runs northward along the right bank of Indus, passing through Larkana, Shikarpur, Jacacobabad and Khandot. At Kashmir, it leaves Sindh and running parallel to Highway No. 55, passes through Dera Ghazi Khan before it crosses Indus from right to left bank (west to east) at Kot-Abbu and then, running northward, touches Leiah, Bakkar, Mianwali, finally terminating at Attock. Pakistan Railways have given it the name ML-2.
3. *Rohri–Chaman Line (ML-3)* – This is a railway line, built very early for strategic reasons. It negotiates the most difficult terrain of Balochistan with Bolan and Khojak Passes en route. It starts from Rohri, a town on the left bank of Indus across Sukkur, and runs northwestward touching Shikarpur and Jacobabad before it crosses into Balochistan at Jhatpal and traverses the Kachhi Plain. Beyond Sibi, the apex of the Kachhi Plain, it is a hilly terrain all the way, starting from the northern end of the Brahui Range, which it crosses through Bolan Pass (1798 m ASL) and touches Quetta (Railway station height 1676 m ASL). Northwest of Quetta, the railway line passes through Qila Abdullah, before it negotiates the steepest Khojak Pass and terminates at Chaman, the frontier town of Pakistan that oversees the Quetta–Qandhar route.

4. *The Quetta–Taftan Line (ML-4)* – This, more than 800 km long route, passes through a desolate terrain of West Balochistan. Sparsely populated, the region does not generate much traffic. Yet, it was built in the late 19th century as a strategic necessity to link Iran at Zahedan. The route has lately gained importance following the discovery of minerals in the Chagai hills area and the copper and gold mines of Reko Diq and Saindak. The route provides quick access to the Iran border and is strategically significant.

Connecting these main lines are a number of branch lines like Khanewal–Wazirabad Line in Punjab, Bannu–Tank Line in Khyber Pakhtunkhwa or Hyderabad–Badin line in the Indus delta region.

14.2.2 International Connection of Pakistan Railways

Pakistan is linked with two adjacent countries – Iran and India – by railways.

Pakistan–Iran link is established by Quetta–Taftan Line, mentioned above. It links Taftan on Pakistan border with Zahedan in Iran. Taftan is situated on Pakistan–Iran border, while Zahedan is about 90 km inside Iran. The distance between Taftan and Zahedan is roughly 100 km.

Pakistan–India link – Immediately after the division of British India into two independent states, India and Pakistan, the pre-existing railway links between the two countries were snapped, The Frontier Express that ran between Bombay and Peshawar was discontinued, so also between the two Punjabs, and between Sindh in Pakistan and Rajasthan in India. Both these routes were revived in 1976 with agreement between Pakistan and India.[2] These were:

1. Jodhpur–Karachi – linked by Thar Express (709 km)
2. Delhi–Lahore – linked by Samjhauta Express

Jodhpur–Karachi link – The route starts from a station called Bhaghat Ki Kothi in the suburbs of Jodhpur (Rajasthan, India) and ends at Karachi Cantonment (Sindh-Pakistan). It is a distance of roughly 700 km, covered by Thar Express in over 13 hours, with four or five halts on the way. The route was first opened in February 2006 but again closed in 2019, following an abrupt change in relations between the two countries.

Delhi–Lahore Link – The Northern Route taken by the Frontier Mail of the colonial period always existed, only the running of trains was discontinued. Following the Simla Agreement and a better understanding between the two neighbours, an express train called Samjhauta Express (*samjhauta*, a vernacular word meaning agreement) was started in 1976. It is a weekly train that runs between Old Delhi Railway Station and Lahore Junction, Punjab, a distance of 502 km. These Pakistan–India links are quite uncertain, and their operation depends on the flow and ebb of relations between the two countries.

14.2.3 Operation of Pakistan Railways

The management of Pakistan Railways is divided into eight divisions. These are (1) Karachi, (2) Lahore, (3) Multan, (4) Peshawar, (5) Quetta, (6) Rawalpindi, (7) Sukkur and (8) Gwadar (not yet fully operational). On a network of 7,791 km route, the Pakistan Railways carries 70 million passengers and 8 million tonnes of freight annually. The railways, though presenting a system of mass transport used by all sections of the society, are the principal means of transport in the country.

The Pakistan Railways run 228 trains, each train named after its origin and destination. Some others are named after some great personage, while some others are named after the territory through which they run. The most prestigious train of Pakistan is the Jinnah Express, a super-luxury train, started in 2019 that runs between Lahore and Karachi and covers a distance of over 1,200 km in 16 hours. The most important train covering the longest route is the Khyber Mail. Started in 1920, it is one of Pakistan's oldest and cross-country train that runs between Karachi and Peshawar, a distance of 1,760 km, and takes over 30 hours to cover the distance.

14.2.3.1 Development of Railways Infrastructure

Pakistan Railways have their own factories and workshops to produce locomotives, coaches and freight wagons. The most important factory is the "Pakistan Locomotive Factory" located at Risalpur, a station halfway between Nowshera and Mardan, in Khyber Pakhtunkhwa. The unit produces 20–25 diesel-electric locomotives annually. Similarly, a passenger coach factory was established near Islamabad in 1970, with a capacity to manufacture 150 passenger coaches annually. Besides, Pakistan Railways own "Moghalpura Railway Workshop" at Moghalpura Junction, on Lahore–Wagah Line. Built in 1904, this workshop has developed over the years into a modern workshop to repair and maintain passenger coaches and goods wagons.

14.2.3.2 Plans for Future Development

Though the network of railways is relatively well developed, yet the country plans further expansion of the railway network, especially in Balochistan and the northern part of Pakistan, along the Karakoram Highway. The proposed Karachi–Gwadar Railway Line (to be named Makran Railway) running along the Makran coast, and Gwadar–Mastung Railway Line, cutting across the entire stretch of western Balochistan, are the two ambitious and costly proposals in the pipeline. More ambitious and technologically more complex is the Karakoram Railway, as yet at an exploratory stage. The proposal involves linking the main Lahore–Peshawar route to Kashgar in the Xinjiang Province of China. The 750 km stretch, between Havelian and Khunjerab Pass, is proposed to be built by Pakistan, linking Havelian and Khunjerab (4,733 m ASL) on the northern border of Pakistan. Beyond Khunjerab, China is to build a 350 km railway track to Kashgar in the Xinjiang Province. The Karakoram Highway already exists, and to lay a railway route parallel to it, traversing a high altitude terrain is a difficult proposition. It is a question of priority. Would Pakistan opt for it as a priority project?

Under the CPEC agreement (China–Pakistan Economic Corridor), China is deeply involved in the expansion and improvement of Pakistan Railways, providing high-speed coaches, wagons and locomotives. The Chinese have already built the urban metro Orange Line in Lahore and are partners in its operation. Chinese companies are also engaged in building the light-rail mass transit system in Karachi. With financial and technological support from external sources, Pakistan Railways should be able to build and maintain a far more elaborate and efficient railway system.

14.2.3.3 The Economy of Railway Transport

Pakistan Railways are a fully state-owned enterprise, managed through a railway board appointed by the Government of Pakistan. As an economic venture, Pakistan Railways don't make any profit and is a loss-making enterprise. The annual operating revenue of Pakistan Railways in the year 2018–2019 was PKR 49.56 billion, whereas the operating expenses amounted to PKR 52.07 billion. One may, observe, that though the railways have incurred losses in their operation, their contribution to the nation and the society far outweighs the economic losses. Railways are, by far, the cheapest mode of transport, and a vast majority of people depend on it for their travel. The lower income group of people depend almost exclusively on the railways for travel and especially for long journeys. As mentioned in the Economic Survey of Pakistan,[3] Pakistan Railways (PR) is a single major mode of transport in public life contributing to economic growth and providing national integration. They are more of a public service enterprise than a commercial enterprise and fall in the category of essential services to the people and have to be looked upon as such.

Development Prospects – In a recent document running into 121 pages, Pakistan Railways has outlined a plan of turning the railway into a profitable entity. It mentioned that

50 % of the train engines are old, besides a shortage of funds is affecting the repair work of the engines. According to this document, Pakistan Railways need to enter into joint venture international agreements to become a profitable entity, and passengers will be provided with secure, comfortable and affordable facilities.[4]

14.3 Roadways

Historically, one of the earliest known routes in South Asia was the "*Uttarapath*", the Northern Route, built during the Mauryan period, a couple of centuries before Christ. The Northern Route was greatly improved during the medieval period and rebuilt by Sher Shah Suri, who ruled India for a period of 15 years (1540–1555). The route is, hence, also known as "Sher Shah Suri Road". The British gave it the title "Grand Trunk Road", that ran from Dhaka to Peshawar. Part of this route lies in Pakistan from Lahore to Peshawar, now transformed into National Highway No. 5. Besides having one of the earliest routes in history, Pakistan also inherited a network of roads, developed during the British period. During the last 75 years of the existence of Pakistan, the road network has been made more dense and efficient by widening and paving the existing roads, building bridges and making long thoroughfares by linking together isolated roads.

Roads are the dominant mode of inland transport in Pakistan, carrying 91 percent of passenger traffic and 96 percent of freight traffic. It may appear unrealistic in view of visible dominance of railways. The reason is that for short distances, daily or hourly travel, and transport requiring haulage from pick-up-point to destination, road transport is the preferred mode of transport. The railways are cost-effective for long distance mass transport. It may be noted, by way of contrast, that while the railway routes have a length of 7,791 km (2017–2018), the road network length is 259,618 km, 33 times larger than the railways. This explains why road transport is so important in the daily life of people.

14.3.1 Classification of Roads

The basic division of Pakistan roads is two-fold: high type and low type.[5] The total length of roads in Pakistan is grouped in these two categories, according to the state and status of specific roads (Table 14.1).

During the ten-year period (2008–2018), there has been a meagre increase of 3.8 percent, in the road- length, but what is significant is considerable improvement and conversion of low-type roads into high-type. During this decade, 18 percent of the total low-type roads have been converted into a high-type. From the point of their importance as thoroughfares, traffic density, and design and maintenance, and their ownership and

TABLE 14.1 Length of High- and Low-Type Roads in Pakistan (in km) for 2008 and 2018

Year	*High Type*	*Low Type*	*Total Length (in km)*
2008	175,000	84,038	259,038
2018	197,452	71,483	268,835

TABLE 14.2 Administrative Categories of Roads in Pakistan and Their Length

Type of Roads	*Controlling and Maintaining Authority*	*Length (in km)*
1. National highways, motorways and strategic roads	National Highway Authority	9,000
2. Provincial roads	(C. and W. D.) Communication and Works Department of provinces and Frontier Highway Authority (FHA)	101,000
3. District roads	District Administration	94,000
4. Municipal and Cantonment roads	Municipal Administration and Army	54,000
Total		258,000

administration, the roadways in Pakistan are grouped into the following categories (Table 14.2).

The most important of these roads are the National Highways and Motorways. The length of these two categories has been extended during the last decade. In 2019–2020, the National Highways and Motorways recorded a length of 13,000 km, an increase of 4,000 km over their length in 2006 (Table 14.3).

14.3.2 National Highways and Motorways

National Highways and Motorways are the two principal types of roads in Pakistan (Figure 14.1). While National Highways always existed, Motorways have been introduced recently. These are the express ways with multiple high-speed lanes. The number of both, the highways and the motorways, has increased. While some of the provincial roads are converted into National Highways, some important highways are transformed into Motorways, upgrading them, with more lanes, ensuring free flow of traffic.

14.3.3 Principal Highways

There are four important national highways. These are 1. Grand Trunk Road, 2. Indus Highway, 3. Karakoram Highway and 4. Makran Highway.

TABLE 14.3 Status and Distribution of Roads in the Provinces[6]

	Pakistan	*Provinces*			
		Punjab	*Sindh*	*Khyber Pakhtunkhwa*	*Balochistan*
Road length (km)	258,214	106,140	79,834	30,049	42,191
Percentage of paved roads	63	78	69	46	13
Area (km^2)	796,096	206,250	140,914	101,741	347,190
Road density (km/km^2)	0.32	0.51	0.57	0.30	0.12
Road length (km) per million population	1736	1244	2420	1292	5909
No. of registered vehicles	4,974,000	2,920,984	1,457,323	430,429	165,264

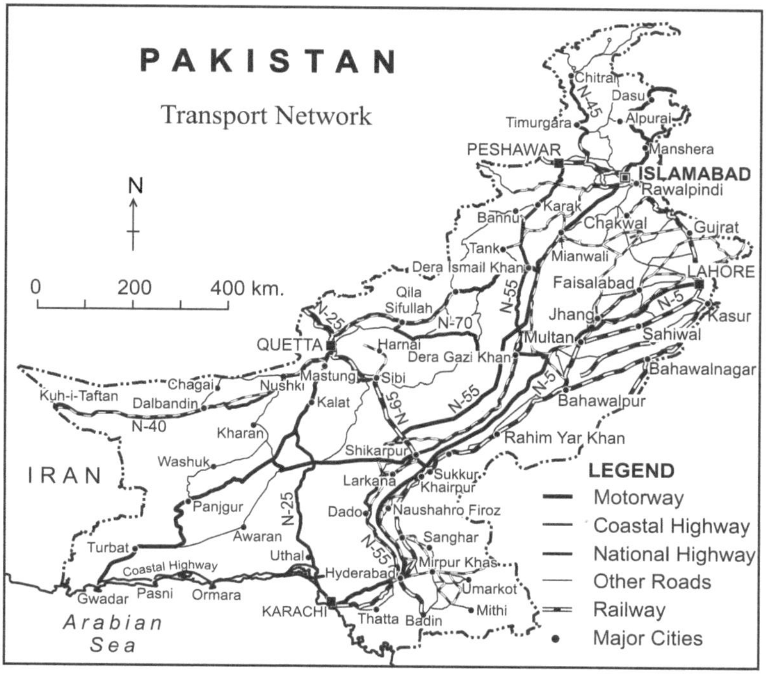

FIGURE 14.1 Transport Network of Pakistan

During the last couple of decades, there has been considerable improvement in these highways by technical up-gradation. Some of these highways are changed into motorways, changing their names and even dividing them into several sectors. The four National Highways are:

1. *National Highway (NH No. 5)* – This is the longest highway with a length of 1,819 km, extending from Karachi to Torkham, divided into three sectors, in the three provinces of Sindh, Punjab and Khyber Pakhtunkhwa. The length of the highway, in the three provinces mentioned above, is 671 km in Sindh, 1,021 km in Punjab, and 165 km in Khyber Pakhtunkhwa. Completed in 1952, the highway runs from Karachi through Hyderabad, Moro, Khairpur (in Sindh); Lahore, Gujranwala, Gujrat, Jhelum, Rawalpindi and Attock in Punjab, and then crossing the Indus, the highway reaches Torkham border, through Nowshera and Peshawar in Khyber Pakhtunkhwa. This is the largest and the most travelled highway. Parts of this highway are converted into motorways.

2. *Indus Highway (NH No. 55)* – With a length of 1,264 km, this is the second longest highway in Pakistan, running along the west bank of Indus. Remodelled in 1980, it runs from Kotri in Sindh to Peshawar, passing through Dadu, Larkana and Shikarpur in Sindh; Rajanpur and Dera Ghazi Khan in Punjab and touches Dera Ismail Khan, Bannu and Kohat in Khyber Pakhtunkhwa before reaching Peshawar.
3. *Makran Highway (NH No. 10) 653 km* – Makran Highway is relatively a new highway in Pakistan, completed in 2004, in a record time of two years, by Pakistan Frontier Works Organisation (PWO). The highway originates at Karachi, and passing through the coastal cities of Ormara and Pasni in Balochistan, terminates at Gwadar. In view of the development of Gwadar as a deep sea port under CPEC, and linking it to Karakoram Highway to open the route to China through Khunjerab pass, the opening of Makran Highway has a special significance. Besides, the coastal highway also provides some impetus for industrial development of the Makran Coast.
4. *Karakoram Highway (NH No. 35)* – This is the most high profile highway in the history of transport and road building in Pakistan. It links the Grand Trunk Road (now Highway No. 5) with Kashgar in Xinjiang, the Uyghur Autonomous region of China. The highway passes through Khunjerab Pass, at 4,693 m ASL (36°- 51 N & 75°30' E) in the Karakoram range. Since it is built jointly by China and Pakistan, it is also called China–Pakistan–Friendship Highway. The length of this highway is 1,300 km, of which 887 km lies in Pakistan and the remaining 413 km in China, terminating at Upat (Kashgar). In a geopolitical context, Karakoram Highway is considered an integral part of Asian Highway No. 4.

In Pakistan, the Karakoram Highway takes off from the Grand Trunk Road (NH-5) at Hasan Abdal, around 35 km northwest of Rawalpindi and traverses the entire length of Khyber Pakhtunkhwa. It, then, turns eastward near Kotgala , and running for around 130 km due east. Following the Indus Valley, the highway takes a northward turn and passes through a number of towns on the way, notably Gilgit, before touching the northernmost point of Pakistan at Khunjerab. From south to north, the highway passes through the following towns:

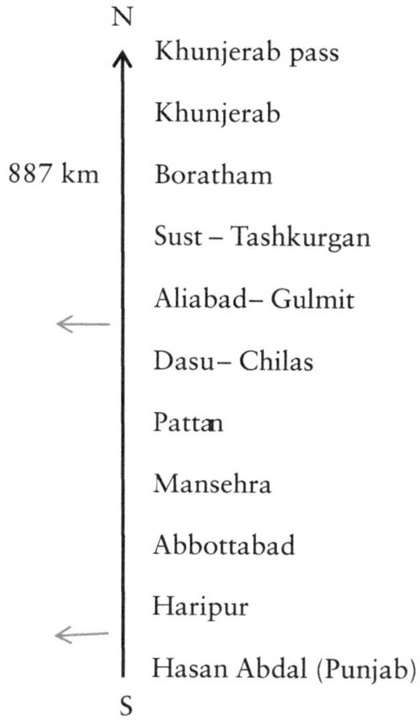

14.3.3.1 Building of Karakoram Highway

Built jointly by Pakistan and China, construction works, the project began in 1959, and it took 20 years to complete the construction of this 1,300 km long road. It was completed in 1979 and opened to the public in 1986. The most dangerous aspect in the construction was the occurrence of frequent landslides. Occasional flooding was another risk as much as glacial avalanches. Over 800 Pakistani and 200 Chinese died during the construction period, mostly from landslides. The Frontier Works Organisation of Pakistan, under the direction of the Army Corps of Engineers, completed this feat. The road covers five long tunnels, ranging in length from 195 m to 3,360 m, and the most notable technical feat is the Shikasht bridge on the Hunza River with a length of 480 m.

The Karakoram Highway, though completed, is not fully in use, as the link with China is not fully established. A petroleum and gas pipeline along the route, connecting Gwadar to Kashgar, will complete this complex project, thus, providing China direct access to the Arabian Sea.

14.3.4 Motorways of Pakistan

Motorways of Pakistan are multiple-lane high-speed freeways having a total length of 3,736 km. These are high-speed freeways, controlled and managed

TABLE 14.4 Motorways in Pakistan, Length in km and the Route Followed

No. of Motorway	*Route Followed*	*Length in km*	*No. of Lanes*
M-1	Peshawar–Islamabad	155	6
M-2	Islamabad–Lahore	344	6
M-3	Lahore–Abdul Hakeem	230	6
M-4	Pindi–Bhattian–Multan	300	4–6
M-5	Multan–Sukkur	392	6
M-6	Sukkur–Hyderabad	296	6
M-7	Dadu–Hub	270	-
M-8	Ratodero–Gwadar*	892	2
M-9	Hyderabad–Karachi	136	6
M-10	Karachi–N. Bypass	57	2
M-11	Lahore Sialkot	105	4
M-12	Sialkot–Kharian	46	6
M-13	Kharian–Rawalpindi	115	6
M-14	Islamabad–Dera Ismail Khan	285	4
M-15	Hassan Abdal–Thakot	180	6–4
M-16	Swabi–Chakadara	169	4

*This is the most important link in the route under construction between Gwadar and Kashgar.

by the National Highway Authority. The following are the motorways of Pakistan as reported in 2020 (Table 14.4).

These motorways, some intra- and some interprovincial in their reach, have smoothened the road transport. Fast, efficient and in a sound state of repair, these have added greatly to the economy of the country. Some like M-2, Islamabad–Lahore, with six lanes are the busiest roads of Pakistan. Some along the Indus like Karachi–Hyderabad, or Islamabad–Multan (M-4) or Islamabad–Peshawar (M-1) pass through densely peopled areas of the country and have rendered transport in these areas trouble free.

14.3.4.1 Road Projects Under the Aegis of China–Pakistan Economic Corridor (CPEC)

Many of the highways and some motorways under CPEC have been rebuilt or are under construction, while others, already built, are operational.

M-14 Islamabad (Hakla)–Yank (Dera Ismail Khan), 255 km (under construction)

M- 5 Multan–Sukkur, completed and operating

Havelian–Thakot (a part of Karakoram Highway)

NH 50 D. I. Khan–Zhob (Fort Sandeman) (235 km) and Zhob–Quetta (331 km).

This 566 km long route, from Dera Ismail Khan to Quetta, passes through difficult terrain. The route (NH 50) does exist, but in a state of disrepair. Its improvement is undertaken under CPEC. Much of the terrain, through which the route passes, is over 2,000 m ASL, and the highway has to negotiate the Sulaiman Range, first traversing through southern Khyber Pakhtunkhwa and then the mountainous northern Balochistan, passing through Qila Saifullah before terminating at Quetta.

The Gwadar–Ratodero Road, as mentioned earlier, is a major initiative under CPEC co-ordinated programme. This Highway, likely to be transformed into M-8, is a major highway that will link Gwadar with the main highway at Ratodora station, located in northern Sindh, between Larkana and Shikarpur. The highway passes from Quetta through Awaran in dry south-central Balochistan, and further on to Khuzdar, and then via Shahdatkot to Ratodero. The route will further link the two principal routes No. 55 and 5 leading further north to join the Karakoram Highway at Hasan Abdul on GT road.

During the last two decades, the road network is considerably improved. Existing roads have been streamlined and even transformed into motorways, and new routes are opened. Most significant is the improvement in the road network of Khyber Pakhtunkhwa and Balochistan. The CPEC initiative has greatly contributed to the infrastructure development in the country including the road network, besides developing other infrastructures.

14.4 Air Transport

The development of airlines in Pakistan coincided with the emergence of the country in 1947, when Orient Airlines, the predecessor of the Pakistan International Airlines Corporation, shifted its headquarters from Calcutta to Karachi. The Airlines was merged into Pakistan International Airlines in 1955 and became an aviation company fully controlled and managed by the Government of Pakistan.

The Pakistan International Airlines (PIA), with its hub at Jinnah International Airport, Karachi, is the flag bearer of Pakistan. Besides PIA, the country has a few more airlines, principally the Air Blue, Serene Air and Air Sial. All the airlines are based at Jinnah International Airport, Karachi. There are a few others, which don't run scheduled flights but charter planes to individuals and companies. These include Akshai Air, ASSL, Princely Jets, Vision International and Air Falcon. Undoubtedly Pakistan International Airlines is the leading airlines. It has a fleet of over 32 aircrafts and operates over 100 flights daily connecting 70 domestic and 34 international terminals in over two dozen countries (Table 14.5).

TABLE 14.5 The Number of Passengers Handled at Civil Airports in Pakistan in 2017-2018

	Embarked	*Disembarked*	*Transit*	*Total*
Domestic	3,646,805	3,845,802	23,965	7,317,569
International	7,164,715	7,46,532	67,174	14,708,421
Domestic and International	10,811,517	11,123,334	91,139	22,025,990

Source: Pakistan Statistical Year Book 2018, table 20–21, p. 481.

TABLE 14.6 Domestic and International Cargo Handled at the Airports in 2017–2018 (Cargo Handled in Tonnes)

Cargo	*Loaded*	*Unloaded*	*Transit*	*Total*
Domestic	38,119	38,119		76,238
International	166,845	71,233	300.881	538,959
Total	204,964	109,352	300.881	615,197

All the airlines, together, carry 6 to 8 million domestic passengers annually, increasing from about 1.5 million in the 1970s to over seven million in 2017–2018. The total number of passengers handled by all the airlines at civil airports in Pakistan is 22 million, of which over 7 million are domestic passengers and the number of international passengers is over 14 million. The role of aviation in the transport system of Pakistan is evident from the number of international passengers, which is twice the number of domestic passengers.

Obviously, the role of air transport in promoting tourism, especially in attracting foreign tourists is important. Over 90 percent of the tourists arrive by air. The last peak year for tourism in Pakistan was the year 2009, when of the total number of 485,558 tourists (about 93 percent) arrived in Pakistan by air. It may be emphasised that a well-organised air service attracts foreign tourists and is a valuable source of earning foreign exchange. This doubly emphasises the importance of well-oiled civil aviation, with domestic and international airlines operating smoothly in the country.

14.4.1 Cargo Handled by Airlines and Civil Airports

Besides the passengers, the airlines handle a large volume of high value cargo. The total cargo handled at the airports is over 300 million tonnes annually (Table 14.6).

The airlines, thus, handled over 300,000 tonnes of cargo annually, which consists predominantly of international cargo that amounts to 75 percent of the total cargo handled. It may be noted that over the last ten years, there has been no significant increase in the volume of cargo handled. In 2009, the tonnage of air cargo was 306,322 tonnes with a marginal increase leading to 315,132 tonnes in 2017–2018.

14.4.2 Airports and Their Distribution

Pakistan has around 150 airports, all of which are not functional. Some do not have paved runways, and some others don't have a proper landing system. There are, however, more than 80 airports, which have proper runways, and some even have 5 km long runways. There are around 50 airports that are operational. These include international as well as domestic airports. Many of these civil airports are also used as air bases, while some others are labelled as military airports. A typical example would be Lahore, which is home to "Allama Iqbal International Airport" and is also privileged to have an air base.

International Airports – There are three principal international airports viz. Karachi, Lahore and Islamabad. Besides, there are second-order international airports.

Principal International Airports: (1) Islamabad, (2) Karachi and (3) Lahore

Second-Order International Airports: (1) Faisalabad, (2) Multan, (3) Rahim Yar Khan, (4) Dera Ghazi Khan and (5) Sialkot, all in Punjab, (6) Quetta, (7) Gwadar and Turbat in Balochistan, and (8) Peshawar in Khyber Pakhtunkhwa.

Domestic Airports – Besides the above airports, which also function as domestic airports, there are others, which function exclusively as domestic airports. These include Bahawalpur, Gujrat, Chashma, and Mangala in Punjab; Dalbandin, Jiwani, Khuzdar, Ormara, Panjgur, Pasni City, Zhob, Sibi and Sui in Balochistan; Sukkur, Sehwan Sharif, Sawan Gas field, Nawabshah, Larkana, Jacobabad, and Hyderabad in Sindh; and Chitral, Dera Ismail Khan and Parachinar in Khyber Pakhtunkhwa.

It has to be emphasised that as a matter of policy, Pakistan appears to have linked far-off places, having minimal surface transport, by air. For example, places like Parachinar or Chitral, both in Khyber Pakhtunkhwa are highly inaccessible, especially in winter, and to establish air links to these centres is very apt, strategically and from the socio-economic point of view. Similarly, many parts of Balochistan have poor accessibility involving long road transport. Linking by air places like Zhob or Dalbandin, Khuzdar, Jiwani or Turbat may not be economically viable; yet, it is socially and even strategically important to have an airport in the area. Many of these cities have a small population; yet, they have the privilege of an airport.

Airports as International Landmarks in the Socio-economic Landscape of a Country – Often the importance of a region is judged by the presence of an airport. Airports are, unlike railway stations, a costly proposition, requiring adequate space clearance for landing and take-off for modern planes. Besides, the establishment of an airport requires a great deal of well-planned supporting structure, like the air terminal with all its constituent functional units, warehouses, security establishments and connectivity of the airport with the hinterland by other means of transport.

14.4.3 Some Important Airports of Pakistan

The three most important international airports of Pakistan:

1. *Jinnah International Airport, Karachi* – Named after Quaid-e-Azam Muhammad Ali Jinnah, the founder of Pakistan, Karachi Airport is by far the oldest and the busiest airport of Pakistan. Managed by PCAA (Pakistan Civil Aviation Authority), the airport is the hub of almost all airlines of Pakistan, including the national carrier, Pakistan International Airways (PIA). With a runway measuring over 3 km (3,360), the airport handles 7 million passengers (7,267,582) and 140,000 metric tonnes of cargo, annually. It has to be noted that the number of international flights coming to Karachi has slightly declined in recent years, as a result of the coming up of Islamabad International Airport, the latter being the federal capital of Pakistan. However, being a sea port, and the foremost industrial, commercial and financial centre of Pakistan, the Jinnah International Airport at Karachi still retains its importance.
2. *Allama Iqbal International Airport, Lahore* – This is the third largest airport of Pakistan. Named after the poet laureate of Pakistan, Muhammad Iqbal, the airport is located east of the city of Lahore, a few kilometre further east of Lahore Cantonment. It handles over 5 million passengers and over a hundred thousand metric tonnes of cargo annually. With a runway of 3,360 km, the airport was built in 1962, replacing the old Walton port of Lahore. The old airport is used as a terminal for Hajj passengers. The airport handles quite a few international airlines especially from the Middle East countries. Though the location of Lahore is highly skewed, the airport has developed because of the importance of Lahore as the capital of Punjab, the most prosperous province of Pakistan. To cope up with the burgeoning air traffic, a new international airport is planned for Lahore, with far greater capacity to handle the increasing number of flights and the passenger load. Reportedly, the construction work is being carried out by a Chinese engineering firm.
3. *Islamabad International Airport* – The airport was built to replace the old Benazir Bhutto Airport and became operational in May 2018. The airport has acquired a pre-eminent position as an international airport,

handling the largest number of international flights and passengers in the country. Operated by the Pakistan Civil Aviation Authority (PCAA), the airport is spacious and handles over 5 million passengers annually, though this number is fast increasing, and the PCAA has plans to expand the capacity of the airport to handle 25 million passengers annually. Presently, Islamabad is the only airport in Pakistan with a capacity to handle the airbus 380, and has attracted most of the international airlines. With the operationalisation of Islamabad Airport, the old Benazir Bhutto International Airport at Rawalpindi is closed and is currently used by Pakistan Air Force.

14.4.4 Economies of Air Transport in Pakistan

Though there are a few domestic airlines and several international airlines operating in Pakistan, the focus here is on the Pakistan International Airlines Corporation, owned and managed by the Ministry of Civil Aviation, Government of Pakistan. Employing roughly 14,500 persons, PIA is perpetually a loss-making enterprise. This is not surprising and is typical of public enterprises in many countries. The airlines began making losses around 2010–2011, and at the end of 2018, the Airlines had a debt of US $3.3 billion. Though the performance of the Airlines has lately improved, it is far from adequate to pull it out of the decade-old indebtedness. It is beyond the scope of this text to look at the financial audit of PIA. A number of factors have combined to bring these reputed airlines to a sorry financial state.

At the same time, one has to observe that in a country like Pakistan where many areas are not accessible, and an air service to areas like the interior Balochistan though not always economic is indispensable to ensure accessibility to all areas and all sections of the society.

14.5 International Trade of Pakistan

The regional or local trade, in countries of South Asia, has been conducted traditionally through weekly markets and strategically located *mandis* (collection and warehousing centres). This kind of regional trade is conducted largely in agricultural commodities. Urban centres of different orders also carry out trade in a variety of goods, raw materials, manufactured commodities or even imported goods, involving multistage trading. Such trade has only local or regional significance and follows the laws of equilibrium between supply and demand. A large number of private operators, with moderate capital investment, participate in this kind of regional trade that forms a link in the flow of commodities at the national level.

Far more important than the local or regional is the international trade that determines the economic health of a country. Commonly known as

TABLE 14.7 Value of Imports and Exports for the Decade (2006–2017), Value in Billion PKR (Pakistan Rupees)

Year	*Imports in Billion PKR*	*Exports in Billion PKR*	*Excess of Imports over Exports*
2008–2009	2,724	1,384	1,340
2009–2010	2,911	1,617	1,294
2010–2011	3,455	2,121	1,334
2011–2012	4,009	2,111	1,898
2012–2013	4,350	2,368	1,982
3013–2014	4,631	2,583	2,048
2014–2015	4,644	2,398	2,246
2015–2016	4,659	2,167	2,492
2016–2017	5,540	2,138	3,402
2017–2018	6,694	2,555	4,139

Source: Pakistan Statistical Year Book-2018, table IX.1 p. 187.

foreign trade, international trade is composed of export and import of commodities from or in a country. It is the trade transaction with foreign countries to which the Pakistani goods are exported, or from where Pakistan imports certain commodities. The latest available data for 2017–2018 show that there is an imbalance between the value of goods imported in and exported from Pakistan (Table 14.7).

The foreign trade data for the last decade, as is clear from the above table, show that there has been a constant increase in the value of imports, from 2,724 billion PKR in 2008–2009 to 6,694 billion PKR in the year 2017–2018, a two and half times increase, as normally expected in a growing economy. The increase in the value of exports, however, is very moderate. This has risen from 1,384 billion PKR in 2008–2009 to 2,555 billion PKR in 2017–2018, a moderate increase of 85 percent. A much higher increase in imports, as compared to a moderate increase in exports, has affected Pakistan's external finances, creating an adverse balance of payment situation.

14.5.1 Export and Import by Commodities

Pakistan exports a variety of goods, under different major heads, and each having a number of sub-heads.

This categorisation conceals the importance of individual commodities, which are important in the export trade of Pakistan. For example, in the category of food material, rice is an important commodity with an export value of 225 billion PKR, accounting for 9 percent of the total export of Pakistan. Meat and fish products together, with an export value

TABLE 14.8 Monetary Value of Export and Import by Commodities (in Billion PKR)

	Major Heads of Exported and Imported Commodities	*Monetary Value in Billion PKR of Exported Commodities*	*Monetary Value in Billion PKR of Imported Commodities*
1.	Food and live animals	523	282
2.	Beverages and tobacco	3	2
3.	Crude materials inedible except fuels	78	644
4.	Mineral fuels, lubricants and related materials	43	1,755
5.	Animal and vegetable oils and fat	4	249
6.	Chemicals	115	983
7.	Manufactured goods classified chiefly by materials	1012	806
8.	Machinery and transport equipment	23	1,743
9.	Miscellaneous manufactured articles	755	224
10.	Commodities and transactions – negligible not classified according to kind		7
	Total	2,556	6,695

Source: Pakistan Statistical Year Book 2018, pp.210–238.

of 70 billion PKR, account for roughly 1.5 percent of the total export. Sugar, honey and associated products and fruits and nuts are important items of export. The most important commodity that earns for Pakistan a significant amount of foreign exchange is textiles. Textiles are exported in different forms, unfinished, semi-finished and finished products. These include raw cotton, cotton waste, cotton yarn, cotton thread, cotton cloth, readymade garments and hosiery and synthetic textiles. Together, they add export valued at roughly 1,000 billion PKR that is roughly 40 percent of the total export earning of Pakistan. Food material like fish and meat, milk products, fruit and dry fruits, sugar and honey are other items of export (Table 14.8).

14.5.2 Commodities of Import

As can be seen from the table above, the principal imported commodities are largely mineral fuel, machinery and transport equipment, chemicals and raw materials for basic industries like iron and steel scrap and ores and concentrate of ferrous metal and non-ferrous metal scrap. This is indispensable, as in the absence of substantial iron ore deposits in Pakistan, import is necessary to feed its steel mills. Mineral fuels, that is, coal and petroleum account for 1,755 billion PKR, which is 27 percent of the total import of

Pakistan. Equally important as the mineral fuels which account for a quarter of imports, are the goods included in the "Machinery and transport equipment" category. These, together, claim over 26 percent of the total value of imports. These include power-generating and transport equipment and machines of all descriptions including those required for railways, textile industry, ships, boats and aircrafts. The import of iron ore or other metals as raw material for basic industries, mineral oil to run the economy and all kinds of machinery are the major import goods. These can't be dispensed with. Dependence on the import of these items may be minimised or altogether done away with, once the country develops its own capacity to substitute these items.

14.5.3 Overseas Countries Trading with Pakistan: Source of Imports and Destinations for Exports

There is hardly a country in the world with which Pakistan does not have trade relations. With some countries, it has well established trade partnership exporting and importing large volume of goods, while with some others, it may be just nominal. The foreign trade value (imports and exports) of Pakistan with other countries, grouped continent-wise, is reproduced here (Table 14.9) (Figure 14.2).

Pakistan's foreign trade relations have a wide network. The important aspect is a wide divergence in the selection and preference for countries for export from those chosen for import. While the import is largely from

TABLE 14.9 Continent-wise Import and Export Value of Pakistan (2017–2018)

	Continent/Areas	*Value of Export from Pakistan (in billion PKR)*	*Value of Import in Pakistan (in billion PKR)*
1.	North America	429	394
2.	Latin and Central America	19	17
3.	South America	31	99
4.	Western Europe (EEC + EFTA + others)	847	735
5.	Other Europe (excl. Turkey)	22	85
6.	Eastern Europe	57	92
7.	Middle East	258	938
8.	African countries	25	21
9.	Other Africa	148	283
10.	Other Asia	735	4083
11.	Oceania	32	58
	Total	2,603	6,713

Source: Pakistan Statistical Year Book, 2018:188-208.

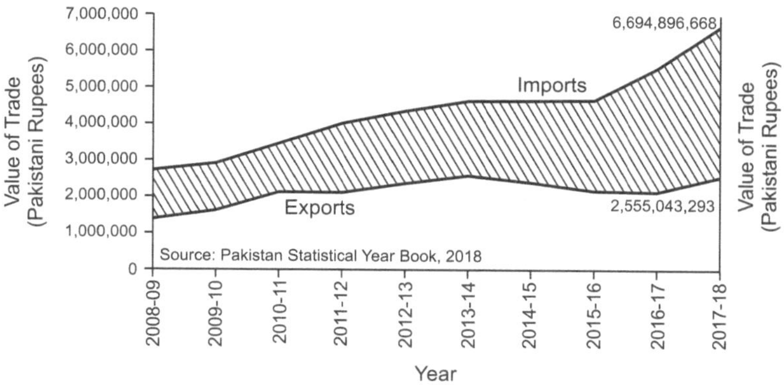

FIGURE 14.2 Value of Pakistan's Foreign Trade (2007–2008 to 2017–2018)

Asian countries, export is more oriented to North America, Europe and the Middle East.

Western Europe and North America together account for more than 50 percent of the total export of Pakistan. The United States, alone, accounts for over 15 percent of the total export of Pakistan, while Western Europe buys one-third (33 percent) of the total goods exported from Pakistan. In Europe, the United Kingdom is the largest importer of goods from Pakistan followed by the Federal Republic of Germany. While the total export of goods to the United Kingdom in 2018 amounted to 187 billion PKR, for Germany, this value was 146 PKR. The Middle East countries account for 16 percent of the total export of Pakistan. Thus, over 60 percent of Pakistan's export is absorbed by the United States, Western Europe and the Middle East. The only country in Asia, that imports substantial volume of goods from Pakistan, is the People's Republic of China. The latter country imported goods worth more than 250 billion PKR, from Pakistan, in 2018.

Import Trade: The pattern of import is quite different. The United States, which received 16 percent of the total exported goods from Pakistan in value terms, figures marginally in the countries preferred for import. Europe does a little better from where Pakistan receives just about 10 percent of the total imports. But in terms of absolute value of trade, both the United States and Western Europe, import more from Pakistan than export to it. Thus, Pakistan has a trade surplus with these countries and a sound current account balance. Leaving aside the United States, the United Kingdom and Western Europe, Pakistan has a trade deficit with many other countries of the Middle East and several Asian countries. The People's Republic of China, Indonesia, Japan, Thailand, Malaysia and Singapore are the principal countries from where Pakistan imports substantial volume of goods.

TABLE 14.10 Trade Surplus and Deficit of Pakistan (2017–2018) with Some Major Trade Partners (Value in Billion PKR)

Destinations to/from	*Values of Pakistan's Export*	*Values of Pakistan's Import*	*Trade Surplus or Deficit +/– of Pakistan*
USA	400	316	+84
Western Europe	847	735	+12
UK	188	86	+102
Germany	147	146	+001
Middle East	259	939	–680
China	186	1731	–1545
Indonesia	35	278	–243
Singapore	10	114	–104
Malaysia	16	132	–116

Source: Pakistan Statistical Year Book 2018, Govt. of Pakistan, Ministry of Planning and Development and Special Initiatives, Pakistan Bureau of Statistics, Islamabad, pp. 187-238.

China, in fact, is the major supplier of goods to Pakistan and accounts for more than a quarter of the total import of Pakistan. Against an import of goods worth 1,732 billion PKR, the value of exports to China is just around 185 billion, showing a huge trade deficit (Table 14.10).

Foreign Trade: Trends and Balance of Trade – During the last decade (2008–2018), there has been a substantial increase in the total volume of trade in Pakistan. In monetary terms, the total value of foreign trade (export and import) has jumped from 4,107 billion PKR to 9,250 billion PKR, an increase of 2.25 times. During the same period, while the exports have grown only 1.84 times, from 1,983 billion PKR to 2,655 billion PKR, the imports have grown much more, jumping from 2,724 billion PKR in 2008 to 6,695 billion PKR in 2018, an increase of 2.5 times. This imbalance in trade, imports exceeding the exports in monetary terms, persists even today.

Shifts in the Source and Destination of Imports and Exports – There has been, of late, a radical shift in the source of imports. While the Western countries like the United States and Western Europe have doubled their trade with Pakistan; and both, export and import trade with Pakistan, have more than doubled, there is a phenomenal increase in imports from China. While the exports to China have increased more than three times, the imports from China have increased almost six times.

The Reality of the Geo-spatial Situation and the Impact of China – Pakistan Economic Corridor are clearly visible in an unprecedented boost in the volume of trade between the two countries. In the beginning of the decade, Pakistan's import from the Middle East was almost three times that from China. Similarly, imports from Western Europe were higher than the imports from China. The situation is completely reversed today, and China has become a major trade partner of Pakistan.

The story with exports is different. To each of the countries mentioned above – the United States, Western Europe and Middle East – the exports from Pakistan are several times higher than the export to China. It is evident that while the exports to Western countries earn valuable foreign exchange for Pakistan, exports to China suffice only partially to compensate for the imports from that country. There is, no doubt, a radical shift in Pakistan's preference of countries for imports. Friendly countries, like China, offer long-term credit and even accept payment in kind.

The Impact of China–Pakistan Economic Corridor – The most easily accessible source of machinery and other goods for Pakistan is the People's Republic of China. Many of the development projects in Pakistan fall under the purview of the CPEC; and hence, a large volume of imports from China does not entail immediate payment in dollar terms, as it comes under the umbrella of the CPEC (China–Pakistan Economic Corridor) that stipulates an investment of the around US $ 50 billion in Pakistan. The terms of the agreement may be partially opaque, but CPEC is decidedly a financial boost to the country in its development.

The Foreign Trade with Neighbouring India – It is ironical that trade between Pakistan and India is riddled with a lot of problems, not the least of which is the non-too friendly relation between these two neighbours. A saner approach would decidedly be to the benefit of both the countries.

Appendix: Some Details about the Operation of Pakistan Railways

1. Route length	7,791 km (2017–2018)
2. Track length	11,881 km (2017–2018)
3. Locomotives owned by Pakistan Railways	478 (2017–2018)
4. Number of coaches owned by Pakistan Railways	1,743 (2017–18)
Passenger coaches	1,460
Other coaches	*283*
Total	1,743
5. Number of trains	228
6. Number of Passengers carried	70 million (2018–2019)
7. No. of Employees	72,078 (2017–2018)
8. No. of freight wagons	16,159
9. Tonnage of freight carried	8.355 million tonnes (2017–2018)
Revenue	
10. Total operating revenue	49.5 billion PKR
11. Total operating expenses	52.07 billion PKR

Source: Pakistan Statistical Year Book, 2018, Government of Pakistan, Ministry of Planning Development and Initiatives, Pakistan Bureau of Statistics, Islamabad, pp. 470–478.

Notes

1 The companies that merged together were
   1. Scinde, Punjab and Delhi Railway
   2. Indus valley State Railway
   3. Punjab Northern State Railway
   4. Sind–Pishin State Railway
   5. Trans–Baluchistan Railways
   6. Kandahar State Railways

   Between 1885 and 1947 these were merged to North-Western State Railways, and after 1947, the entire network of railways came to be known as Pakistan Railways.

2 Simla Agreement was executed between India and Pakistan, following the 1971 war, which led to the separation of Bangladesh.

3 Government of Pakistan, Finance Survey (2020) *Economic Survey 2019–20*, June, Islamabad, p. 257.

4 Government of Pakistan, Finance Survey (2020) *Economic Survey 2019–20*, June, Islamabad, p. 257.

5 This classification appears in Pakistan Year Book 2018–2019, p. 479.

6 National Transport Research Centre – 2006. The figures in the table are of the year 2006 and are only indicative and don't represent the actual status.

15

ECONOMY OF PAKISTAN

Changing Public Policy and Its Impact on the Economy – An Overview

Like some of its South Asian neighbours, Pakistan is one of the developing economies of the world. With an annual Gross Domestic Product (GDP) of around US $ 383 billion (nominal–2023 estimate) and a per capita income of US $1,798, it ranks 44th in the world, but if evaluated on a Purchasing Power Parity (PPP) scale, the country's GDP rises to over US $1,512 billion (2023), and it occupies a respectable 24th rank, leaving 170 countries behind. In nominal terms, the GDP of Pakistan has grown from a mere US $3 billion in 1950 to US $383 billion in 2022.

During the 70-year period of 1952 to 2022, the country's economy has grown at an average annual rate of 4.84 percent, with occasional negative as well as positive fluctuations. The total GDP of a country, however, reflects only the size of its economy, and a more appropriate picture of a country's prosperity emerges only when the GDP is weighted against its population, usually expressed as per capita GDP. The per capita income for Pakistan has risen from US $88 in 1950 to US $1,788 in 2022, a 20-fold increase that is not a mean achievement.

There has been, no doubt, an overall growth in all sectors of the economy. Agriculture, the main contributor to the primary sector showed remarkable growth, raising the production of wheat from 3.4 million tonnes in 1948 to 26 million tonnes in 2022 and that of rice from 0.7 million tonnes in 1948 to 9.4 million tonnes in 2022. Similar growth is observed in the production of all other agricultural commodities. A major change has occurred in the manufacturing sector that claims 65 percent of the value of the total economy rising from 15 percent in 1950. In fact, in the 1950s, there was not much to narrate as the industrial base of Pakistan,

DOI: 10.4324/9781003510147-18

but during the last 70 years, the country has been able to develop multiple industries like textiles, occupying the apex of Pakistan's industrial landscape, and many other industries like cement, sugar, automobiles, chemicals and fertilisers.

15.1 Pakistan Economy on a Comparative Scale

With a population reaching 220 million – the official estimate is 207 million for 2017 – the per capita GDP of Pakistan, as estimated by the IMF, is US $1543 (2023) However, seen on a PPP scale, the per capita GDP of Pakistan rises to International $6,836, claiming a much higher rank among the countries of the world. In this respect, the economic situation of Pakistan is not very dissimilar from several of its South Asian neighbours. A comparison with some of the economically advanced, some not so very advanced and some struggling economies, especially of South Asia, as quoted below, will prove more enlightening (Table 15.1).

TABLE 15.1 Per Capita GNP on a Comparative Scale of Different Countries

Countries	*GDP in US $ (billion) Nominal (2023)* 1	*Per Capita GDP in US $ (2023 estimate)* 2	*Per Capita GDP (PPP) in International $ (2023 estimate)* 3
A. South Asian Countries			
1. Pakistan	377	1,543	6,836
2. Bangladesh	421	2,450	8,663
4. Bhutan	2.7	3,063	14,170
5. India	3,737	2,274	9,073
6. Maldives	7	10,366	36,358
7. Nepal	41	1,159	5,101
8. Sri Lanka	74	3,362	14,223
B. Developed Countries			
9. Germany	4,309	51,073	66,132
10. Japan	4,410	39,650	51,809
11. UK	3.159	45,603.3	56,471
12. USA	26,856	69,542	80,035
C. Better Developed Asian Countries			
13. China	19,374	12,472	23,382
14. Indonesia	1,392	3,514	15,855
15. Malaysia	447	11,101	36,847

Source: Col 1, C Col.2, IMF estimate of per capita GDP (nominal) in US$, 2023: Col.3, IMF estimate of Per capita GDP (PPP) in International $ from Wikipedia, accessed on June 17, 2023.

In terms of the sheer size of the economy, as expressed as total GDP, Pakistan is the third largest economy in South India, next to India and Bangladesh. But when compared with the developed countries of the world, its total GDP is only 1.4 percent that of the United States and 1.95 percent that of China, 5 times larger than Sri Lanka and 12 times greater than its next-door neighbour, Afghanistan. The relative position of a country, in terms of the per capita GDP vis-à-vis other countries of the world, calculated either nominally or on a PPP scale, doesn't change substantially but only appears to ameliorate the relative position of less developed or undeveloped countries because of their wage and price structure. To take an example, the per capita GDP of Pakistan is only 12 percent that of China, but when seen on a PPP scale, it rises to 29 percent of the Chinese per capita GDP.

15.2 Growth Trends of Pakistan's Economy During the Last Seven Decades (1950–2022)

Pakistan's economy has grown at an estimated average growth rate of 4.92 percent since its inception in 1947. During the period of over six decades, from 1947 to 2014, the economic growth was characterised by phases of "boom and bust" and virtual stagnation interspersed with violent fluctuations The teething troubles of early years, marked by domestic disturbances, episodes of wars, natural catastrophes, secession of East Pakistan leading to the emergence of an independent state of Bangladesh and several coups d'état, have contributed to the unsteady economic progress of the country. The decadal variation in growth, often quoted, is as follows (Table 15.2).

As can be seen, there has never been, in the history of Pakistan, a sustained growth of 6 percent or above. Usually, a bad decade or a bad patch in economic development is followed by a higher growth rate, reflecting the

TABLE 15.2 Average Annual Growth of Pakistan During 1950–2023

Decades	*Average Annual Growth (in percentage)*	*Decades*	*Average Annual Growth (in percentage)*
1950s	3.1	2015	4.73
1960s	6.8	2018	6.15
1970s	4.8	2019	2.5
1980s	6.5	2020	–1.7
1990s	4.6	2021	6.49
2000	4.26	2022	5.97
2005	6.52	2023	1.9(estimated)
2010	1.61		

Source: Economic History of Pakistan 1. (https.//en.wikipedia.org/wiki/ accessed on 21/6/2015; 2. Pakistan Economic Survey 2014–15, p. 2, Ministry of Finance, Government of Pakistan; 3. Data for 2005 to 2022 is from World Bank open data http://data World Bank.o rg accessed on July 19, 2023.

effect of a lower base in the previous period. The sixties and the eighties of the last century were the periods of a maximum growth of over 6 percent followed by a yearly growth of around 6 percent in 2005. Ironical though it may appear, these were the years of military rule in the country. It may be just a coincidence or the result of a more disciplined approach of the much-maligned rule of the army. More probably the higher growth rate during army rule resulted from financial support from external sources, especially the United States, in return for the support extended by Pakistan in strategic matters.

The fluctuating figures of economic growth give a confusing picture, though an overall growth of 4.7 percent growth is not good enough for Pakistan or for that matter other poor and developing countries of South Asia. Seen in a larger context of the welfare of the society, even this moderate growth of the economy gets neutralised by a concurrent high growth rate of population, especially of Pakistan where the average annual growth rate of population is as high as 2.5 percent (2023 Census).

15.3 Pakistan's Economy through the Decades – A historical Perspective

15.3.1 The Early Years

Emerging as an independent country, carved out of British India, Pakistan had to contend with the difficulties and the disadvantage of having to negotiate a transition, from a colonial administration to one of self-rule. It had to reorganise an administrative structure and streamline the functioning of the government, while undertaking, simultaneously, the enormous task of rehabilitating seven million refugees who had migrated from India in the wake of partition. The immigration stream continued for several years, adding to the ever-increasing burden on the administration and the economy of Pakistan.

Development of Governance Structure – Unlike India, which inherited the administrative infrastructure including the parliament building, the central secretariat, the Vice-regal palace and, in fact, the entire New Delhi, designed and built just around four decades ago. Pakistan had to start from ground zero. It had to install its own infrastructure, streamline governance, commission its own judiciary and defence services, and a whole lot of organs of governance. A silver lining in the functioning of the tangled administrative machinery was the presence of that colossus of a leader, Quaid-i-Azam Muhammad Ali Jinnah, the first Governor General of Pakistan, at the helm of Pakistan's affairs. He commanded the trust and confidence of every Pakistani; his words were law and guided the destiny of a newborn nation for the remaining 13 months of his life. His address to the Constituent Assembly of Pakistan on August 11, 1947, where he laid out the spirit of

Pakistan's political philosophy, will always remain a *Magna Charta,* and a beacon of hope and guidance for the people of Pakistan.

Thus, the first decade of Pakistan's existence was completely utilised in providing the country with a streamlined administration, promoting the idea of a unified Pakistan and trying to reconcile the internal socio-cultural and religious differences in the newly formed state.

The untimely demise of Jinnah on September 11, 1948, much before the Constitution of the country was framed under his guidance and in consonance with his ideas, was a death blow to the future of Pakistan. The political fallout of Jinnah's departure from the scene was mutual distrust and one-upmanship among the top leaders and bureaucrats. A struggle for power ensued and Pakistan faced a decade of instability, which ended with a coup d'ètat and the establishment of a military rule, barely a decade after its birth.

The political instability of the 1950s is clearly reflected in the dismal economic growth of the country, limited to an annual increase of 3.5 percent in its GDP. During the 1950–1960 decade, Pakistan experienced very modest economic growth and, occasionally, even a decline, as in 1952, when the country recorded a negative growth of –0.18 percent. The decline, obviously, was the result of socio-political disturbances that rocked Pakistan. The Bengali language movement of 1952–1953 in East Pakistan, as much as the Anti-Ahmadiya movement in West Pakistan, not only muddled the political waters but even stymied the economic growth of the country.

15.3.2 The 1960s of the Last Century

The agenda for economic growth received focused attention for the first time in the 1960s. Ironical though it might appear, this was a period of a military regime headed by General Ayub Khan. Added to this was the 1965 Pakistan–India war that should have adversely affected the economic growth of the country. The most significant political and economic event of the 1960s, however, was the Indus Water Treaty, signed between India and Pakistan in 1960. Though much maligned in certain quarters, it has stood the test of times and still holds good ensuring an adequate supply of water to Pakistan from the Indus and its three tributaries Ravi, Chenab and Jhelum. It ensured an uninterrupted agricultural development of Pakistan and supply of water to its growing industries and urban centres. In the absence of this treaty, the none-too-cordial relations between India and Pakistan would have one more reason to be perpetually at war of words, between the two countries, over the distribution of water. This was the period that witnessed the construction of the Mangla Dam on Jhelum that added to the overall supply of energy and water to the plains of Punjab. The most remarkable achievement of the 1960s was the establishment of Islamabad, the planned

Federal capital of Pakistan, shifting it from Karachi. This was not an economic event, yet an important event in the life of Pakistan. It provided Pakistan not only the necessary infrastructure but also a grand capital with all its avenues and architectural marvels that provided the Government of Pakistan and its administration a glorious new phase in their life, signifying order and dignity (Figure 15.1).

15.3.3 Economic Achievements of 1960s

Development of Agriculture – Agriculture, contributing around 50 percent of the national GDP in the 1960s, was an important segment of the economy that needed reforms. A limit was imposed on the holdings of big landlords, limiting these to 200 acres of irrigated and 400 acres of un-irrigated land. Besides expediting irrigation projects and the introduction of modern technology, the procurement prices of agricultural products, especially cereals, were raised as an incentive to farmers to increase agricultural productivity. All these measures ushered in a phase of Green Revolution and food self-sufficiency.

Economic Reforms – Economic reforms in the non-agricultural sector were not very spectacular. Growth of the economy, creation of wealth and enlargement of the economy were the prime objectives. As emphasised by a Harvard Advisory Group, headed by Gustav Fritz Papanek that the "creation of wealth was of prime importance ignoring a more equitable distribution of resources".[1] In fact, the advisory group of General Ayub Khan labelled the economic disparity as "functional inequality" and considered

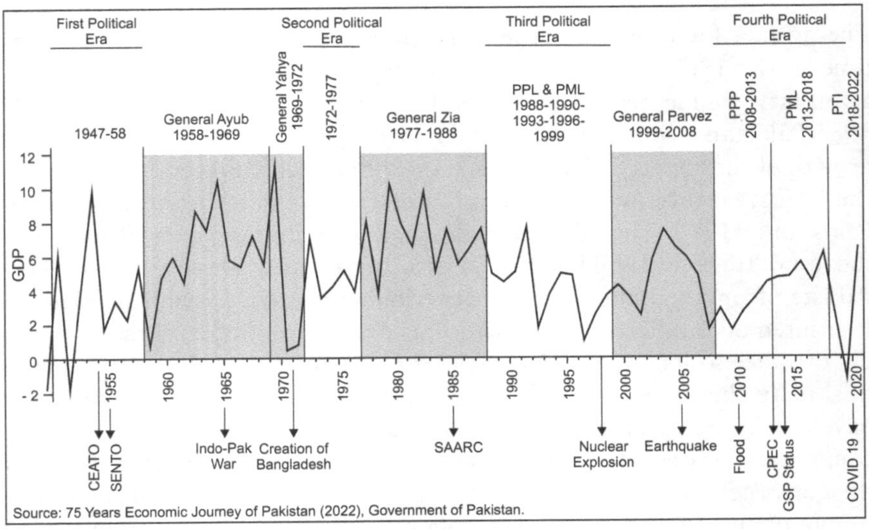

FIGURE 15.1 Annual Growth of Pakistan's GDP (1951 – 2021)

it as a pre-condition for rapid economic growth. It was theorised that the increasing income of the wealthy would lead to greater saving and investment in the economy, which would, in turn, produce more wealth that would be transmitted, down the line, to other sections of the society by Myrdal's trickle-down effect. The economy grew at a reasonable pace, and there was visible all-round progress, though without a concurrent diffusion of economic gains to all sections of the society.

Adverse Features of Economic Growth during the Ayub Era – While every issue related to the economy was going *hunky dory*, a single statement from the Chief Economic Advisor seems to have virtually trivialised all the gains of the Ayub era. The statement of the Chief Economist of the Planning Commission that "twenty-two families controlled 70 per cent of manufacturing and 90 per cent of banking and insurance business in the country" jolted his regime to its foundations. It was a supreme irony as Ayub Khan was following the growth philosophy as espoused by the Chief Economist of the Planning Commission who in his Cambridge Ph.D. thesis turned book *The Strategy of Economic Planning* argued that

> *in the initial stages of economic development some concentration of income and wealth in a few hands was necessary and appropriate to stimulate animal spirits and foster the conditions for rapid growth, Considerations of equity and income distribution could be tackled at a later stage of development.*[2]

Reaction and Aftermath of Mahbub-ul-Haq's Statement – The reaction to the statement of Mahbub-ul-Haq, lamenting the concentration of wealth in a few hands, was hostile and swift. Pakistan's industrialists came to be called "Robber Barons" who had earned monopolistic profit behind high walls of protection, subsidy and Government patronage. An empirical study by Professor Lawrence White of New York University lent credence to this proposition.

In hindsight, one finds that Ayub Khan was the victim of his own success. The Government of President Ayub Khan believed in the effectiveness of this process as did most of the economists, and economic progress did take place, but it also led to the concentration of wealth in a few hands. The problem of a very skewed development, leading to the concentration of wealth in a small section of the society, though considered indispensable by many economists, in the early stages of economic growth, remained unaddressed. This caused a deep resentment in the Pakistani middle class and they agitated for better economic reform.

Despite the flaws in the policy and its execution during General Ayub Khan's regime, some visible progress was noticed. The discovery of the Sui gas field in Balochistan, made in the 1950s, close to the tri-junction of the three provinces of Punjab, Sindh and Balochistan, was a major boost to the country's fuel economy. Its full potential was realised only in the 1960s, after the necessary infrastructure for transport was developed, and natural gas began to be utilised as a fuel for transport and other industrial purposes. In fact, Sui gas proved a boon to take care of the energy crisis, the country faced then. Excessive use of the newly discovered gas, as if it is inexhaustible, has led to a slow depletion of this energy resource. The energy crisis of Pakistan still remains one of the major the constraint in Pakistan's economic growth.

Large deposits of limestone, the principal raw material for cement, were exploited, a number of cement factories were established, and the domestic production of cement sets the trend for the multiplication of a large number of industrial units. The decade was particularly significant for the beginning of a renewed exploration of petroleum in Pakistan. The development of the Toot Oil field in the Potwar Basin was the direct result of the mineral development initiative of the government. Toot oil field produced 2,400 barrels of oil per day, at its peak in 1966. This, together with the supply of gas from the Sui gas field kept the economic wheels of Pakistan running. Textiles, the principal industry of Pakistan today, was invisible in the 1950s, but made a spectacular rise in the 1960s. In the pre-1947 era, Punjab and Sindh were integral to the cotton textile industry of colonial India, but chiefly as the supplier of raw cotton for Ahmedabad and Mumbai mills. At partition, the bulk of cotton textiles needs were met by imports chiefly from India. The trade war with India in 1949, however, required Pakistan to develop its own production capabilities. Thus, the textile industry, the pride of Pakistan's industrial landscape, began to take shape in the late fifties and the sixties of the last century. Pakistan recorded a respectable annual growth rate of 6.8 percent in the 1960s, though in the closing years of the decade, a negative per capita growth was recorded.

Apart from the economy, Pakistan had very unstable political condition. The military regime of General Ayub Khan, who wanted to introduce a progressive regime comparable to that of Ataturk Kemal Pasha, ended up in a series of debacles, starting with the failure of economic reforms, the 1965 war with India over Kashmir, which ended in a stalemate, and a series of cascading events during the subsequent regime of General Yahya Khan, which finally led to the secession of East Pakistan in 1971, leaving the country truncated, confined to its western wing.

15.3.4 Economy in the Regime of a Constitutionally Elected Government

The next stage in Pakistan's economic history began in the early seventies with the resumption of civilian rule under Pakistan People's Party chief Zulfikar Ali Bhutto, who acted first as President and then as the Prime Minister of Pakistan.

Though possessing an aristocratic life style and bearing, as head of the Pakistan People's Party, he pretended to be a leader of the masses and tried to inject an element of socialism in the organisation and working of the economy. Through his ability to articulate the public sentiments for economic and social reforms, he gathered support from a wide constituency – the peasantry, the factory workers, the urban lower middle class, the proletariat and the landlords from Sindh.[3]

Bhutto described the economic strategy under Ayub as "a monstrous economic system of loot and plunder which the regime branded as free enterprise".[4]

In west Pakistan the revelation of income and wealth provided a spring board for the rise of Z.A. Bhutto's party and its socialist agenda. The 1971 war placed pressure on Government spending and imports and the dismemberment of the country gave pragmatic argument for nationalisation of the financial sector ... Bhutto's nationalisation programme was a shock to the system and a grievous blow to private sector confidence that would take years to rebuild. New private sector investment came to a virtual halt. There was much (concealed) capital flight as businessmen took their money out of the country.[5]

That the private sector survived at all in the wake of this calamity with the additional ideological socialisation imposed by Bhutto is, as some feel, nothing short of a miracle. But, by and large, the nationalisation programme was well received by the people who saw it as a means of redressing the evils of concentration of wealth and wide disparities in income. "The economy was kept afloat and on a reasonable even keel thanks to Bhutto's diplomatic success in securing financing from friendly Islamic countries ... and the remittances from abroad."[6]

The most spectacular symbol of Bhutto's socialist economy was the establishment of the Pakistan Steel Mill in the public sector in the early seventies with technical and financial assistance from the USSR. The public ownership of the steel plant symbolised a socialist approach, though Bhutto was only partly responsible for this, as the proposal was conceived earlier. The Pakistan Steel Mill, discussed in the sequel, has turned into an economic disaster, a white elephant with accumulated losses, and is now being considered for disinvestment. Issuing of licenses for exploration of energy resources, both onshore and offshore, and enactment of land reforms were other measures that the Government took. Bhutto's land reforms remained symbolic and could not be fully implemented, though on the eve of the general election, believed to be highly rigged to favour the PPP, he reduced the size of holdings of un-irrigated land to 300 acres and irrigated land to 150 acres. This was only a marginal reduction in the size of agricultural holdings decreed by General Ayub Khan. In 1975, the land tax was abolished on holdings with an area of less than 25 acres and was progressively increased on those with larger holdings. It is doubtful, if these measures brought the desired result, except in the case of Khyber Pakhtunkhwa where a large area was redistributed to landless tenants.

15.3.5 Bhutto's Nationalisation Scheme

Bhutto's nationalisation scheme, including even small industries like cotton ginning, paddy husking and even flour mills, was not very significant in the national economy. His economic policy was a disincentive to private capital and entrepreneurship. All these measures, ironically labelled as *Islamic socialism*, did not make much impact on the country's economy, nor was there a visible spurt in economic development following the resumption of civilian rule. The disgruntled petty traders and mill owners played an active part in paralysing the government during national strikes. Though nationalisation, introduced in the name of socialism, did stop the monopolies of rich industrial and financial houses, it, almost, always carried the germ of mismanagement and the loss of several million rupees. Many of these nationalised units employed surplus manpower to the extent that in 1976, the total employment of the ten public sector corporations amounted to 61,731.[7] Besides the surplus manpower employed in nationalised industrial units incurring losses, there was also huge expenditure on defence.

One of the drastic measures that Bhutto's Government took was the devaluation of the Pakistan currency in 1972. This induced significant growth in export and in one year "despite the loss of revenue, especially foreign exchange from the East Pakistan's exportable produce, the country doubled its foreign exchange earnings".[8] The annual growth rate of the economy in the 1970s lingered around 4.5 percent. A great deal of the nation's energy was spent up in suppressing the revolt in Balochistan and settling scores with political opponents. Bhutto had the vision of some grand projects that were not only symbolic of rising Pakistan but were also the focus of attention of the entire nation. These included the Karachi Steel Mill (now Pakistan Steel Mill), the Indus Highway, the Nuclear Power Development Programme, the Lowari Pass Tunnel and the Z. A. Bhutto Sports and Culture Institute and Complex at Islamabad. While the Pakistan Steel Mill, the Indus Highway and the Nuclear power project materialised, the Sports complex at Islamabad came up as Jinnah Sports Stadium, much after the regime of Prime Minister Bhutto. The strengthening of economic ties with oil-rich Islamic countries led to the export of labour in increasing number. Starting with a modest figure of 2,262 in 1970, the number of emigrants to the Middle East reached a level of 1.6 million in 1981. The momentum built then has continued uninterrupted, and today, out of over 7 million Pakistani emigrants, over 3 million work in the Middle East, and contribute significantly to the foreign exchange reserve of Pakistan.

Bhutto's style of governance marked by confrontation and not accommodation, his alienation of some of the important members of his own party, including some co-founders, antagonising the feudal elites and hard-core religious leaders of Pakistan cost him his position as the Prime minister of Pakistan. Even the middle class felt cheated. The highly rigged general election of 1977, where he emerged victorious, evoked anger among the people in general and political leaders in particular. The use of the Federal Security Force, an armed unit independent of the army, for the brutal suppression of the revolt in Balochistan and imprisonment of several of his own followers caused widespread anger not only among his followers but even among the rank and file of the regular army. The chaotic situation resulted in a military coup, the arrest of Prime Minister Bhutto, and his execution, following a majority judicial verdict (4–3) in a criminal case, in April 1979.

15.3.6 The Second Coup d'Etat – The Regime of General Zia-ul-Haq and the Rise of Radical Islam

The sheet anchor of the state policy during the military regime of General Zia-ul-Haq (1977–1988) was primarily "the Islamisation of Pakistani society", meant to strengthen his position among the masses.

Economy during Zia Regime – On the economic front, as a result of the *policy of non-interference* of the Government, some of the large corporations and public sector undertakings, started during Bhutto's regime, began performing better after the initial teething trouble. This was particularly true of steel, cement, fertilisers and automobiles. The foreign investment climate also improved as a result of the guarantee provided by the government against future nationalisation. The most significant growth was of cotton textiles, where devoid of any apprehension of nationalisation, private entrepreneurs entered into the fray and started many textile mills, using locally produced cotton, leading to an almost unparalleled expansion of the cotton textile industry in Pakistan.

The most significant factor, however, that propelled the economic growth during the regime of General Zia was foreign aid. In fact, the most important financial support came from the United States and the Arab world to bolster resistance to Soviet forces by the Afghan Mujahedeen who were indoctrinated and trained in the *madrasahs* established on the Afghanistan–Pakistan border by the Pakistani administration. These *madrasahs* played a major role in the recruitment of *jihadists* (religious warriors) who, equipped with American weapons like Springer missiles, fought the Soviet forces. While fighting the Soviets to throw them out of Pakistan, Zia pushed his own agenda of Islamisation by promoting the multiplication of *madrasahs* in Pakistan and introducing Sharia courts and Islamic banking. An additional source of national income, the remittances from abroad, largely from the Middle East, amounted to roughly 10 percent of the GDP in 1982–1983 and formed an important element in budgetary calculations and foreign exchange requirements of the country. Being involved in the war against the Soviet Union, as a frontline state, social development was virtually neglected, and fields like science and technology were ignored. Yet, the economy grew rather well, attributed largely to the initiative of previous governments in areas like agriculture and industry.

While the most noticeable event of Zia's regime has been the forced retreat of the Soviets from Afghanistan, the regime also witnessed the rise of the *Taliban,* a retrograde group of Islamic fundamentalists who believed in, and adopted the *Sharia* law for statecraft. Their assumption of power in Afghanistan in 1996 was not a blessing for Pakistan, as it is being proved now. The indoctrination of Pakistani youth in religious fundamentalism, through the mushrooming growth of *madrasahs* during the regime of General Zia, introduced a new element in Pakistani society. This was, one may call, the primacy of Islam over everything mundane.

The influx of a large mass of Afghan refugees in Pakistan meant an additional burden on the finances of the state. But the most damaging effect of the rise of Taliban in Afghanistan has been its contagious effect in the neighbouring FATA (the Federally Administered Tribal Areas of Pakistan), where a similar fundamentalist organisation called the Tehreek-e-Taliban Pakistan (TTP) emerged. Using violence and terror as a means to achieve their goal, the TTP demands some measure of autonomy from Pakistan and freedom to conduct the affairs in their area, strictly according to Sharia. In their demand to enforce the Islamic system of jurisprudence and social behaviour, they have turned against the state. The rise of *Taliban* in Pakistan as Tehreek-e-Taliban Pakistan (TTP), based in Khyber Pakhtunkhwa, has turned a nightmare for the administrators, as much as for the society in general, in Pakistan.

15.4 The Last Decade of the 20th Century

The last decade, in the history of Pakistan, following the death of General Zia, the then President of Pakistan, was marked politically by a game of *musical chairs*. In less than a decade, half a dozen governments appeared, and they either fell or were dismissed by the President, invested with the power that General Zia had assumed by bringing in a constitutional reform. There was not much progress. It was a period of political instability. Benazir Bhutto and Mian Nawaz Sharif, each was elected Prime Minister twice, and each time they were sacked by the President. During his penultimate tenure as the Prime Minister, Nawaz Sharif restored the primacy of an elected Prime Minister and the legislature by the 13th Constitutional Amendment, terminating the presidential power to dismiss the governments and the prime ministers. The decade (1988–1998) was unstable and uneventful with frequent changes of Governments and Prime Ministers. The 1998 election resulted in the election of Nawaz Sharif as Prime Minister of Pakistan, only to be ousted and exiled by another coup d`état by the then Chief of the Army Staff, General Pervez Musharraf who remained at the helm of the affairs of Pakistan till 2008.

It may be noted that despite the sympathy Mian Nawaz Sharif attracts because of him being deposed and exiled by his Chief of the Army Staff, his economic policy proved detrimental to economic growth. It was percieve to be anti-capital as seen from his treatment of HUBCO[9] "the single largest private foreign investor to acquire stakes in the country until that point of time", in seizing their headquarters and freezing their accounts, alleging corruption in the grant of tariff- This had adverse impact on Pakistan's relationship with its creditors. "The economy was on its knees. Foreign exchange reserves depleted ... strong fear of default plagued the economy and multiple exchange rates were in effect".[10]

15.4.1 Economy During the Military Regime of General Musharraf (1999–2007)

The last military coup in Pakistan happened in 1999 when General Pervez Musharraf, like General Zia-ul-Haq, usurped power from the duly elected Prime Minister Mian Nawaz Sharif and exiled him for an indefinite period.

The military regime of General Musharraf inherited a very unhealthy economy. It took some time for him to mend things. He revived the relationships with the creditors and invited private capital that started flowing in. Debts were restructured and private investment picked up. Despite the fact that the Musharraf regime coincided with a very unstable political phase, there was unexpected economic growth.

Political instability was partly the result of a sudden change in the government regime, but more because of the American military campaign against the Taliban ruled state of Afghanistan. The reason for the American invasion of Afghanistan was the reluctance of the Taliban, the ruling party, to hand over Osama bin Laden, the supreme commander of al-Qaeda, a terrorist organisation, responsible for bombing the Twin Trade Towers in New York on September 11, 2001, in which more than 3,000 persons lost their lives.

The United States invaded Afghanistan to seek out Osama from his hideout. The operation that followed had the reluctant, yet tacit support of Pakistan. During the operation, much of American hardware, especially fuel supply and arms, passed through Pakistan. Though Pakistan forces were not directly involved, but helped the United States through logistical support. In hindsight, it appears that Pakistan suffered disproportionately heavily, possibly more than the benefits it derived. No doubt, the country was a regular recipient of a sum known as "Coalition Support Fund" (CSF). But, this pales into insignificance when seen against the backdrop of a huge refugee problem. Over three to four million refugees, displaced from Afghanistan and having taken shelter in Pakistan, proved an unwelcome burden on Pakistan, straining the country's resources and even proving detrimental to the interests of the local population.

15.4.2 Economy in the First Decade of the 21st Century

Though considered a constitutional pariah, the unwavering military regime of General Musharraf, with its strict disciplinarian attitude, brought forth a measure of stability. It guaranteed safe investment and trade. A decline in corrupt practices in administration generated confidence in the population. The economy unexpectedly appeared to be moving forward.

The military regime eliminated many of the bureaucratic and legal constraints, and the economy picked up after a couple of years. It is reported that in the eight years (1999–2007), the size of Pakistan's economy increased by 50 percent, and the per capita income by 25 percent following the military takeover. Concurrently with the economic growth, the population growth was also moderated, resulting in a declining growth rate of the population (Dawn, Oct. 17, 2007). There was certainly an increase in the level of investment, which rose to 25 percent of the GDP in 2006–2007, at the close of the Musharraf era, presenting a sharp contrast with the pre-Musharraf regimes. It must be admitted that the type of investment that flowed in did not increase the expansion of employment opportunities as it did not always lead to new industrial units and remained limited to the enlargement of the industrial base. It was characterised by the acquisition of some existing industrial units by some foreign industrial houses, or it could be an investment in the equity market. Besides, much of the growth was witnessed either in the service sector or high-investment industrial units, limiting employment to white-collar educated people. The ancillary industries took time to develop, and the dearth of employment was always felt. This is not only true of Pakistan, but this is the tragic corollary of modern industrial growth where the level of employment does not correspond to growth in investment, retarding an equitable distribution.

Besides a flow of investment and loans from international agencies like the World Bank, an invisible flow of money from the diaspora, the remittances from abroad, contributed to the national kitty. The recent spurt in IT services, mobile telephony, and the establishment of a few much-needed pharmaceutical firms added to the GDP. It may be emphasised that the remittances from abroad don't only add to GNP but also form a strong support base for the foreign exchange reserve stability of Pakistan.

An important aspect in the management of Pakistan's economy during the Musharraf era was the appointment of Shaukat Aziz as the Finance Minister and Prime Minister of Pakistan. He is credited with foresight and policy decisions that held Pakistan in stable economic health, despite a severe drought, a global recession and disturbed conditions following the war in Afghanistan. There was an all-round improvement in the productivity of large corporations like oil, gas and telecommunication. Also, small- and medium-scale industries contributed 40 percent to the GDP and employing 80 percent of the non-agricultural labour. The IT sector also received a huge foreign direct investment. The period (2003–2007) is marked by a higher collection of taxes, lower fiscal and current account deficit and a foreign investment of US $ 8 billion, all combining to boost the economy. The economy of Pakistan during this period grew at a rate of 7 percent. It is on record that after 2002, the revenue collection improved. A decline in corruption led to a substantial increase in revenue collection. Privatisation of some industries in the public sector infused efficiency and increased dividend. Tax holiday for newly emerging IT sector, duty reduction on the import of hardware and Government incentive to export provided further impetus to industrial growth.

One wonders if it is the quick and unwavering decision, the orderly conduct of the economy and a decline in corrupt economic practices during the military regime that showed accelerated economic growth, a unique phenomenon for a military regime.

15.5 A Democratically Elected Government Run by Pakistan Peoples' Party, and a Declining Economy

It is ironical, yet true, in the case of Pakistan, that with the arrival of a democratically elected Government of Pakistan People Party (PPP) in 2008, stagnation in the economy of the country set in. It is certainly a reflection on the quality of economic governance by an elected Government.

A pathetic economic situation at the close of the PPP rule in 2013 could be attributed to the rise of Tehreek-e-Taliban Pakistan (TTP), with its core in the Federally Administered Tribal Area but active in several parts of Pakistan and fighting for an Islamic state with "Sharia Laws". Ethnic conflicts and an overall deteriorating law and order situation were the main problems the state had to contend with, resulting in a neglect of the economy The country witnessed a lack of confidence in the federal government resulting in a decline in foreign investment, widening current account deficit and shrinking foreign exchange reserve from US $16 million in 2007 to US $7 million in 2008, and a decline in revenue collection from US $24 billion in 2007–2008 to US $18.3 billion in 2010–2011. The number of income taxpayers declined from 2.8 million in 2006–2007 to less than a million, the actual number being 768,000 in 2013. To make export commodities more competitive, the State Bank of Pakistan devalued the Pakistani rupee to 80 PKR per US dollar.

In a scathing attack on the performance of the Pakistan People Party (PPP), Murtaza Haider, a Pakistani expatriate and an academic, observed that "after five years of rule of the PPP, the nation is poorer, hungrier and more impoverished."[11] In a comparison with the Musharraf regime, he observed that while the Musharraf regime posted a growth rate of 6–8 percent, the Zardari Government could not muster even a 4.4 percent growth of GDP. The foreign debt ballooned from US $4.5 billion in 2008, at the beginning of the PPP regime, to US $65 billion in 2012. This happened despite the strong remittances from abroad, a steady grant of coalition support fund from the United States and a relatively low trade deficit.

15.5.1 The Economy in the 2013–2023 Decade

The two most serious economic problems that the country faced following the PPP (Pakistan Peoples' Party) regime that ended in 2013 were the power shortage, hampering industrial expansion, and power subsidies and debt servicing that ate into the State revenue. In 2013, the growth rate of Pakistan's economy was just about 3.6 percent. The regime witnessed a gradually declining economic situation.

15.5.2 A Period (2013–2017) of Economic Stability During the Regime of Mian Nawaz Sharif as the Prime Minister of Pakistan

The elections to the National Assembly that followed the end of the tenure of the Pakistan People Party in 2013 brought another party – Pakistan Muslim League (PML-N) – to power, headed by its President Mian Nawaz Sharif. The government of Prime Minister Mian Nawaz Sharif inherited an economy suffering from inflation, a low economic growth and a large budgetary deficit.

The government run by Pakistan Muslim League (PML-N), though having a comfortable majority in the National Assembly (NA), had to struggle hard to revive the economy, beleaguered as it was by continuing instability, threat of terrorist activities, ethnic conflicts, and opposition by religious fundamentalists. Added to these woes was the disposition of the opposition parties like Pakistan Tehreek-e-Insaf (PTI), which launched in 2014, a vitriolic attack on the democratically elected PM(N) government, and tried to pull it down by undemocratic and even coercive means. This was a period of a four-month-long *dharna* and a Tsunami March from Lahore to Islamabad. The protest march was later called off following the massacre of about 130 students of the Army Public School in Peshawar in 2014.

The cold-blooded assassination of 149 people including 130 children of the Army Public School at Peshawar led to a national outrage, followed by a punitive military offensive against the perpetrators of the attacks. The operation known as *Zarb-e-Azb* directed against the miscreants, largely in the Federally Administered Tribal Area (FATA) especially, North and South Waziristan, lasted for almost a year, causing death and destruction in the area. While it is difficult to gauge the impact of this operation, what is certain is that it alienated a large mass of people in the area, besides adding to the already existing refugee problem.

The military operation against the terrorists' hide-outs in the frontier areas was followed by another Civil–Military operation known as *Radd-ul-Fasaad* (Ending the conflicts) in 2017 to eliminate the terrorists and terror activities from all parts of Pakistan. It is difficult to assess the impact of these operations on terror activities in Pakistan. But, certainly, it had some impact in subduing the morale of the miscreants and boosting the morale of the civil population. The entire operation diverted the attention of the government from an administration directed largely to public good and economic development to punitive operations and restoration of people's confidence in the ability of the government.

Yet, despite these disturbances, things appeared moving in the right direction. The economy, during the 2014–2015 period, attained a growth rate of 4.4 percent, the highest level after 2008–2009. The country took a turnaround with stability in prices, industrial growth, stable foreign exchange reserve and improved energy supply (though much short of requirement). The Gross Domestic Product of Pakistan for 2014–2015 was US $252 billion (nominal) with a per capita GDP of US $1,513. Seen on a scale of Price Purchasing Parity (PPP), these figures swell to a GDP of US $928 billion and a per capita GDP of US $4,886. An exceptional increase in remittances from abroad further helped strengthen the Foreign Exchange Reserve position.

This may be also partly because of falling crude prices internationally, as Pakistan depended, and depends heavily even today, on the import of fuel. During the year (2014–2015), the inflation in Pakistan was the lowest at 2.1 percent, the interest rate was brought down to 7 percent, the lowest in the last 42 years,[12] and the unemployment had declined marginally to 6 percent. The current account deficit of the country attributed to the decline in crude prices was halved in 2014–2015 and was estimated at US $2.8 billion in the year 2014–2015. A major reason for a decline in the current account deficit was the realisation of US $2 billion from Eurobonds in 2014 and 1 billion from the sovereign sources.[13]

One may observe that despite all the odds, the economic situation in Pakistan during the 2014–2015 period was forward-looking, producing enough food and industrial goods for its population and enough revenue to run the government machinery and generating foreign exchange to meet its international payment obligations. There was no distress situation, things appeared to be moving in the right direction and the economy appeared to be gradually picking up. But, yes, there was an energy problem that caused substantial suffering to the economy.

However, it seems, Prime Minister Nawaz Sharif could not manage his relations with some of the important organs of the government and had to resign prematurely following a judicial verdict by the Supreme Court of Pakistan, which disqualified him from holding the office of the Prime Minister of Pakistan. However, the Muslim League (N) Government, though considerably weakened, remained in office till the end of its tenure in 2018.

15.5.3 The Euphoria of China–Pakistan Economic Corridor

In April 2015, during the PML (N) regime of Nawaz Sharif, as a part of a cooperative undertaking to establish an all-weather strategic partnership, an agreement, known as the *China–Pakistan Economic Corridor (CPEC)*, was signed between Pakistan and China on the background of a declaration of friendship "higher than the mountains and deeper than the oceans." The agreement was widely welcomed in Pakistan, and the country was euphoric about it. The protocol comprised 51 agreements covering almost all sectors of the economy. Under the agreement (CPEC), the corridor meant the route from Gwadar on the Makran Coast in Balochistan to Khunjerab Pass on the Karakoram ranges and beyond, to be linked by a super highway and subsequently by pipelines. The agreement attained a very high profile as the signing of the agreement was accompanied by the state visit of Xi Jinping, the President of the People's Republic of China.

The total finances involved in the 51 agreements were of the order of US $46 billion. Of this amount, about US $28 billion was meant to kick start "early harvest" projects. The remaining projects were to be carried out after feasibility studies and the completion of necessary formalities. Simultaneously, eight projects were inaugurated. These agreements covered a wide range of subjects, like joint marine research, Bio-tech laboratory, communication and infrastructure projects, climate change, Banking, Karachi-Multan Motorway, railways, Sister-cities relationship, Lahore-Orange-Line Metro project, Gwadar-East-Bay expressway and Gwadar International Airport. The energy sector included a number of projects like 870 MW Hydroelectric Power Project, Suki Kinari project, 1,370 MW Port Qasim Coal-fired project, 720 MW Karot Hydropower project, 9x100 MW solar power projects in Punjab, some wind power projects, several coal mining projects (lignite at Thar block 2) and a coal-fired project from block two, amounting to 2×330 MW. The most important of these projects was the north-south highway link.

One is not sure how much of this sum of US $46 billion is a grant and how much it is a loan. The assurance of an investment of US $46 billion was a substantial foreign investment, and it is reported that this amount is raised further. The famous and long-delayed Bhasha-Daimar hydroelectric project is now being implemented with financial and technical support from China.

15.5.4 The Economy of Pakistan During the Last Five Years (2018–2023)

This has been and continued to be the most unstable phase in the political and economic history of Pakistan. The national election of 2018 brought into focus another political party the Pakistan Tehreek-e-Insaf (PTI), headed by its chairman Imran Khan, a cricketer turned politician. With PTI having the largest number of members elected to the National Assembly, Imran Khan staked his claim to form the Government by stitching a majority in the National Assembly, with the support of several smaller parties, and was finally sworn in as the Prime Minister of Pakistan in 2018.

Prime Minister Imran Khan, during the election campaign, promised a New Pakistan, more prosperous, with better law and order situation, and one devoid of any corrupt practices. While the economic situation of Pakistan was deteriorating and needed urgent remedial measures, the PTI government headed by Imran Khan concentrated on uprooting corruption and invested considerable energy in instituting enquiries through the National Accountability Bureau against the opposition leaders, getting them arrested and sent to prison. The opposition parties, notably Pakistan Muslim League (Nawaz), Pakistan Peoples' Party and Zamiat-e-Islam, grouped together and formed an opposition coalition known as Pakistan Democratic Movement (PDM) and thus started a political battle between Pakistan Tehreek-e-Insaf, the ruling party and the opposition represented by PDM. The political battle that started with a war cry led to the fall of the PTI government, following a no-confidence motion, and the dismissal of Imran Khan as the Prime Minister (2018–2021) of Pakistan.

The new coalition government of PDM, headed by a new Prime Minister Mian Shehbaz Sharif completed the remaining one-and-a-half-year term of the National Assembly, following which, it was dissolved in August 2023, paving the way for a fresh election to the National Assembly. In the interim, a Care Taker Government has taken charge of the administration of the country and will continue till a new government is sworn in. The most recent election to the National Assembly was held in February 2024. Pakistan Muslim League (N) secured the largest number of seats and was able to form a coalition government with the support of other parties.

15.5.5 Whither Goes Pakistan's Economy?

The bitter struggle for power between the contending political parties led to a complete neglect of the economy that needed immediate attention and remedial measures for revival. But what happened was an utter disregard of the economy, and the adverse economic situation was allowed to slip further into an unprecedented economic crisis.

15.5.6 The Present State of Economy – Economic Survey Report of Pakistan 2022–2023

To give an idea of the economic situation of Pakistan, an official version of the state of Pakistan economy as it appeared in the Pakistan Economic Survey 2022–2023 is given in a nutshell (Table 15.3). It must be emphasised that this is the version of the coalition government run by PDM, which ran the country for over a year after the fall of the Pakistan Tehreek-e-Insaf (PTI) Government. The Survey observes that

> *inconsistent policies (of the governments), unsustainable fiscal deficit, an exponential rise in public debt, economic uncertainties, and mounting circular debt have been the main ingredients of the previous Government tenure. As such, the last fiscal year of the previous government alone witnessed an overall deficit of 7.9 per cent and a trade deficit of US$ 39.1 billion.*

It observes further that

> *currently the challenge is to reduce the fiscal deficit for fiscal sustainability along with macroeconomic stability. At the same time there is a difficult task to support the vulnerable segments of the society and meet expenditure on rising interest servicing.*

Inflationary pressures emanated from supply disruptions created by flood damages, higher global food prices and broader tariff reforms and weaker exchange rates for electricity and other fuels.

15.5.7 The Present Economic Situation of Pakistan through the Prism of Pakistani Observers

There are two principal issues that worry most of the Pakistani observers. First, the current economic situation, and second, a situation of constitutional entanglement coupled with the manoeuvres of political parties, creating a climate of uncertainty. The latter is as worrying for Pakistan as its current economic situation.

TABLE 15.3 Some Microeconomic Indicators of Pakistan (2022–2023 and 2021–2022)[14]

Item	*2022–23*	*2021–22*	*Increase/Decrease*
Revenue	6,938.2 billion PKR	5,874 billion PKR	18.1 %
Total Expenditure	10,016 billion PKR	8,439 billion PKR	18.7 %
Borrowing	3,043 billion PKR	1,570 billion PKR	51.6 %
Inflation	29.2 %	11.3 %	
Import	46.9 billion US$	65.5 billion US$	-28.4 %
Export	23.2 billion US$	26.2 billion US$	-11.7 %
Current account	3.3 billion US$	26.1 billion US$	78.1 %
Workers' remittances	22.2 billion US$	26,1 billion US$	-13.3 %
Foreign investments *	1.2 billion US$	1.5 billion US$	-23.2 %
China investing	347.6 million US$		29.7 % of total FDI
Japan investing	162.9 million US$		13.9 % of total FDI
Switzerland investing	131.6 million US$		11.2 % of total FDI
UAE investing	113.6 million US$		9.7 % of total FDI
Hong Kong investing	81.3 million US$		6.7 % of total FDI
Foreign exchange reserve	4.6 billion US$		
Exchange rate	283.8 per US$	204.8 per US$	-27.8 %
Indebtedness	59,247 billion PKR		
Domestic Debt	35,078 billion PKR		
External Public Debt	24,121 billion PKR (85.2 billion US$)		

Source: Pakistan Economic Survey 2022–23, Government of Pakistan.
Note. *Investment during July–April period, 2023.

There is a large section of public opinion in Pakistan that believes that the contemporary crisis is the worst ever experienced by Pakistan. As observed by Niaz Murtaza, a political economist, "Pakistan faced many problems since 1947 but the last one year has been very bad given the numerous challenges, their intensity and duration ...There is chaos under the heavens".[15] The state of chaos he attributes to Imran Khan's incompetence, maverick acts and desire for autonomy. An almost similar view was expressed by Ashraf Jehangir Qazi, a senior diplomat and former of ambassador of Pakistan to the United States, India and China when he wrote on March 30, 2023, that

> *Pakistan is imploding. The President and Prime Minister are at war. The Election Commission of Pakistan challenges the Constitution and the Supreme Court. The Prime Minister attacks the Chef Justice of Pakistan. The Government leaders threaten it is either Imran or them. The doctrine of necessity is being revived by a Care Taker Government and its puppeteers pulling the strings.*[16]

There are others who think that "Our country (Pakistan) is in the midst of a major civil war. All the pillars of the State are in conflict with each other, and within each pillar there are conflicts as well".[17]

He attributes this state of affairs to two principal reasons. First, political reason – the political tussle that surfaced around October 2021 morphed into a full-blown political and constitutional crisis by late March 2022. The second reason, according to him, is the conflation of a number of unexpected global crises including the Covid-19 pandemic and the Russia–Ukraine war that caused disruption in the global supply chain and an unprecedented 40-year high inflation, putting massive stress on Pakistan's economy, already struggling with high inflation and rising unemployment.

Unlike other observers, Maleeha Lodhi, a diplomat and former ambassador of Pakistan to the United States and even to the United Nations, looks at this crisis rather holistically. She observes

> *Pakistan's macroeconomic crisis, the most serious in its history, continues to warrant urgent and sustained attention. The roots of this crisis lie in chronic fiscal deficits and external imbalances that are responsible for its perpetual balance of payment problems, high inflation and macroeconomic instability necessitating repeated financial bailouts". "There is another crisis that is worsening and consequential for Pakistan's future.* "This is *the crisis in human development – with most indicators of literacy, education, health and other aspects of social justice and human welfare deteriorating in recent years.*[18]

It appears that the economic situation deteriorated at an unprecedented speed, showing the inadequacy of measures adopted by the Government and allowing the devaluation of rupee.

> Since 2008, Pakistan has been continuously under IMF supervision. Over these 15 years, there have been four programmes (the current standby arrangement included). Out of these 15 years, it has spent 11 years under Fund programme. This doesn't augur well for a large country like Pakistan.

The most troubling aspect of Pakistan's economy is the energy sector. The domestic growth of industries depended a great deal on imported energy sources, petroleum and even LNG, in a situation of gradual decline of domestic gas resources. The import of fuel resources is unavoidable, and this chink in the Pakistani economic armour has to be addressed by finding domestic sources of supply, promoting export and filling the current account deficit.

According to the Pakistan Economic Survey 2018–2019,[19] "Pakistan's economic growth is largely driven by consumption led spending as opposed to investment. This non-productive nature of growth has led to imbalance in current account deficit and an adverse balance of payment". In recent years, the current account deficit rose from a couple of billion US$ in 2000 to around US $50 billion in 2022, when the value of import was over US $80.17 billion as against the export value that declined to US $31.7 billion. The invisible income like foreign remittances amounting to US $31 billion (2022), and other service earnings could not compensate this current account deficit. The foreign direct investment has been miniscule, around US $2 billion. All this led to an increase in external debt and repeated borrowing from international sources, more specifically the International Monetary Fund. The regular fiscal deficit and borrowing from external sources led to rising inflation and devaluation of Pakistani currency and erosion of the real income of the citizens.

The inability to repay the debt led to more borrowing and a situation of circular debt developed. The energy sector's long precarious situation has become acute, while severe liquidity pressures now adding to the continued overaccumulation of unsustainable payment arrears.

> *The stock of circular debt has crossed Pakistani RS. 2.5 trillion … Interest payments have grown by 460 pc since 2008, while tax revenues have grown by 360 pc. No wonder, the debt service bill is larger than net federal revenue, a milestone, and Pakistan is sinking increasingly into financial and economic un-viability.*[20]

The presentation and discussion of public perception and the enlightened view of knowledgeable people of Pakistan about Pakistan's economy is closed here with the quotation of the perception of Pakistan's economic reality by Professor Ishrat Husain, former Governor of the State Bank of Pakistan and a distinguished National Professor of economics and public policy. While the analysis of Professor Husain, outlined below, is not only plausible but largely true, there are certain other aspects of economic crisis that are generally ignored. This is what he has to say:

> *"Most observers and analysts, both inside and outside Pakistan firmly believe that the quality of economic governance and decision making and the capacity of key institutions have gradually deteriorated over time" ... Improvement of economic governance has not been consistently pursued in Pakistan as implementation requires several decades while elected and military governments have short term horizons. Elected governments in their pursuit of winning the next election, and military governments in their attempts to gain legitimacy, get bogged down in ad hoc and occasionally popular measures without addressing the root-cause that is building institutional capacity to deliver improved living standards for majority of the population and setting up a viable governance structure. ... "Personalised decision making according to the whims and caprices of individuals at the helm has therefore displaced well thought out institutionalised processes."*
>
> *The common view is that the centre has assumed too much power and authority and this excessive concentration of power has led to inefficiency, social fragmentation and ethnic divisiveness. Political leaders with the help of a small coterie of loyalists exercise absolute power and take all important decisions. Dissent within the political parties is hardly tolerated and parliaments usually rubberstamp the decision taken by the leader of the party.*[21]

15.5.8 The Way the Countries Have Prospered

A country tends to become economically prosperous if it is technologically very advanced as the West European countries and Japan or even South Korea in Asia. There are other countries that have a huge resource base and a large market by virtue of their population and even cheap and skilled labour with the added benefit of attracting foreign investment. One can put People's Republic of China in this group. India is another country with a substantial base of natural resources, a huge market and equally cheap unskilled and semi-skilled labour. These conditions have to be balanced with a relatively stable population. A runaway growth of population eats into the gains of development.

15.5.9 Pakistan's Advantages and Disadvantages

Pakistan has a huge market and a mass of unskilled, semi-skilled labour and even technically trained manpower. As a country, however, it is not adequately bestowed with natural resources like cultivable land. It has hardly 25 percent of its area under cultivation. The country does not have some of the basic raw materials for its industries like iron nor is it rich in fuel resources. The much-discussed gas fields are nearing exhaustion, and the huge Thar lignite is out of tune with contemporary concerns about pollution, global warming and climate change. The import of crude or fuel resources, in whatever form, is unavoidable, ballooning the import bill. To overcome these perceived deficiencies, the country needs to develop a large industrial base to be able to export or exchange manufactured goods. What is required is more investment and more surplus production of goods to be exported to generate the necessary foreign exchange reserve. This is a difficult proposition in the absence of funds, technical know-how and even raw materials. The much-touted gold and copper reserves, yet to be exploited, cannot adequately boost the Gross Domestic Production (GDP).

15.5.10 Direct Foreign Investment – the Sole Option

Perhaps, the sole option available is to invite/attract foreign investment – not an easy task. The country has to create conditions to attract foreign investment. Most enlightened citizens of Pakistan are familiar with what the conditions should be. Besides the terms and conditions to attract foreign capital are well known, there is more to it that meets the eye. A harmonious social climate and a "law and order situation" free from any violence are key factors in the entrepreneurial calculus of foreign investors, who should feel welcome and at home.

There are examples of countries in South Asia with precarious economic situations but with well-structured economic policy and an accommodative disposition of the Government towards foreign investors, they have been able to secure their economy with increasing growth and a build of substantial foreign exchange reserve. India, the next-door neighbour of Pakistan, had a similar bad batch in its economic history when it had to transfer tonnes of gold from the vaults of the Reserve Bank of India to some Swiss banks, as security against international loans. It was such a humiliating experience. How the Indian economy has turned the corner is for anyone to observe. Pakistan can do still better.

With an enlightened leadership, a genuine concern for the people and economy leading to a concentrated effort from the government, with a host of economic experts planning and directing the country and a welcome arrival of Foreign Capital Investment, the country will bounce back to a flourishing and booming economy. Creation of a climate, suitable for foreign investment, is perhaps the most urgent task the country should embark on. Pakistan has its job cut out, and undoubtedly it will succeed, and happy days will be back again.

Notes

1 Papanek, G. F (1967) *Pakistan's Development: Social Goals and Private Incentives*, Harvard University Press, Cambridge, MA.
2 Meekal Ahmad (2011) An Economic Crisis State, in Maleeha Lodhi (ed.) *Pakistan Beyond the Crisis State*, Rupa, New Delhi, p. 171). The statement of the Chief Economist of the Planning Commission, Mahbub-ul-Haq, also appears in a statement in London Times, 22 March 1973.
3 Zahoor, Muhammad Abrar (2011) A Critical Appraisal of the Economic reforms of Zulfikar Ali Bhutto: An Assessment, *Pakistan Journal of History and Culture*, vol. 32, pp. 145–161.
4 Rizvi, Hassan Askari (1972) PPP-The First Phase (1969–71). Lahore Progressive Publications, p. 7, quoted from Zahoor, M. A. (2011) A Critical Appraisal of the Economic reforms of Zulfikar Ali Bhutto: An Assessment, *Pakistan Journal of History and Culture*, vol. 32, pp. 145–161.
5 Meekal, Ahmad (2011) An Economic Crisis State, in Maleeha Lodhi (ed.) *Pakistan beyond the Crisis State*, Rupa, New Delhi, p. 173). The statement of the Chief Economist of the Planning Commission, Mahbub-ul-Haq, also appears in a statement in London Times, 22 March 1973.
6 Meekal, Ahmad (2011) An Economic Crisis State, in Maleeha Lodhi (ed.) *Pakistan beyond the Crisis State*, Rupa, New Delhi, p. 174). The statement of the Chief Economist of the Planning Commission, Mahbub-ul-Haq, also appears in a statement in London Times, 22 March 1973.
7 Goverment of Pakistan – Pakistan Economic Survey 1978, quoted from Abrar Zahoor-2011.
8 Hussain, Khurram (2023) The Tale without a Plot, *Dawn*, 20 July 2023.
9 Hub Power Company Limited.
10 Hussain, Khurram. (2023) The Tale without a Plot, *Dawn*, 20 July 2023.
11 Haider, Murtaza (2013) Five Years On: A Poorer, Hungrier Nation, *Dawn*, 16 March 2013.
12 Pakistan Economic Survey (2014–15) Finance Division, Government of Pakistan, p. 1.
13 Pakistan Economic Survey (2014–15) Finance Division, Government of Pakistan.
14 Government of Pakistan, Economic Advisor's Wing, Finance Division (2022–23) *Highlights Pakistan Economic Survey, 2022–23,* Islamabad.
15 Niaz, Murtaza (2023) A State in Chaos, *Dawn*, 16 May 2023.
16 Qazi, Ashraf Jehangir (2023) Saving Pakistan, *Dawn*, 30 March 2023.
17 Arif, Hasan (2023) Coordination is Key, *Dawn*, 6 September 2023.
18 Lodhi, Maleeha (2023) Pakistan's other Crisis, Dawn, 18 September 2023.
19 Pakistan Economic Survey (2018–19) Finance Division, Government of Pakistan.
20 Husain, Khurram. (2023) The Tale without a Plot, *Dawn*, 20 July 2023.
21 Husain, Ishrat (2011) Retooling Institutions, in Lodhi Maleeha (ed.) *Pakistan: Beyond the Crisis State*, Rupa Publications Ltd, New Delhi, pp 150 & 152.

BIBLIOGRAPHY

A

Abdul, Majid Kazi (2014) Overview of Water Resources in Pakistan, *Pakistan.com*, 7 April 2014.

Abdul, Rashid (2004) Impact of El-Nino on Summer Monsoon Rainfall of Pakistan, *Pakistan Journal of Meteorology*, vol. 1 (2), pp. 35–43.

Abdullah, Hamid (1973) *The Historical Background of Pakistan and its people*, Tanzeem Publications, Karachi.

Afzal, M., M.A. Haroon, A.S. Rana and A. Imran (2013) Influence of North Atlantic Oscillation and Southern Oscillation on the Winter Precipitation of Northern Pakistan, *Pakistan Journal of Meteorology*, vol. 9 (18), pp. 1–8.

Agricultural Census Organization (2008) *Mouza Statistics*, Government of Pakistan, Statistics Division.

Agricultural Statistics, Pakistan Bureau of Statistics (2011) Government of Pakistan, Islamabad.

Ahmed, Ayaz, Henna Ifthikar, and G. M. Chaudhry (2007) Water Resources and Conservation, Strategy of Pakistan, *The Pakistan Development Review*, vol. 46 (4) pt. II (winter) pp. 997–1009, DOI:10.30541/v46i4IIpp.997-1009.

Ahmad, Meekal (2011) An Economic Crisis State, in Maleeha Lodhi (ed.) *Pakistan Beyond the Crisis State*, Rupa, New Delhi, p. 171.

Ahmad, Nafis (1947) *The Basis of Pakistan*, Thacker Spink & Co., Calcutta.

Ahmad, Nafis (1958) *Economic Geography of East Pakistan*, Oxford University Press, London and Karachi.

Ahmed, Usman (2011) Abstract of the Ph. D, thesis "Social Stratification in Punjabi Village in Pakistan: The Dynamics between Caste, Gender and Violence", University of Leeds.

Ahmed, Usman (2016) Marginalised Voters and Supporters: Biradari System, Caste Hierarchy and Rights to Political Participation in Rural Punjab, *Journal of Political Studies*, vol. 23 (2), pp. 515–24.

Akhtar, Shamshad and Mohammad Sharif (2007) Population Growth and Land Use Pattern of Karachi City, *Pakistan Geographical Review*, vol. 62, pp. 63–71.

Al Biruni (1030) *Kitab-ul-Hind,* Translated by Edward C. Sachau as 'India- An Account of the Religion, Philosophy, Literature, Geography, Chronology, Astronomy, Customs, Laws and Astronomy of India (English edition with notes and indices) in two volumes, Kegan Paul, Trench, Trubner & Co., London, 1910.

Ali, Akhtar (2013) *Indus Basin Floods, Mechanism, Impacts and Management*, Asian Development Bank, Philippines.

Ali, R. (2008) Understanding Urbanisation, in Arif Hasan (ed.) *Migration and Small Towns in Pakistan*, JIED Workshop, London, 18–19 November 2008.

Ali, Tariq (1983) *Can Pakistan Survive? The Death of a State*, Penguin Books Ltd., London.

Amin, Mohamed, Duncan Welletts and Graham Hancok (1982) *Journey through Pakistan*, Vikas Publishing House, New Delhi.

AQUASAT – FAO (UNO), http://www.fao.org/nr/water/aquasat/basins (accessed on 6-8-2016).

Arif, Hasan (2023) Coordination Is Key, *Dawn*, 6 September 2023.

Arshad, Muhammad (2015) Pakistan, Soil Resources, Issues, Threats, On-going Activities and their Sustainable Management – Lecture delivered on 13–15 May 2015, in 'Asian Soil Partnership Consultation Workshop on 'Sustainable Management and Protection of Soil Resources' Bangkok.

Asian Development Bank (2002) *Water Resources Sector Strategy, Natural Water Sector Profile*, Asian Development Bank, Philippines.

Ayesha, Siddiqa (2016) Pakistan's Elite, *The Express Tribune*, p. 24.

Azad, Azra Parveen and Razzaq Ahmed (2006) A Geographical Study of Land Use in the Commercial Heart of Karachi (Saddar), *Pakistan Geographical Review*, vol. 61, pp. 64–82.

Azmat, Hayat Khan (2004) The Influence of La-Nina Phenomena on Pakistan Precipitation, *Pakistan Journal of Meteoroloy*, vol. 1 (2), p. 23.

B

Baker, Aryn (2008) Landowner Power in Pakistan Election, *Time Magazine*, 18 February 2008.

Brinkman, A. (1971) *Soil Genesis in West Pakistan*, Pakistan Soil Bulletin No. 4, Central Soil Research Institute, Lahore, p. 5.

Bureau of Statistics, Final Results of Census (2017).

Bureau of Statistics, Government of Pakistan, Ministry of Petroleum and Natural Resources & HDIP (Hydrocarbon Development Institute of Pakistan). Pakistan Economic Survey 2020–21.

Bureau of Statistics of Pakistan (2014).

Burrard and Hayden (1933) *A Sketch of Geography and Geology of Himalaya*, Manager of Publication, Delhi.

Burrard and Hayden (1933) *A Sketch of Geography and Geology of Himalaya*, Manager of Publication, Delhi.

C

CDPR (2115) Urban Development in Punjab: A Political Economy Analysis.

Census of India (1941) vol. I, pt. I, Tables, p. 91.

Census of Pakistan (1951) vol. I and VI.

Census of Pakistan (2017).

Census Organisation 1951, by E.H. Slade, vol. I, Report and tables.

Central Water Commission India – World – Land and water Resources at a glance in http://www.cwc.nic.in/main/webpages/statistics. html (accessed 16-3-2016).
Choudhry, M. Bashir (1978) *Soil Map of Punjab*, Soil Survey of Pakistan, Lahore.

D

Dames, M. L. (1904) *An Historical and Ethnological Sketch of the Baloch Race*, quoted from Imperial Gazetteer of India, 1908, vol. VI, p. 289.
Dames, M.L. (1904) *The Baloch Race: An Historical and Ethnological Sketch, Asiatic Society Monograph*, Vol. IV, The Royal Asiatic Society, London. Quoted from Imperial Gazetteer of India, 1908, Vol. VI, The Clarendon Press, Oxford, p. 289
Davis, Kingsley (1951) *The Population of India and Pakistan*, Princeton University Press, Princeton.
Directorate of Crop Reporting Services, Balochistan.
Directorate of Crop Reporting Services, Khyber Pakhtunkhwa, Peshawar.
Douie James (1916) *The Punjab, North-West Frontier Province and Kashmir*, Cambridge University Press, Cambridge, 373 p., classified as a rare book.

E

Encyclopaedia Britannica (1985) vol. 4 and 11.
Encyclopaedia Britannica (2016).
Encyclopaedia Britannica, Book of the Year (2014).

F

Faisal, N. and A. Gaffar (2012) Development of Pakistan's New Area-weighted Rainfall using Thiessen Polygon Method, *Pakistan Journal of Meteorology*, vol. 9, July, pp. 107–16.
FAO, UNO Aquastat Data base, http://www.fao.org/nr/water/aquastat/data/query/results - (accessed on 1-3-2016).
FAO/UNO (2011) *AQUASAT – Indus Basin Regional Report*, Water Report 37.
FAO/UNO -AQUASTAT (2011) Indus Basin, http//www.fao.org./nr/water/aquastat /Basins (accessed on 6-8-2016).
Finance Survey (2020) *Economic Survey 2019-20*, Government of Pakistan, June.
Forestrypedia – Classification of Forests and Forest Types in Pakistan (detailed note), https:// forestpedia.com/classification of forests and – fores/# Forest-Types of Pakistan – (accessed on 5-9-2020).

G

G.M. Arif and Sabha Ibrahim (1998).The Process of Urbanisation in Pakistan 1951–81, *The Pakistan Development Review*, vol. 37 (4), pp. 507–22.
Geological Survey of Pakistan (1964) Geological Map of Pakistan.
Ghazala, Naheed and Ghulam Rasul (2011) Investigation of Rainfall Variability in Pakistan, *Pakistan Journal of Meteorology*, vol. 7 (14), pp. 25–32.
Government of Pakistan – Ministry of Environment (2009) Land Use Atlas of Pakistan.
Government of Pakistan, Economic Advisor`s Wing, Finance Division (2022–23) Highlights Pakistan Economic Survey, 2022–23.
Government of Pakistan, Finance Survey (2020) Economic Survey 2019–20, June.

Government of Pakistan - Pakistan Economic Survey (1978).
Governmet of Pakistan, Pakistan Economic Survey 2014–15, table 4, table 3–1.

H

Haider, Murtaza (2013) Five Years On: A Poorer, Hungrier Nation, *Dawn*, 16 March 2013.
Halsey, A. H. (ed.) (1988) British Statistical Trends, since 1900. *A Guide to Changing Social Structure of Britain*, Macmillan, London, pp. 322–356.
Hanif, Muhammad (2002) Management of Water Resources in South Asia, *Science Vision*, vol. 7 (3–4), pp. 20–32.
Hasan, Arif (2023) Coordination Is Key, *Dawn*, 6 September 2023.
Hazala, Naheed and Ghulam Rasul (2011) Investigation of Rainfall Variability in Pakistan, *Pakistan Journal of Meteorology*, vol. 4 (14), pp. 25–32.
Hewitt, K., C.P. Wake, G.J. Young, and C. David (1989) Hydrological investigations at Biafo Glacier, Karakoram Himalaya: An important source of water for the Indus River, *Annals of Glaciology*, vol. 13, pp. 103–108.
Hunter, (Sir) W.W. (1881) *Imperial Gazetteers of India, in Multiple Series*, Government of India, Oxford University Press, Oxford.
Husain, Ishrat (2011) Retooling Institutions, in Lodhi Maleeha (ed.) *Pakistan: Beyond the Crisis State*, Rupa Publications Ltd, New Delhi, pp 150 & 152.
Husain, Ishrat (2016) Messy and Hidden Urbanisation, *Dawn* 16 May 2016.
Hussain Danish (2014) Slum Survey, Over 80,000 People Living in Capital's Katchi Abadi Says Report, *The Express Tribune*, Pakistan, 2 February 2014.
Hussain, Khurram (2023) The Tale without a Plot, *Dawn*, 20 July 2023.

I

Iftikar, Ahmad, Sun Zhaobo, Deng Weitao, and Romana Ambreen (2010) Trend Analysis of January Temperature in Pakistan over the Period of 1961–2006: Geographical Perspective, *Pakistan Journal of Meteorology*, vol. 7 (13), pp. 11–22.
Imperial Gazetteer of India, vol. 13.
Imperial Gazetteer of India 1908, vol. VI.
Imperial Gazetteer of India, Provincial Series, Punjab, vol. I, Calcutta 1908.
Imperial Gazetteer of India, Punjab (1908) vol. I. p, 50, Government Printing Press, Calcutta.
Imperial Gazetteer, Provincial Series North-west Frontier Province (1908) Government Press, Calcutta.
Indus River System Authority Act. No. XXII 1992 – Water Apportionment Accord 1991.

J

Johnson, B.L.C. (1979) *Pakistan*, London, Heinemann Educational Books Series, 214 p. (The book was finally placed on-line by Cambridge University Press in 2011).

K

Kahlown, Muhammad Akram and Abdul Majeed (2002) Water Resources Situation in *Pakistan: Challenges and Future Strategies, Science Vision*, vol. 7 (3–4), pp. 33–45.

Ayaz Ahmed, Henna Iftikhar (2007), Water Resources and Conservation Strategy of Pakistan, *The Pakistan Development Review*, vol. 46 (4), pp. 997–1009.

Kazi, Said-Uddin Ahmad (1964) *A Geography of Pakistan*. Originally published from the University of California, the book went into a second edition, and was published by (Oxford University Press, Karachi, 1968).

Kazmi, A.H. and M. Qasim Jan (1997) *Geology and Tectonics of Pakistan*, Graphic Publishers, Karachi.

Khan, A.H. (2004) The Influence of La-Nina Phenomena on Pakistan Precipitation, *Pakistan Journal of Meteoroloy*, vol. 1 (2), p. 23.

Khan, Jarwar Attar (Director, Soils Survey of Punjab, Pakistan) (2012) Soil Survey of Pakistan: History, Achievement and Impact on Agriculture – Paper presented at the Regional Conference on the Asian Soils Partnership, held at Nanjing, China, 8–11 February, 2012. Sponsored by Food and Agricultural Organisation of United Nations.

Khan, Fazle Karim (1991) *A Geography of Pakistan: Environment, People and Economy*, Oxford University Press, Karachi.

Khan, Fazle Karim (2015) *Pakistan: Geography, Economy and People*, Karachi, Oxford University Press, Karachi.

Khan, Imran Ahmad and Muddassar Hassan Arsalan (2018) Satellite based Assessment of Soil Moisture and associated Factors for Vegetation Cover. A Case Study of Pakistan and adjoining Region, *Pakistan Journal of Botany*, vol. 50 (2), pp. 699–709.

Khan, M.A., M. Ahmad and H.S. Hashmi (2012) *Review of Available Knowledge on Land Degradation in Pakistan*, OASIS Country Report 3, ICARDA (International Centre for Agricultural Research in Dry Areas), p. 4, quoted from Lal Rattan (2018).

Khan, Sanaullah and Mukhtar Ahmad (2003) High Altitude Pastoral Systems in Malakand Division, Pakistan, in J. M. Shuttle and S. G. Reynold (eds) *Transhumant Grazing Systems in Temperate Asia, Chapter XII – Pakistan Case Study* 2, Plant Production and Protection Series no. 31, FAO, Italy, pp. 239–248.

Khurram, Hussain (2023) The Tale without a Plot, *Dawn* e-paper, 20 July 2023.

Kronstadt, K. Alan, Parvez A. Sheikh and Vaughn Bruce (2010) *Flooding in Pakistan: Overview and Issues for Congress*, Congressional Research Service, Washington, November, p. 3.

Kugelman, Michael (2013a) Pakistan`s Fixation with Feudalism, *Dawn*, 3 January 2013.

Kugelman, Michael (2013b) Urbanisation in Pakistan. Causes and Consequences. Norwegian Peace Building Research Centre, *NOREE Expert Analysis*, June, p. 4.

Kureshi, K. U. (1977) *A Geography of Pakistan*, Oxford University Press, London and Karachi.

Kureshi, K. U. (1988) *A Geography of Pakistan* (Revised edition), National Book Service, Karachi.

L

Lal, Rattan (2018) Managing Agricultural Soils of Pakistan for Food and Climate, *Soil and Environment*, vol. 37 (1), Islamabad, pp. 1–10.

Lancaster, J. (2003) Pakistan's Modern Feudal Lords, *Washington Post*, 8 April 2003.

Land use – Land cover Atlas of Pakistan's Provinces, based on remotely sensed data GCP/PAK/125/USA, the Govt. of Pakistan and FAO of the UNO, and the US Dept. of Agriculture.

Lieven, Anatol (2011) *Pakistan: A Hard Country*, Allen Lane, London.
Lodhi, Maleeha (2023) Pakistan's other Crisis, *Dawn*, 18 September 2023.

M

Magazine Ravi (2015) Sugar Industry in Pakistan, an Academic Report, 5 May, http://www.ravimagazine.com/sugar-industry.
Malkani, M.S. (2012) A Review of Coal and Water Resources of Pakistan, *Journal of Science, Technology and Development*, vol. 31, pp. 202–218.
Marcus, Nusser and Sussane Schmidt (2017) Nanga Parbat Revisited: Evolution and Dynamics of Socio-Hydrological Interactions in the Northwestern Himalaya. *Annals of the American Association of Geographers*, vol. 107 (2), pp. 403–415.
Meekal Ahmad (2011) An Economic Crisis State, in Maleeha Lodhi (ed.) *Pakistan Beyond the Crisis State*, Rupa, New Delhi, p. 171. The Statement of the Chief Economist of the Planning Commission, Mahbub-ul-Haq, also appears in a statement in London Times, 22 March 1973.
Mian, M. Alim and M. Yasin Javed. (1993) *Pakistan's Soil Resources.* A Pakistan National Conservation Strategy Paper 4, IUCN, Karachi.
Michel Morrish (2011) Geofile on line, September 2011 (Pakistan Floods, 2010 – Causes, Impacts and Aid).
Ministry of Environment (2009) *Land Use Atlas of Pakistan*, Government of Pakistan, Islamabad.
Mohamed Amin, Duncan Welletts and Graham Hancok (1982) *Journey through Pakistan*, Vikas Publishing House, New Delhi.
Mohammad Hanif – Management of Water Resources in South Asia – a Note Agricultural Development Commissioner.
Muhammad, Akram Kahlown and Abdul Majeed (2007) Water Resources and Conservation Strategy of Pakistan, *The Pakistan Development Review,* vol. 46 (4), pp. 997–1009.

N

NASA Earth Observation System, Oak Ridge National Laboratory Distributed Active Archive Centre (ORNLDAAC), http://daac.ornl.gov/data/rivdis/Stations (accessed on 13-8-2016).
Nasrullah Khadim and Kandhkot (2022) Feuadalism in Sindh, *Pakistan Today*, 22 June 2022.
National Transport Research Centre (2006) The figures in the table are of the year 2006 and are only indicative and don't represent the actual status: Freshwater.
National Water Policy (2018) Ministry of Water Resources, Government of Pakistan, April 2018, Preamble.
Nazir, Ahmad (1993) *Water Resources of Pakistan*, Lahore, Pakistan.
Niaz, Murtaza (2023) A State in Chaos, *Dawn*, 16 May 2023.

O

Ojeh E. (2006) Hydrology of the Indus Basin, Pakistan, quoted from AQUASAT, FAO (UBO) 2016.
Olaf Caroe (1958) *The Pathans- 550 BC – AD 1956*, Macmillan, London.
Oldham, T. (1883) Catalogue of Indian Earthquakes from the Earliest Times to the end of AD 1869, *Mem. GSI*, vol. 19, p. 3.

P

Pakistan Bureau of Statistics (2015) *Compendium of Environment Statistics of Pakistan*, Pakistan Bureau of Statistics, Islamabad.
Pakistan Bureau of Statistics (2017) Final Census results.
Pakistan Bureau of Statistics (2018) Pakistan Statistical Year Book, Government of Pakistan.
Pakistan Bureau of Statistics (2023) Digital Census, July.
Pakistan Economic Review (2011–12).
Pakistan Economic Survey (1978) Finance Division, Government of Pakistan.
Pakistan Economic Survey (2012) Finance Division, Government of Pakistan.
Pakistan Economic Survey (2013–14) Finance Division, Government of Pakistan.
Pakistan Economic Survey (2014–15) Finance Division, Government of Pakistan.
Pakistan Economic Survey (2018–19) Finance Division, Government of Pakistan.
Pakistan Economic Survey (2019–20) Finance Division, Government of Pakistan.
Pakistan Economic Survey (2021–22) Finance Division, Government of Pakistan.
Pakistan: Generalized Soil Map (1993) Soil Survey of Pakistan, Lahore.
Pakistan Statistical Yearbook (2018) Government of Pakistan.
Government of Pakistan, Economic Advisor`s Wing, Finance Division (2022–23) Highlights Pakistan Economic Survey, 2022–23.
Pakistan Statistical Year Book (2018) (Provisional) Government of Pakistan, Pakistan Bureau of Statistics, Islamabad.
Pakistan Statistical Year Book (2019) Government of Pakistan, Pakistan Bureau of Statistics, Islamabad.
Pakistan Tourism Development Corporation, http//www. tourism.gov.pk. peoples-of-Pakistan (accessed on 29-12-2016).
Papanek, G.F (1967) *Pakistan`s Development: Social Goals and Private Incentives*, Harvard University Press, Cambridge, MA.
Population Census Organization, Government of Pakistan, http//www.census.gov .pk/religion.htm (accessed on 24-6-2013).
Punjab Irrigation and Power Department (2013) *Water Resources of Pakistan*, Government of Punjab, Pakistan.

Q

Qazi, Ashraf Jehangir (2023) Saving Pakistan, *Dawn*, 30 March 2023.
Qureshi, A.S., P.G. McCornick, M. Qadir, and Z. Aslam (2008) Managing Salinity and Waterlogging in the Indus Basin of Pakistan, *Agricultural Water Management*, vol. 95 (1), pp. 1–10. doi:10.1016/j.agwat.2007.09.014.

R

Report of the Pakistan Sugar Mills Association (2015).
Rizvi, Hassan Askari (1972): PPP - *The First Phase* (*1969–71*). Lahore Progressive Publications, p. 7, quoted from Zahoor (2011).

S

Safir Rammah (2017) Punjabi in Pakistan, http://www.Punjabics.com/exclusive article (accessed on 2-5-2017).
Salma S., S. Rehman and M. A. Shah (2012) Rainfall Trends in Different Climatic Zones of Pakistan, *Pakistan Journal of Meteorology*, vol. 9 (17). pp. 37–47.

Satterthwaite, David (2002) *Scale of Urban Change Worldwide (1950–2000) and its Underpinnings*, Royal Institute of Chartered Surveyors, London.
Searle, M. P. and B. R. Hacker (2018) 1 Structural and Metamorphic of the Karakoram and Pamir following India-Karakoram-Asia collision from Trelor, P. J. and Searle, M. P. (eds–2019) *Himalayan Tectonics-A Modern Synthesis*, Geological Society, London,, spl publication No. 483.
Shehryar Nabi (2024) *Urban Development in Punjab: A Political Economic Analysis, A Note*, Consortium for Development and Policy Research, p. 2.
Sheikh, H. Y. (2022) *Seventy-five Years of Economic Journey of Pakistan; Towards a Vibrant Pakistan*, Ministry of Finance and Revenue, Government of Pakistan, Islamabad.
Sheikh, Muhammd Munir (2002) Drought Management and Prevention in Pakistan, *Science Vision*, vol. 7, pp. 117–31.
Sibte, Nabi Naqvi and M. Rahamatullah (1962) Weather and Climate of Pakistan, *Pakistan Geographical Review*, vol. 17 (1), pp. 1–18.
Siddiqua Ayesha (2012) Urbanisation in Pakistan and Its Consequences, *Tribune*, 21 November 2012.
Siddiqua, Ayesha (2016) Pakistan's Elite, *The Express Tribune*.
Siddiqui, K. M. (1997) 'Forests of Pakistan' – Working paper no. 11, Asia Pacific Forestry Sector Outlook Study Working Paper series, 'Asia-Pacific Forestry towards 2010,' FAO, Rome.
Social Development in Pakistan, Annual Review 2014–15, Social Policy and Development Centre (SPDC).
Social Policy and Development Centre, Annual Review (2014–15) Social Development in Pakistan, ch.1.
Soil Map of Punjab (1978) *1:1,000,000 Prepared Under the Direction of Dr. M. Bashir Choudhry*, Director Soil Survey of Pakistan, Lahore.
Soil Survey of Pakistan, Lahore (1993) Pakistan: Generalized Soil Map, 7th Draft June 1993, scale 1:2,000,000.
Sorkhabi, Rasoul (2010) Geologic formation of the Himalayas, *Himalayan Jour*, vol. 66, pp. 87–102.
Spate, O. H. K. (1954) *India and Pakistan, A General and Regional Geography*, Methuen Publishing, London.
Stamp Sir Dudley (1929) *Asia: An Economic and Regional Geography*, Methuen Publishing, London..
Syed Sajidin Hussain, Muhammad Mudasser, Muhammad Munir Sheikh, Naeem Manzoor (2005) Climate Change and Variability in Mountain Regions of Pakistan Implications for Water and Agriculture, *Pakistan Journal of Meteorology*, vol. 2(4), p. 75.

T

Tayyeb, A. (1966) *Pakistan: A Political Geography*, Oxford University Press, Karachi.
Thar Coal Energy Board, Government of Sindh, httpt://www//sindhcoal.gos.pk/thar-coalfield/ (accessed on 30-5-2015).

U

United States, Science Education Administration (1975) *Soil Taxonomy: A Basic System of Soil Classification for Making and Interpreting Soil Survey or Soil Conservation Service*, United States Department of Agriculture, Washington, DC, Agricultural Handbook 436.

V

Valdiya, K. S. (1998) *Dynamic Himalaya,* Universities Press, Hyderabad.

W

Wasim Wagha (2012) *Country Technical Note on Indigenous People's Issues Islamic Republic of Pakistan*, International Fund for Agricultural Development, Rome.

Wissmann, Von. H. (1959) Die heutige Vergletscherung und Schneegrenze in Hochasien mit Hinweisen die Vergletscherung der letzten Eiszeit. *Akademie der Wissenchaften und der Literatur, Mainz*, vol. 14, pp. 1101–1407.

World Bank (2015) *World Development Indicators 3.5 Environment: Freshwater.*

World Nuclear Association – Nuclear Power in Pakistan updated April 2015, http://www.world-nuclear.org/country profiles (accessed on 3-5-2015).

World Urbanisation Prospects, 2018, Dept. of Economic & Social Affairs, Population Division, United Nations, New York.

Y

Yeats, M.W.M. (1941) *Census of India,* vol. I, pt. I table 1943, Govt. of India Press, Simla.

Z

Zahoor, Muhammad Abrar (2011) A Critical Appraisal of the Economic reforms of Zulfikar Ali Bhutto: An Assessment, *Pakistan Journal of History and Culture*, vol. 32 pp.145–161.

Zuabair Ahmed Siddiqui and Abdul Rashid (2008) Thunderstorm Frequency Over Pakistan (1961–1990), *Pakistan Journal of Meteorology*, vol. 5 (9), pp. 39–63.

Weblinks.

http://www.fao.org/nr/water/aquasat/data/query results. html (accessed on 1-3-2016).

Indus Basin; http//www.fao.org./nr/water/aquastat/Basins (accessed on 6-8-2016).

http://daac.ornl.gov/data/rivdis/Stations (accessed on 13-8-2016).

http.//www. tourism.gov.pk. peoples-of-Pakistan (accessed on 29-12-2016).

http//www.census.gov.pk/religion.htm (accessed on 24-6-2013).

http://www.Punjabics.com/exclusivearticle (accessed on 2-5-2017).

INDEX

MIX
Papier | Fördert gute Waldnutzung
FSC® C083411

Kolibri 360 GmbH
Ferdinand-Jühlke-Straße 7
99095 Erfurt, Deutschland
produktsicherheit@kolibri360.de